MYCORRHIZAL TECHNOLOGY

Managing Plant Stress and Mitigating Climate Change Using Mycorrhizae

Current Advances in Biodiversity, Conservation and Environmental Sciences

MYCORRHIZAL TECHNOLOGY

Managing Plant Stress and Mitigating Climate Change Using Mycorrhizae

Edited by

Jeyabalan Sangeetha

Abdel Rahman Mohammad Al-Tawaha

Devarajan Thangadurai

AAP APPLE ACADEMIC PRESS

First edition published 2024

Apple Academic Press Inc.
1265 Goldenrod Circle, NE,
Palm Bay, FL 32905 USA

760 Laurentian Drive, Unit 19,
Burlington, ON L7N 0A4, CANADA

CRC Press
2385 NW Executive Center Drive,
Suite 320, Boca Raton FL 33431

4 Park Square, Milton Park,
Abingdon, Oxon, OX14 4RN UK

Library and Archives Canada Cataloguing in Publication

CIP data on file with Canada Library and Archives

Library of Congress Cataloging-in-Publication Data

CIP data on file with US Library of Congress

ISBN: 978-1-77491-510-3 (hbk)
ISBN: 978-1-77491-511-0 (pbk)
ISBN: 978-1-00342-970-8 (ebk)

BOOK SERIES

Current Advances in Biodiversity, Conservation and Environmental Sciences

Series Editor
Jeyabalan Sangeetha, PhD
Assistant Professor, Central University of Kerala, Kasaragod, India
Devarajan Thangadurai, PhD
Professor, Karnatak University, Dharwad, India

Books in the Series

- **Biodiversity and Conservation: Characterization and Utilization of Plants, Microbes, and Natural Resources for Sustainable Development and Ecosystem Management**
  Editors: Jeyabalan Sangeetha, PhD, Devarajan Thangadurai, PhD, Hong Ching Goh, PhD, and Saher Islam, PhD
- **Beneficial Microbes for Sustainable Agriculture and Environmental Management**
  Editors: Jeyabalan Sangeetha, PhD, Devarajan Thangadurai, PhD, and Saher Islam, PhD
- **Algal Genetic Resources: Cosmeceuticals, Nutraceuticals, and Pharmaceuticals from Algae**
  Editors: Jeyabalan Sangeetha, PhD and Devarajan Thangadurai, PhD
- **Organic Farming for Sustainable Development**
  Editors: Jeyabalan Sangeetha, PhD, Kasem Soytong, PhD, Devarajan Thangadurai, PhD, and Abdel Rahman Mohammad Al–Tawaha, PhD
- **Mycorrhizal Technology: Managing Plant Stress and Mitigating Climate Change using Mycorrhizae**
  Editors: Jeyabalan Sangeetha, PhD, Abdel Rahman Mohammad Al–Tawaha, PhD, and Devarajan Thangadurai, PhD
- **Algal Farming Systems: From Production to Application for a Sustainable Future**
  Editors: Jeyabalan Sangeetha, PhD and Devarajan Thangadurai, PhD
- **Microbial Fertilizer Technology for Sustainable Crop Production**
  Editors: Jeyabalan Sangeetha, PhD, Abdel Rahman Mohammad Al-Tawaha, PhD, and Devarajan Thangadurai, PhD

ABOUT THE EDITORS

Jeyabalan Sangeetha, PhD

Assistant Professor, Central University of Kerala, Kasaragod, India

Jeyabalan Sangeetha, PhD, is Assistant Professor at Central University of Kerala, Kasaragod, India. She has edited/co-edited several books in her research areas, which include environmental toxicology, environmental microbiology, environmental biotechnology, and environmental nanotechnology. She earned her BSc in Microbiology and PhD in Environmental Science from Bharathidasan University, Tiruchirappalli, Tamil Nadu, India. She holds an MSc in Environmental Science from Bharathiar University, Coimbatore, Tamil Nadu, India. She is the recipient of a Tamil Nadu Government Scholarship and a Rajiv Gandhi National Fellowship of the University Grants Commission, Government of India for her doctoral studies. She served as Dr. D.S. Kothari Postdoctoral Fellow and UGC Postdoctoral Fellow at Karnatak University, Dharwad, India, during 2012–2016 with funding from the University Grants Commission, Government of India, New Delhi.

Abdel Rahman Mohammad Al-Tawaha, PhD

Professor, Al Hussein Bin Talal University, Jordan

Abdel Rahman Mohammad Al-Tawaha, PhD, is a full Professor of Plant Science, former Head of the Department of Biological Sciences, and former Director of Planning, Information and Quality Unit at Al Hussein Bin Talal University, Jordan. He obtained his PhD from McGill University, Montreal, Canada in 2006. He is the author/coauthor of more than 200 publications in plant science in leading peer-reviewed journals, chapters in edited volumes, and books on a broad range of development issues. He is the Founder and Editor-in-Chief of *Advances in Environmental Biology*. He has also participated in many international conferences as chair or member of the scientific committee.

Devarajan Thangadurai, PhD

Professor, Karnatak University, Dharwad, India

Devarajan Thangadurai, PhD, is Professor at Karnatak University in India. He has authored/edited over 35 books with national and international publishers and has visited 24 countries in Asia, Europe, Africa, and the Middle East for academic visits, scientific meetings, and international collaborations. He received his PhD in Botany from Sri Krishnadevaraya University in South India as a CSIR Senior Research Fellow with funding from the Ministry of Science and Technology, Government of India. He served as a Postdoctoral Fellow at the University of Madeira, Portugal; University of Delhi, India; and ICAR National Research Centre for Banana, India. He is the recipient of a Best Young Scientist Award with a Gold Medal from Acharya Nagarjuna University, India, and the VGST-SMYSR Young Scientist Award of the Government of Karnataka, Republic of India.

CONTENTS

CONTRIBUTORS

Yousef Abu-Zaitoon
Department of Biological Sciences, Al Hussein Bin Talal University, Maan, Jordan

Hiba Alatrash
General Commission for Scientific Agricultural Research, Syria

Abdel Rahman Mohammad Al-Tawaha
Department of Biological Sciences, Al Hussein Bin Talal, Maan, Jordan

Nujoud Alimad
Damascus University, Faculty of Agriculture, Damascus, Syria

Abdel Razzaq Al-Tawaha
Department of Crop Science, Faculty of Agriculture, Universitiy Putra Malaysia, Selangor, Malaysia

Amanullah
Department of Agronomy, The University of Agriculture, Peshawar, Pakistan

Mina Bayanati
Department of Horticultural Science, Ferdowsi University of Mashhad, Mashhad, Iran

Vedavyas Shivanand Chavan
Department of Botany, Karnatak University, Dharwad, Karnataka, India

Pavitra Chippalakatti
Department of Botany, Karnatak University, Dharwad, Karnataka, India

Paulami Dam
Department of Sericulture, Raiganj University, Raiganj, Uttar Dinajpur, West Bengal, India

Abhijit Dey
Department of Life Sciences, Presidency University, Kolkata, India

Maha El-Kady
Department of Self-Pollinated Vegetables Crops, Horticulture Research Institute (HRI), Agriculture Research Centre (ARC), Giza, Egypt

Elif Gunal
Department of Soil Science and Plant Nutrition, Faculty of Agriculture, Harran University, Sanliurfa, Turkey

Imran
Department of Agronomy, The University of Agriculture, Peshawar, Pakistan

Saher Islam
Institute of Biochemistry and Biotechnology, University of Veterinary and Animal Sciences, Lahore, Pakistan

Arun Karnwal
Department of Microbiology, School of Bioengineering and BioSciences, Lovely Professional University, Phagwara, Punjab, India

Shah Khalid
Department of Agronomy, The University of Agriculture, Peshawar, Pakistan

Samia Khanum
Department of Botany, University of the Punjab, Lahore, Pakistan

Poojashree Nagappa Kummur
Department of Botany, Karnatak University, Dharwad, Karnataka, India

Mariyana Lyubenova
Sofia University, Department of Ecology and Environmental Protection, Sofia, Bulgaria

Bhaswatimayee Mahakur
Department of Botany and Biotechnology, Ravenshaw University, Cuttack, Odisha, India

Debasis Mitra
Department of Microbiology, Raiganj University, Raiganj, Uttar Dinajpur, West Bengal, India

Rittick Mondal
Department of Sericulture, Raiganj University, Raiganj, Uttar Dinajpur, West Bengal, India

Simran Panigatti
Department of Botany, Karnatak University, Dharwad, Karnataka, India

Abdur Rauf
Department of Chemistry, University of Swabi, Anbar, Khyber Pakhtunkhwa, Pakistan

Rohan Sachan
Department of Microbiology, School of Bioengineering and Biosciences,
Lovely Professional University, Phagwara, Punjab, India

Aveek Samanta
Department of Botany, Prabhat Kumar College, Contai, West Bengal, India

Jeyabalan Sangeetha
Department of Environmental Science, Central University of Kerala, Kasaragod, Kerala, India

Palani Saranraj
Department of Microbiology, Sacred Heart College (Autonomous), Tirupattur, Tamil Nadu, India

Mohamad Shatnawi
Department of Biotechnology, Faculty of Agricultural Technology, Al-Balqa Applied University,
Al-Salt, Jordan

Devarajan Thangadurai
Department of Botany, Karnatak University, Dharwad, Karnataka, India

ABBREVIATIONS

ABA	abscisic acid
AM	arbuscular mycorrhiza
AMF	arbuscular mycorrhizal fungi
APX	ascorbate peroxidases
Arg^+	arginine
ASC	ascorbate
CAT	catalase
CNN	common mycorrhizal network
ERM	extraradical hyphae
Gln	glutamine
G-POD	guaiacol peroxidase
GPX	glutathione peroxidase
GR	glutathione reductase
GRSP	glomalin-related soil protein
GSH	glutathione
H_2O_2	hydrogen peroxide
Hc	*Hebeloma cylindrosporum*
HRGP	hydroxyproline-rich glycoproteins
IRM	intraradical mycelium
IWUE	intrinsic water use efficiency
JA	jasmonic acid
LBC	lignin biochar
Lj	*Lotus japonicus*
MAMPs	microassociated molecular patterns
MAPK	mitogen-activated protein kinase
MGR	mycorrhizal growth response
MHB	mycorrhization helper bacteria
N	nitrogen
NH_4^+	ammonium
NO_3^-	nitrate
OH	hydroxyl radicals
P	phosphorus
Pe	*Populus euphratica*
PI	phosphatidylinositol

PNP	plant natriuretic peptides
PR	pathogensis-related
Ps	*Pinus sylvestris*
PS	photosystem
Pt	*Pisolithus tinctorius*
Pt	*Populus tremula*
PtKUP	KUP transporter of *Populus trichocarpa*
Ri	*Rhizophagus irregularis*
ROS	reactive oxygen species
RWC	relative water content
SOD	superoxide dismutase
SOS	salt overly sensitive
SSU	small subunit
VAM	vesicular arbuscular mycorrhiza
VOCs	volatile organic compounds

PREFACE

Gradual climate change causes severe consequences on ecosystems and instigates various abiotic stressors that adversely affect the plant developmental process. So, there is a progressively growing grasp of the elemental importance of soil for agricultural sustainability that embraces plant–fungal symbiotic associations. Among these kinds of symbioses, mycorrhiza is a leading player. A mycorrhizal symbiosis is a fundamental component of a terrestrial ecosystem as hyphal networks significantly improve the utilization of water and essential nutrients with colonized roots. Developing mycorrhizal technology for sustainable agriculture includes ecological engineering of mycorrhizal fungi, production of mycorrhizal inoculum, and agricultural management. Moreover, a series of research studies based on mycorrhizal technology has explored the positive impact of root symbiosis on plant adaptation to climate-associated abiotic stressors such as salinity, drought, flooding, and high temperatures.

Research in mycorrhizal technology is significantly evolving, and cutting-edge findings have provided fair evidence to understand how complex plant–mycorrhiza associations work. Thus, this current volume implies an integrated description of mycorrhizal symbiosis as an essential component to ensure plant growth and health and topsoil quality. The book confers the prospects of transfer of mycorrhizal technology into commercial regularities that make exclusive emphasis on challenges pertaining to quality of inoculum production and accompanies successful results of introduction of mycorrhizal communities into plants production systems.

The prime objective of this book is to deliver the remarkable theoretic and practical findings to the scientific community, emphasizing outstanding applications of mycorrhizal fungi in recovering plant health, producing the mycorrhizal edible forest trees, and conserving natural resources.

The book comprises 13 chapters, partitioned into four different parts, authored by international experts and researchers from different domains of plant–mycorrhiza research. Part I encompasses three chapters that explain the roles of mycorrhiza in plant growth, minerals metabolism, and biological control. Part II of this book also covers three chapters, presenting the extensive commercial applications of mycorrhizal fungi in various industrial sectors, including agriculture, horticulture, and biotechnology. Part III consists of four chapters that investigate the diverse role of mycorrhizal fungi in managing different stress conditions. Part IV provides in-depth analyses and explores the ways to maintain global environmental sustainability and food security with commercialized mycorrhizal technology.

Contributions from different academic circles, case reports, and industrial applications make this book a broad review of recent developments in mycorrhizal technology. So, this book seeks the great attention of researchers, academicians, students, and professionals in the field of mycology, botany, biotechnology, bioengineering, and environmental sciences for sustainable crop production by managing abiotic stressors with mass production of mycorrhizae.

—***Jeyabalan Sangeetha***
Abdel Rahman Mohammad Al–Tawaha
Devarajan Thangadurai

PART I

MYCORRHIZAL PRODUCTION FOR PLANT GROWTH, PHOSPHORUS METABOLISM, AND BIOLOGICAL CONTROL

CHAPTER 1

TYPES, IMPORTANCE, AND FACTORS AFFECTING MYCORRHIZA PRODUCTION FOR SUSTAINABLE PLANT GROWTH

AVEEK SAMANTA[1], ABDEL RAHMAN MOHAMMAD AL-TAWAHA[2], JEYABALAN SANGEETHA[3], DEVARAJAN THANGADURAI[4], and SIMRAN PANIGATTI[4]

[1]*Department of Botany, Prabhat Kumar College, Contai, West Bengal, India*

[2]*Department of Biological Sciences, Al Hussein Bin Talal University, Maan, Jordan*

[3]*Department of Environmental Science, Central University of Kerala, Kasaragod, Kerala, India*

[4]*Department of Botany, Karnatak University, Dharwad, Karnataka, India*

ABSTRACT

Abiotic constraints on crop productivity have been exacerbated by climate change and agricultural malpractices such as overuse of fertilizers and pesticides, which has degraded the ecosystem. There is a critical need for environmentally friendly crop management strategies, such as the utilization of arbuscular mycorrhizal fungi (AMF), to increase agricultural productivity. AMF are frequently referred to as bio-fertilizers. Additionally, it is widely considered that inoculating host plants with AMF confers resistance on

Mycorrhizal Technology: Managing Plant Stress and Mitigating Climate Change using Mycorrhizae.
Jeyabalan Sangeetha, Abdel Rahman Mohammad Al-Tawaha, & Devarajan Thangadurai (Eds.)

them against a variety of stressful conditions such as heat, salinity, drought, metals, and severe temperatures. The mycorrhizal association between plants and fungi is an important symbiotic relationship for plant growth. There are various types of mycorrhizal types found in nature. These types are found in diverse types of plant groups. The symbiotic association helps the host plants to survive better way than nonassociate plants. Increased nutrient supplies, carbon transfer, water uptake, the establishment of seedlings, pest resistance are the most important benefit for plants. This association has many ecological roles also. Many abiotic and biotic factors like soil temperature, acidity, fertilizer, crop rotation, cultivation of nonhost plants affect the mycorrhizal association. The sustainability of plant growth and development critically depends on this symbiotic association. Mycorrhizal technology could be a more eco-friendly alternative for agriculture practices.

1.1 INTRODUCTION

The symbiotic association between higher plant root and fungi is called Mycorrhiza. This is an obligate type of symbiosis (Smith and Read, 1997; Brundrett, 2004; Barman et al., 2016). Two associated partners are benefited by each other for their yield and reproduction (Clapperton and Read, 1992; Gange et al., 1993). Compared with the fungal partner the host plants are more benefitted by this association (Fogel and Hunt, 1979; Vogt et al., 1982; Hogberg and Hogberg, 2002). Host plants greatly depend for nutrition on associated fungi (Brundrett and Abbott, 2002), and they are more efficient in nutrient uptake than nonhost plants (Janos, 1980; Wilson and Hartnett, 1998, Siqueira and Saggin-Junior, 2001). The uptake of nutrients depends on the growth pattern of root and soil hyphal activity (Hepper, 1985; Lussenhop and Fogel, 1999; Wallander et al., 2001). Besides the role in nutrient uptake mycorrhiza have much other beneficial activity to the associated plants as well as to the ecosystem (Koske and Gemma, 1990; Douds et al., 2000). According to the groups of fungi and host plant involved there are seven types of mycorrhiza. The groups are Ectomycorrhiza, Endomycorrhiza, Ectendomycorrhiza, Ericoid mycorrhiza, Orchid mycorrhiza, Arbutoid mycorrhiza, and Monotropoid mycorrhiza (Read, 1983; Hadley, 1989) (Fig. 1.1). Among the seven groups arbuscular and ectomycorrhiza are common and most abundant (Smith and Read, 1997; Allen et al., 2003). Several abiotic and biotic factors affect growth and development of mycorrhiza. Among these factors, natural calamities, soil pH, nutrient level, soil moisture, salinity, temperature, and crop rotations are most important (Morton, 1988; Abott

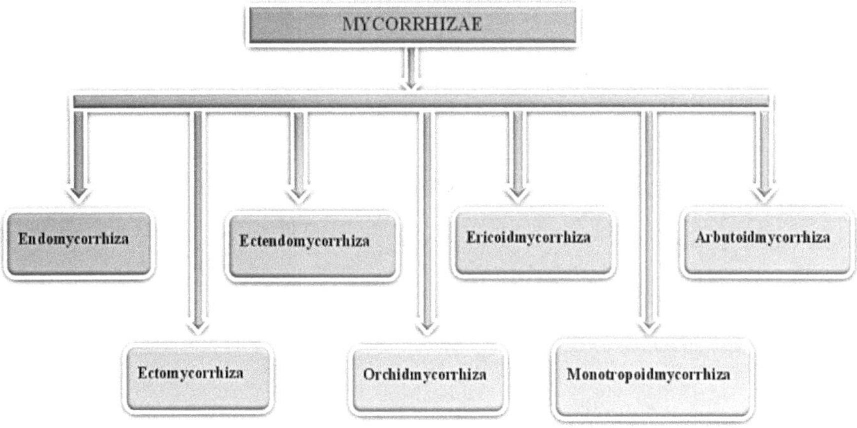

FIGURE 1.1 Types of mycorrhizae.

and Robson, 1991). In this chapter, a brief account of mycorrhizal types and factors affecting its growth and development is described (Fig. 1.2).

1.2 ECTOMYCORRHIZA

In this type of mycorrhiza, the fungal partner forms a sheath around the host root surface (Parke, 1983; Pearson and Jacobson, 1993). This association causes metabolic energy input from the host plant to fungus (Smith et al., 1994; Douds et al., 2000). On the other hand, the fungal partner provides mineral nutrition (Marschner, 1995; Miller and Kling, 2000). The amount of nutrient fluxes depends upon the area of the symbiotic interface (Burgess et al., 1994; Dikson et al., 1999). The ectomycorrhizal fungi belong to the family Basidiomycetes, Ascomycetes, Zygomycetes on gymnosperm and angiosperms (Pathan et al., 2020). These fungi can be grown on pure culture (Brundrett, 2004).

1.3 ENDOMYCORRHIZA

The endomycorrhizal fungi forms branch (arbuscular) and vesicle structure inside plant root. The arbuscular are the site of nutrient exchange and vesicles are the reproductive structure (Brundrett et al., 1996). The plant taxa involved in this type of association can be Bryophytes, Pteridophytes, Gymnosperm, and Angiosperms and fungal partner belongs to the family zygomycetes and cannot be grown on pure culture (Smith and Read, 1997) (Fig. 1.3).

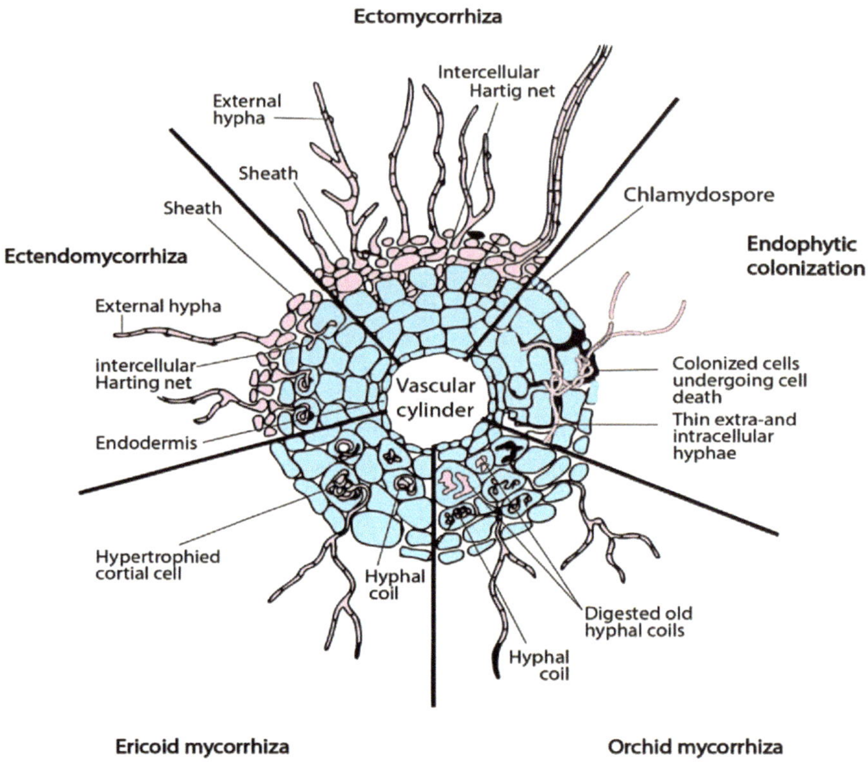

FIGURE 1.2 Various types of mycorrhizal associations.

Source: Reproduced with permission from Dhanker et al. (2020), Copyright © Springer Nature (2020).

1.4 ECTENDOMYCORRHIZA

This mycorrhizal type shows both the characters of ectomycorrhiza and endomycorrhiza. They form a hyphal net around and inside the root. The fungi belong to the family Basidiomycetes and Ascomycetes. Plants belong to Gymnosperm and Angiosperm (Bratek et al., 1996).

1.5 ARBUTOID MYCORRHIZA

This group of mycorrhiza forms a sheath around the root and penetrates the outer cortical cells of the root. The plant taxa are Ericaceae and fungi are Basidiomycetes (Newsham et al., 2009).

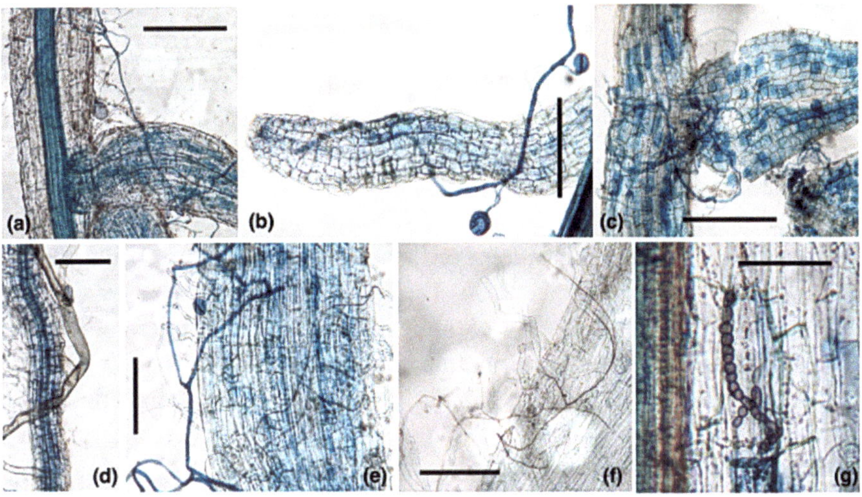

FIGURE 1.3 Root colonization by arbuscular mycorrhizal fungi. (a–e) Arbuscular mycorrhizal fungi (AMF), (e–g) dark septate endophytes (DSE), and other endophytic fungi inhabiting roots of grass species. (a) *A. purpurea* (bare soil)—intracidal and extraradical AMF hypha; (b) *P. jamesii* (bare soil)—AMF extraradical hypha and terminal vesicles in the rhizodermis; (c) *P. jamesii* (bare soil)—AMF intraradical hypha and vesicles in the root cortex; (d) *B. gracilis* (biocrust)—AMF intercidal hypha and extraradical fruiting body; (e) *Z. mays* (biocrust)—AMF extraradical hypha and terminal vesicles and fine, melanised intracidal DSE hypha; (f) *Z. mays* (bare soil)—infection and growth of septate, melanised hypha (DSE); (g) *A. purpurea* (biocrust)—microsclerotium-like structure formed by dark septate fungal endophytes. Bar = 100 µm.

Source: Reproduced with permission from Havrilla et al. (2020), Copyright © Springer Nature (2019).

1.6 MONOTROPOID MYCORRHIZA

The mycorrhizae are found in the plant taxa Monotropoideae. The plants depend for organic and inorganic nutrients on their fungal partner. The fungi belongs to Basidiomycetes (Lutz and Sjolund, 1973).

1.7 ERICOID MYCORRHIZA

Mycorrhizal fungi form a loose network inside the epidermal cells of root hair. The plants belong to the family Ericaceae and fungi under Basidiomycetes (Read, 1996).

1.8 ORCHID MYCORRHIZA

This mycorrhizal association is only found exclusively in orchids. The associated fungi infect host plant just after the seed germination and provide carbohydrate for the host plant. The fungi are belonging to Basidiomycetes. The excessive intracellular growth of fungus inside the plant is controlled by antifungal substances that digest the hypha (Rasmussen, 2002).

The mycorrhizal association is a beneficial relationship to the plant as well as to the fungal partners also. The most important role of mycorrhizae is nutrient supply for the host plant. Phosphorus and other micronutrients are the one available to the plants by this relationship. The hyphal development is proportional to the scarcity of nutrient in soil (Miller and Kling, 2000; Wallander et al., 2001). Another role is the increase in root surface area so more soil is accessible for root (Conn and Dighton, 2000). These associations can protect the host plant from other parasitic fungi (Sanders, 2002). Agricultural weeds growth is suppressed in mycorrhizal association (Rinaudo et al., 2010). The amount of nitrogen fixation also increases in the presence of more phosphorus to the plant (Hayman, 1986). In a drought-prone area, a plant associated with mycorrhiza grows better than nonassociated plants (Parke et al., 1983; Coleman, 1985). Plants associated with mycorrhiza survive better in saline soil than plants not associated with fungal partner (Pond, 1984).

Apart from the role of direct benefit to the plants mycorrhiza has a major impact on the ecosystem. Growth of host plants outside its natural habitat requires mycorrhizal association (Krikun, 1983; Brundrett, 1991). The early phase of plant succession also requires this association (Koske and Gemma, 1990). Nutrient transfer from dead material to living plants ultimately leads to the biochemical cycle (Read and Perez-Moreno, 2003). Mycorrhiza also influences other soil microflora which enriches rhizosphere (Ames et al., 1984; Linderman, 1991). Another importance is the establishment of tree plant seedling (Nara, 2006).

There are several abiotic and biotic factors that are responsible for mycorrhizal growth and development. Some of the important abiotic factors are salinity and temperature (Daniels, 1984; Trappe and Molina, 1986; Morton, 1988). Some of the factors like soil nutrient level, soil moisture, soil pH, and other symbiotic microorganisms affect the association positively, whereas presence of pesticides, fertilizers, drought, and crop rotation harm the mycorrhizal association.

Natural calamities like a volcanic eruption, glaciation, mining also affect mycorrhiza growth. Low phosphate level in soil influences mycorrhizal colony formation (Thomson et al., 1986; Henkel et al., 1989). Availability of calcium is also a nutritional factor for colony formation of mycorrhiza (Lapeyrie and Bruchet, 1986). Soil temperature stimulates more inoculum production that ultimately influences root morphology and nutrient uptake (Dodd and Jeffries, 1986; Sieverding, 1988; Raju et al., 1990). Soil pH above seven has a great influence on arbuscular mycorrhizal growth inside plant (Evans and Miller, 1988; Juniper and Abbott, 1993; Soulas et al., 1997). Sometimes, low pH level also influences mycorrhizal association (Hayman and Tavares, 1985; Koomen et al., 1987; Porter et al., 1987). Pesticide, herbicide, and chemical fertilizers hamper the biochemical activity and reduce the level of association (Termorshuizen and Schaffers, 1987; Sidhu and Chakravarti, 1990; Brundrett and Abbott, 2002). Presence of other soil microbes like actinomycetes enhances mycorrhizal growth. Cultivation of non-host species in a field reduces spore formation of mycorrhizal fungi (Pearson and Schweiger, 1993; Lamhamedi et al., 1994).

1.9 CONCLUSION

The symbiotic event of mycorrhiza is an important factor for soil and plant health. Plants are greatly benefited from this association. In recent years, many works have been done to get benefit from mycorrhiza in agriculture, horticulture, and forestry. Efforts are also done to reclaim unfertile soil. Present-day practice of agriculture like overuse of inorganic fertilizer, pesticides, leads to the destruction of this association. Many anthropogenic activities like mining, waste disposal, fly ass have detrimental effects on the symbiosis. So, it urgently needs to develop some technique on mycorrhizal technologies that can maintain better plant growth as well as sustainable development.

KEYWORDS

- **arbuscular mycorrhizal fungi**
- **biotic factors**
- **climate change**
- **orchid mycorrhiza**
- **abiotic constraints**

REFERENCES

Abbott, L. K.; Robson, A. D. Factors influencing the occurrence of vesicular arbuscular mycorrhizas. *J. Agric. Ecos. Environ.* **1991**, *35*, 121–150.

Allen, M. F.; Swenson, W.; Querejeta, J. I.; Egerton-Warburton, L. M.; Treseder, K. K. Ecology of mycorrhizae: A conceptual framework for complex interactions among plants and fungi. *Ann. Rev. Phytopathol.* **2003**, *41*, 271–303.

Ames, R. N.; Reid, C. P. P.; Ingham, E. R. Rhizosphere bacterial population responses to root colonization by a vesicular-arbuscular mycorrhizal fungus. *New Phytol.* **1984**, *96*, 555–563.

Barman, J.; Samanta, A.; Saha, B.; Datt, S. Mycorrhiza: The oldest association between plant and fungi. *Resonance* **2016**, *21*(12) 1093–1104.

Bratek, Z.; Jakucs, E.; Bóka, K.; Szedlay, G. Mycorrhizae between black locust (*Robinia pseudoacacia*) and *Terfezia terfezioides*. *Mycorrhiza* **1996**, *6*(4) 271–274.

Brundrett, M. Diversity and classification of mycorrhizal associations. *Biol. Rev.* **2004**, *79*(3), 473–495.

Brundrett, M.; Bougher, N.; Dell, B.; Grove, T.; Malajczuk, N. *Working with mycorrhizas in forestry and agriculture*; Australian Centre for International Agricultural Research: Canberra, 1996.

Brundrett, M. C. Mycorrhizas in natural ecosystems. *Adv. Ecol. Res.* **1991**, *21*, 171–313.

Brundrett, M. C.; Abbott, L. K. Arbuscular mycorrhizas in plant communities. In: *Microorganisms in plant conservation and biodiversity*; Sivasithamparam, K., Dixon, K. W., Barrett, R. L., Eds.; Kluwer Academic Publishers: Dordrecht, 2002; pp 151–193.

Burgess, T.; Dell, B.; Malajczuk, N. Variations in mycorrhizal development and growth stimulation by 20 *Pisolithus* isolates inoculated on to *Eucalyptus grandis* W. Hill ex Maiden. *New Phytol.* **1994**, *127*, 731–739.

Clapperton, M. J.; Read, D. J. A relationship between plant growth and increasing VAM mycorrhizal inoculum density. *New Phytol.* **1992**, *120*, 227–234.

Coleman, D. C. Through a ped darkly; an ecological assessment of root-soil microbial-faunal interactions. In: *Ecological interactions in soil, plants, microbes and animals*; Fitter, A. H., Atkinson, D., Read, D. J., Usher, M. B., Eds.; Blackwell: Oxford, 1985; pp 1–21.

Conn, C.; Dighton, J. Litter quality influences on decomposition, ectomycorrhizal community structure and mycorrhizal root surface acid phosphatase activity. *Soil Biol. Biochem.* **2000**, *32*(4), 489–496.

Daniels, B. A. The ecology of VA mycorrhizal fungi. In: *VA mycorrhiza*; Conway, L., Powell, L., Bagyaraj, D.J., Eds.; CRC Press: Boca Raton, FL, 1984; pp 35–36.

Dhanker, R.; Chaudhary, S.; Kumari, A.; Kumar, R.; Goyal, S. Symbiotic signaling: Insights from arbuscular mycorrhizal symbiosis. In: *Plant microbe symbiosis*; Varma, A., Tripathi, S., Prasad, R., Eds.; Springer: Cham, 2020; pp 75–103. https://doi.org/10.1007/978-3-030-36248-5_5

Dickson, S.; Smith, S. E.; Smith, F. A. Characterization of two arbuscular mycorrhizal fungi in symbiosis with *Allium porrum*: Inflow and flux of phosphate across the symbiotic interface. *New Phytol.* **1999**, *144*, 173–181.

Dodd, J. C.; Jeffries, J. C. Early development of vesicular-arbuscular mycorrhizas in autumn-sown cereals. *Soil Biol. Biochem.* **1986**, *18*, 149–154.

Douds, D. D. Jr.; Pfeffer, P. E.; Schachar-Hill, Y. Application of in vitro methods to study carbon uptake and transport by AM fungi. *Plant Soil* **2000**, *226*, 255–261.

Evans, D. G.; Miller, M. H. Vesicular-arbuscular mycorrhizas and the soil-disturbance-induced reduction of nutrient absorption in maize. I: Causal relations. *New Phytol.* **1988**, *110*, 67–74.

Fogel, R.; Hunt, G. Fungal and arboreal biomass in a western Oregon Douglas-fir ecosystem: distribution patterns and turnover. *Can. J. Forest Res.* **1979**, *9*, 245–256.

Gange, A. C.; Brown, V. K.; Sinclair, G. S. Vesicular arbuscular mycorrhizal fungi: A determinant of plant community structure in early succession. *Funct. Ecol.* **1993**, *7*, 616–622.

Harley, J. L. The significance of mycorrhiza. *Mycol. Res.* **1989**, *92*, 129–139.

Havrilla, C.; Leslie, A. D.; Di Biase, J. L. et al. Biocrusts are associated with increased plant biomass and nutrition at seedling stage independently of root-associated fungal colonization. *Plant Soil* **2020**, *446*, 331–342. https://doi.org/10.1007/s11104–019–04306–4.

Hayman, D. S. Mycorrhizae of nitrogen-fixing legumes. *MIRCEN J. Appl. Microbiol. Biotechnol.* **1986**, *2*(1), 121–145.

Hayman, D. S.; Tavares, M. Plant growth responses to vesicular-arbuscular mycorrhiza XV: Influence of soil pH on the symbiotic efficiency of different endophytes. *New Phytol.* **1985**, *100*, 367–379.

Henkel, T. W.; Smith, W. K.; Christensen, M. Infectivity and effectivity of indigenous vesicular-arbuscular mycorrhizal fungi from contiguous soils in southwestern Wyoming, USA. *New Phytol.* **1989**, *112*, 205–214.

Hepper, C. M. Influence of age of roots on the pattern of vesicular-arbuscular mycorrhizal infection in leek and clover. *New Phytol.* **1985**, *10*, 685–693.

Hogberg, M. N.; Hogberg, P. Extramatrical ectomycorrhizal mycelium contributes one-third of microbial biomass and produces, together with associated roots, half the dissolved organic carbon in a forest soil. *New Phytol.* **2002**, *154*, 791–795.

Janos, D. P. Mycorrhizae influence tropical succession. *Biotropica* **1980**, 12, 56–64.

Juniper, S.; Abbott, L. Vesicular-arbuscular mycorrhizas and soil salinity. *Mycorrhiza* **1993**, *4*, 45–57.

Koomen, I.; Grace, C.; Hayman, D. S. Effectiveness of single and multiple mycorrhizal inocula on growth of clover and strawberry plants at two soil pHs. *Soil Biol. Biochem.* **1987**, *19*, 539–544.

Koske, R. E.; Gemma, J. N. VA mycorrhizae in strand vegetation of Hawaii: Evidence for long-distance codispersal of plants and fungi. *Am. J. Bot.* **1990**, *77*, 466–474.

Krikun, J. Edaphic and climatic factors in the evolution of vesicular-arbuscular mycorrhizal fungi. In: *Developments in ecological and environmental quality*; Ecological Society, Hebrew University of Jerusalem: Jerusalem, 1983; pp 31–36.

Lamhamedi, M. S.; Godbout, C.; Fortin, J. A. Dependence of *Laccaria bicolor* basidiome development on current photosynthesis of *Pinus strobus* seedlings. *Can. J. Forest Res.* **1994**, *24*, 1797–1804.

Lapeyrie, F. F.; Bruchet, G. Calcium accumulation by two strains, calcicole and calcifuge, of the mycorrhizal fungus *Paxulus involutus*. *New Phytol.* **1986,** *103*, 133–141.

Linderman, R. G. Mycorrhizal interactions in the rhizosphere. In: *The rhizosphere and plant growth*; Springer: Dordrecht, 1991; pp 343–348.

Lussenhop, J.; Fogel, R. Seasonal change in phosphorus content of *Pinus strobus—Cenococcum geophilum* ectomycorrhizae. *Mycologia* **1999**, *91*, 742–746.

Lutz, R. W.; Sjolund, R. D. *Monotropa uniflora*: Ultrastructural details of its mycorrhizal habit. *Am. J. Bot.* **1973**, *60*(4), 339–345.

Marschner, H. *Mineral nutrition of higher plants*; Academic Press: London, 1995.
Miller, R. M.; Kling, M. The importance of integration and scale in the arbuscular mycorrhizal symbiosis. *Plant Soil* **2000**, *226*, 295–309.
Morton, J. B. Taxonomy of VA mycorrhizal fungi: Classification, nomenclature and identification. *Mycotaxon* **1988**, *32*, 267–324.
Nara, K. Ectomycorrhizal networks and seedling establishment during early primary succession. *New Phytol.* **2006**, *169*(1), 169–178.
Newsham, K. K.; Upson, R.; Read, D. J. Mycorrhizas and dark septate root endophytes in polar regions. *Fungal Ecol.* **2009**, *2*(1), 10–20.
Parke, J. L.; Linderman, R. G.; Black, C. H. The role of ectomycorrhizas in drought tolerance of douglas-fir seedlings. *New Phytol.* **1983**, *95*, 83–95.
Pathan S. I.; Ceccherini M. T.; Sunseri F.; Lupini A. Rhizosphere as hotspot for plant-soil-microbe interaction. In: *Carbon and nitrogen cycling in soil*; Datta, R., Meena, R., Pathan, S., Ceccherini, M., Eds.; Springer: Singapore, 2020.
Pearson, J. N.; Jakobsen, I. Symbiotic exchange of carbon and phosphorus between cucumber and three arbuscular mycorrhizal fungi. *New Phytol.* **1993**, *124*, 481–488.
Pearson, J. N.; Schweiger, P. *Scutellospora calospora* (Nicol. Gerd.) Walker Sanders associated with subterranean clover: Dynamics of colonization, sporulation and soluble carbohydrates. *New Phytol.* **1993**, *124*, 215–219.
Pond, E. C.; Menge, J. A.; Jarrell, W. M. Improved growth of tomato in salinized soil by vesicular-arbuscular mycorrhizal fungi collected from saline soils. *Mycologia* **1984**, *76*, 74–84.
Porter, W. M.; Robson, A. D.; Abbott, L. K. Field survey of the distribution of vesicular-arbuscular mycorrhizal fungi in relation to soil pH. *J. App. Ecol.* **1987**, *24*, 659–662.
Raju, P. S.; Clark, R. B.; Ellis, J. R.; Maramille, J. W. Effects of species of VA mycorrhizal fungi on growth and mineral uptake of sorghum at different temperatures. *Plant Soil* **1990**, *121*, 165–170.
Rasmussen, H.N. Recent developments in the study of orchid mycorrhiza. *Plant Soil* **2002**, *244*(1–2), 149–163.
Read, D. J. The biology of mycorrhiza in the Ericales. *Can. J. Bot.* **1983**, *61*, 985–1004.
Read, D. J. The structure and function of the ericoid mycorrhizal root. *Ann. Bot.* **1996**, *77*(4), 365–374.
Read, D. J.; Perez-Moreno, J. Mycorrhizas and nutrient cycling in ecosystems—A journey towards relevance? *New Phytol.* **2003**, *157*(3), 475–492.
Rinaudo, V.; Bàrberi, P.; Giovannetti, M.; van der Heijden, M. G. Mycorrhizal fungi suppress aggressive agricultural weeds. *Plant Soil* **2010**, *333*(1–2), 7–20.
Sanders, I. R. Specificity in the arbuscular mycorrhizal symbiosis. In: *Mycorrhizal ecology*; van der Heijden, M. G. A., Sanders, I., Eds.; Springer: Berlin, Heidelberg, 2002; 415–437.
Sidhu, S. S.; Chakravarty, P. Effect of selected forestry herbicides on ectomycorrhizal development and seedling growth of lodgepole pine and white spruce under controlled and field environment. *Eur. J. Forest Pathol.* **1990**, *20*, 77–94.
Sieverding, E.; Toro, T. T. Influence of soil water regimes on VA mycorrhiza. V. Performance of different VAM fungal species with cassava. *J. Agron. Crop Sci.* **1988**, *161*, 322–332.
Siqueira, J. O.; Saggin-Junior, O. J. Dependency on arbuscular mycorrhizal fungi and responsiveness of some Brazilian native woody species. *Mycorrhiza* **2001**, *11*, 245–255.

Smith, S. E.; Dickson, S.; Morris, C.; Smith, F. A. Transfer of phosphate from fungus to plant in VA mycorrhizas: Calculation of the area of symbiotic interface and of fluxes of P from two different fungi to *Allium porrum* L. *New Phytol.* **1994**, *127*, 93–99.

Smith, S. E.; Read, D. J. *Mycorrhizal symbiosis*; Academic Press: London, 1997.

Soulas, M. L.; Bihan, B. L.; Camporota, P.; Jarosz, C.; Salerno, M. I.; Perrin, R. Solarization in a forest nursery: effect on ectomycorrhizal soil infectivity and soil receptiveness to inoculation with *Laccaria bicolor*. *Mycorrhiza* **1997**, *7*, 95–100.

Termorshuizen, A. J.; Schaffers, A. P. Occurrence of carpophores of ectomycorrhizal fungi in selected stands of *Pinus sylvestris* in the Netherlands in relation to stand vitality and air pollution. *Plant and Soil* **1987**, *104*, 209–217.

Thomson, B. D.; Robson, A. D.; Abbott, L. K. Effects of phosphorus on the formation of mycorrhizas by *Gigaspora calospora* and *Glomus fasciculatum* in relation to root carbohydrates. *New Phytol.* **1986**, *103*, 751–765.

Trappe, J. M.; Molina, R. Taxonomy and genetics of mycorrhizal fungi: Their interactions and relevance. In: *Physiology and genetical aspects of mycorrhizae*; Gianinazzi-Pearson, V.; Gianinazzi, S.; Eds.; INRA: Paris, 1986; pp 133–146.

Vogt, K. A.; Grier, C. C.; Meier, C. E.; Edmonds, R. L. Mycorrhizal role in net primary production and nutrient cycling in *Abies amabilis* ecosystems in Eastern Washington. *Ecology* **1982**, *63*, 370–380.

Wallander, H.; Nilsson, L. O.; Hagerberg, D.; Baath, E. Estimation of the biomass and seasonal growth of external mycelium of ectomycorrhizal fungi in the field. *New Phytol.* **2001**, *151*, 753–760.

Wilson, G. W. T.; Hartnett, D. C. Interspecific variation in plant responses to mycorrhizal colonization in tallgrass prairie. *Am. J. Bot.* **1998**, *85*, 1732–1738.

CHAPTER 2

MYCORRHIZAL ROLE IN PHOSPHORUS METABOLISM

SAMIA KHANUM[1], ABDEL RAHMAN MOHAMMAD AL-TAWAHA[2], YOUSEF ABU-ZAITOON[2], ABDEL RAZZAQ AL-TAWAHA[3], HIBA ALATRASH[4], ABDUR RAUF[5], ARUN KARNWAL[6], ABHIJIT DEY[7], MOHAMAD SHATNAWI[8], DEVARAJAN THANGADURAI[9], JEYABALAN SANGEETHA[10], SAHER ISLAM[11], IMRAN[12], AMANULLAH[12], SHAH KHALID[12], PALANI SARANRAJ[13], and ELIF GUNAL[14]

[1]*Department of Botany, University of the Punjab, Lahore, Pakistan*

[2]*Department of Biological Sciences, Al Hussein Bin Talal University, Maan, Jordan*

[3]*Department of Crop Science, Faculty of Agriculture, University Putra Malaysia, Selangor, Malaysia*

[4]*General Commission for Scientific Agricultural Research, Syria*

[5]*Department of Chemistry, University of Swabi, Anbar, Khyber Pakhtunkhwa, Pakistan*

[6]*Department of Microbiology, School of Bioengineering and BioSciences, Lovely Professional University, Phagwara, Punjab, India*

[7]*Department of Life Sciences, Presidency University, Kolkata, India*

[8]*Department of Biotechnology, Faculty of Agricultural Technology, Al-Balqa Applied University, Jordan*

[9]*Department of Botany, Karnatak University, Dharwad, Karnataka, India*

Mycorrhizal Technology: Managing Plant Stress and Mitigating Climate Change using Mycorrhizae. Jeyabalan Sangeetha, Abdel Rahman Mohammad Al-Tawaha, & Devarajan Thangadurai (Eds.)

[10]*Department of Environmental Science, Central University of Kerala, Kasaragod, Kerala, India*

[11]*Institute of Biochemistry and Biotechnology, University of Veterinary and Animal Sciences, Lahore, Pakistan*

[12]*Department of Agronomy, The University of Agriculture, Peshawar, Pakistan*

[13]*Department of Microbiology, Sacred Heart College (Autonomous), Tirupattur, Tamil Nadu, India*

[14]*Department of Soil Science and Plant Nutrition, Faculty of Agriculture, Harran University, Sanliurfa, Turkey*

ABSTRACT

The mycorrhizal association between plant and fungi under certain circumstances is beneficial for both partners. The phosphorous present in the soil is the component of many important compounds like nucleic acids, phospholipids, inositol phosphates, and other metabolic intermediates. Instead of so many different forms of phosphorous present in the soil, the only form uptake by hyphae and root is orthophosphate in the form of H_2PO_4 ions. There are many ways for the acquisition of P from the soil. The direct uptake consists of Pi acquisition from the soil solution by plant root hairs and indirect pathway involves uptake via hyphae, its translocation via extra radical mycelium, release of P within roots by hyphae, and uptake of these P by root cells. The direct involvement of AMF and other soil microbes in the mineralization of organic phosphorous under nonsterile soil situations are still not clear.

2.1 INTRODUCTION

The mycorrhizal association under certain circumstances is beneficial for both partners. The AM association first time reported between terrestrial plants and fungi was some 400 to 500 million years ago (Redecker et al., 2000). The association of AMF is reported as the most widely spread association with all important groups of plants like bryophytes, gymnosperms, angiosperms, and ferns and can grow in all kinds of ecosystems of the world (Wang and Qiu, 2006). The fungal partner diversity is very limited and only few hundred species take part in this association which shows that the association is species unspecific. According to some studies, there is some specificity between plant and fungal species that have also been reported (Opik et al., 2009). The ECM, ericoid, and orchid are the type of mycorrhizal association limited to some plant taxa and belong to phyla Basidiomycota, Ascomycota, and Zygomycota and also the species specification reported in these groups (Smith and Read, 2008). Some plants are having both AM and ECM associations and when colonized by mycorrhizal fungi they give rise to their root system simultaneously (Adams et al., 2006).

The response of plants toward mycorrhizal association is depending on their environmental conditions and ecosystems. Plants belonging to Polar Regions or from higher altitudes are not mycorrhizal. The roots of these plants are colonized by some dark-colored septate endophytes, but the function of this colonization is unclear (Newsham et al., 2009). The endophytes spend their entire life inside the plant body, but on the other hand mycorrhizal fungi inhabit roots and soil around the plant roots (Faeth and Fagan, 2002; Jansa and Gryndler, 2010). The plants from grassland ecosystem and from tropical forests have AM association (Castillo et al., 2006).

Plant root and fungus combined, results in a single structure through which exchange of nutrients and carbon takes place. Plants are conferred many characteristics by mycorrhiza such as increase in the growth rate due to increase in uptake of minerals and nutrients, improve the survival in harsh conditions as well as reduced the risk of diseases. The survival of both participants (plant and fungus) is required in the mycorrhizal symbiosis. Plants when growing in highly weathered and severely phosphorous needed soil are not in the mutualistic association and they fulfill their P requirements by clustering of roots (Lamber et al., 2008). When in the field, plants formed the mycorrhizal association with different fungal species (Burke et al., 2005). Such diversity has the important significant part in the uptake of nutrients and also for the maximum benefits for both partners (Jansa et al., 2008).

2.2 THE ROLE OF MYCORRHIZA

The role of mycorrhiza is not explained clearly yet because under the natural condition association is absence or rare (Merryweather and Fitter, 1995). The establishment of non mycorrhizal association in a container or field is difficult challenge because certain treatments of soil like steaming and autoclaving change the chemical structure of the soil and also disturb the other soil organisms also make it more difficult (Serrasolses et al., 2008). The mycorrhizal association improved the nutrient's uptake, especially the elements having low mobility like P, Zn, Cu, etc., and also helped the plants to cope up the stresses and thus help plants to grow better (Sikes et al., 2009). The mycorrhizal association results in the expansion and complementation of the function of the root system in plants. The AMF colonized the inner root area as well as the soil around the root surface, thus uptake the nutrients from the zones that are far from the roots and root hairs of a plant, because of the ability of hyphae to grow several centimeters in the soil (Jansa et al., 2005). The increased uptake and translocation of Phosphorous up to 40 cm via fungal hyphae have been reported (Timonen et al., 1996). Some mycelium growing pattern is very dense and only extends to few millimeters in the soil (Smith et al., 2004). Through hyphae and mycelium nutrient's uptake has been improved in plants. The ECM fungi also play an important role in uptake of water from the soil (Allen, 2007). This colonization of roots also has an impact on the interaction of the plant with soil-borne pathogens because of several reasons like plant nutrition level changed, struggle between fungi and pathogens for resources, and changed microbial communities in the rhizosphere (Toljander et al., 2007). The mineralization of certain organic nutrients and bio-weathering of inorganic nutrients by ECM from carbonates, micas, and apatite have also been reported (Blum et al., 2002). Other soil-borne organisms also involve in the mineralization of the nutrients.

The direct involvement of AMF and other soil microbes in the mineralization of organic phosphorous under nonsterile soil situations are still not clear (Finlay, 2008). The AMF also help in the modification of soil-like aggregation, wettability, and biological activities in the root zone (Rillig and Mummey, 2006). Mycorrhizal fungi are heterotrophic and get organic carbon directly from their host plants required for growth, respiration, and biological maintenance. Some mycorrhizal fungi considered limited saprophytic growth (Koide et al., 2008). Mycorrhizal fungi dependence on the host plant for organic carbon based on environmental circumstances like the presence of nutrients, light, and CO_2, thus this association varies from beneficial to parasitic (Schroeder and Janos, 2004). The net acquisition of P is less by mycorrhizal association when the limited soil P is available because of low concentration of P or high sorption in weathered soil than the soil in which small amount of P is added (Bolan, 1991). When the P level is very low plants prefer mobilization of P and uptake rather than mycorrhizal association (Lamber et al., 2008). Similarly when the high amount of soluble P fertilizers was added in the soil then plant can directly gain P from the soil results in no mycorrhizal association (Sorensen et al., 2003). Low or high availability of P, plant species, soil condition, environmental factors have the negative effect on the growth of plant when the mycorrhizal association is observed (Morgan et al., 2005). Nutrient's uptake is limited when the availability of P is limited, but mycorrhizal association benefits the plant and improves plant growth by increasing nutrients uptake (Li et al., 2005). The mycorrhizal association is not always specifically beneficial for the plants and required proper isotopic labeling for better understanding of the mechanism. The negative impact of mycorrhizal association has been reported (Smith et al., 2009), but it is studied that the mycorrhizal association is beneficial for the plant growth mostly (Choi et al., 2005). The benefits and the degree of colonization of roots with different fungi mostly depend on relevant environmental circumstances rather than the accessibility of P (Hoeksema et al., 2010) (Fig. 2.1).

2.3 FORMS OF P AND ITS ACCESSIBILITY TO MYCORRHIZA

2.3.1 TYPES OF PHOSPHOROUS

Different types of phosphorous are present in the soil. The inorganic types of P differ in water solubility and by chemical reactivity and include calcium phosphates, potassium phosphates, iron phosphates, aluminum phosphates, crystalline apatites, inorganic polyphosphate, and orthophosphate (Dou et al., 2009). The phosphorous present in the soil is the component of many important compounds like nucleic acids, phospholipids, inositol phosphates, and other metabolic intermediates. Instead of so many different forms of phosphorous present in the soil, the only form uptake by hyphae and root is orthophosphate in the form of H_2PO_4 ions (Smith, 2002). Microorganisms utilized many different forms of phosphorous, but the enzymatic cleavage of Pi occurs in the close area of the microbial cell before the cross membrane uptake (Heath, 2005). The presence of inorganic phosphate in the soil solution is the main reason for direct or indirect uptake of P by the plants (Smith et al., 2004). The direct uptake consists of Pi acquisition from the soil solution by plant root hairs and indirect pathway involves uptake via hyphae, its translocation via extra radical mycelium, release of P within roots by hyphae, and uptake of these P by root cells (Table 2.1).

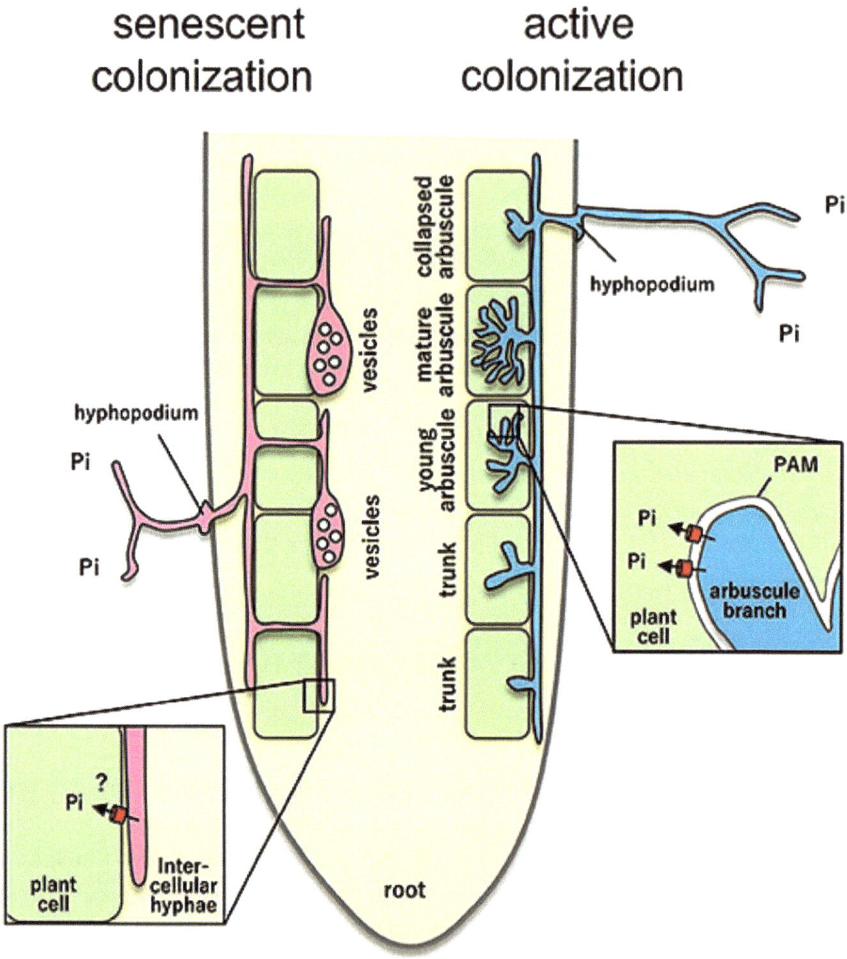

FIGURE 2.1 Lifecycle of intracellularly (Paris-type; not shown) in cortical cells. In these processes, arbuscules that are involved in nutrient exchange are formed in cortical cells in order from the initial penetration site (hyphopodium). Therefore, the arbuscules that are closest to the penetration site are the oldest, and the arbuscule close to the tip of the growing hyphae is the newest. Since the arbuscules have short lifespan (1–2 days), they collapse in order from the oldest one. Arbuscules are surrounded by plant membrane (periarbuscular membrane). Plant phosphate transporter proteins are localized on this membrane and are responsible for the uptake of phosphate from the arbuscules. In the senescent colonization stages of infection unit, vesicles and intercellular hyphae are often observed. The role of these intraradical AMF structures is still unclear.

Source: Reprinted from Yoshihiro et al. (2019), https://creativecommons.org/licenses/by/4.0/

TABLE 2.1 Example of Plant Growth-Promoting Substances Released by Phosphate-Solubilizing Bacteria.

Phosphate-solubilizing bacteria	Plant growth-promoting substances	Reference
Pseudomonas fluorescens, *P. putida*	IAA, siderophore, ACC deaminase	Zabihi et al. (2011)
Serratia nematodiphila	IAA, siderophore, HCN	Dastager et al. (2011a)
Pontibacter niistensis	IAA, HCN, ACC deaminase, and siderophore	Dastager et al. (2011b)
Klebsiella spp.	IAA, siderophore, HCN, ammonia, EPS	Ahemad and Khan (2011)
Pantoea agglomerans	IAA	Mishra et al. (2011)
Arthrobacter, *Bacillus*	IAA, antifungal activity, HCN, NH_3	Banerjee et al. (2010)
Pantoea	IAA, siderophore, antifungal activity	Taurian et al. (2010)
Pseudomonas aeruginosa	IAA, siderophore, antifungal activity, HCN, EPS	Ahemad and Khan (2010)
P. mendocina, *P. stutzeri* and *P. putida*	IAA, gibberellic acid, trans-zeatin riboside, and abscisic acid	Naz and Bano (2010)
Enterobacter aerogenes, *E. cloacae*, *E. asburiae*	IAA, siderophore, HCN	Deepa et al. (2010)
P. alvei	IAA	Hassen and Labuschagne (2010)
Pseudomonas sp.	ACC deaminase, IAA, siderophore	Poonguzhali et al. (2008)
Serratia marcescens	IAA, siderophore, HCN	Selvakumar et al. (2008)
Acinetobacter sp., *Pseudomonas* sp.	ACC deaminase, IAA, antifungal activity, N_2-fixation	Indiragandhi et al. (2008)
Enterobacter sp.	ACC deaminase, IAA, siderophore	Kumar et al. (2008)
Burkholderia	ACC deaminase, IAA, siderophore	Jiang et al. (2008)
Pseudomonas jessenii	ACC deaminase, IAA, siderophore	Rajkumar and Freitas (2008)

Source: Reproduced with permission from Mohammad et al. (2013), Copyright © Springer-Verlag (2013).

2.3.2 *KINETICS OF ACQUISITION OF P BY HYPHAE*

The phosphorous uptake from the soil is mediated by Pi transporters (Tatry et al., 2009). Pi transporters consist of large-sized protein molecules and transmembrane domains. These transporters are the reason for the symport of proton and sodium of phosphate against the electrochemical gradient across the membranes (Smith, 2002). Mostly these transporters are reported in AMF, but recently two transporters are also reported in ECM (Tatry et al., 2009). From the cDNA libraries, the first Pi transporter was recognized by using hybridization of a probe derivative from the Pi transporter of yeast. The expression of Pi transporter of (pho84)-mutant Pi transporter in yeast shows Michaelis-Menten kinetic with the maximum Km value of 18 mM which is one or else two orders higher in magnitude than the estimated value of P transporters of AMF (Smith et al., 2001). This difference is because of heterologous expression of the gene (Smith et al., 2001). But current studies show that the Km values of two Pi transporters of ECM are 4 and 55 mM which are closer to the estimated values of hydroponic experiment (Tatry et al., 2009). The Pi transporters that are considered high-infinity transporter family play an important role for the uptake of Pi from the soil when the concentration of Pi is less than 10 mM in soil solution (Harrison, 1999). The Pi transporters of ECM and AM fungi are with the high rate of structural conservation and other similarities with the other organism's Pi transporters (Karandashov and Bucher, 2005). The acquisition of high-infinity Pi is not clear yet, but it is estimated about 2–4 protons per molecule of P when the H^+-ATPase generated the proton gradient (Smith and Read, 2008). The other kinetics of uptake of Pi involves its per unit inflow of hyphal biomass. The interspecies variation of both ECM and AMF groups linked to the kinetic and expression parameters of various Pi transporters in the mycelium of fungi in the soil (Smith and Read, 2008). Further modulation of Pi uptake by hyphae depends on the difference of the proliferation of hyphae and dynamics of the hyphal network between fungal species and genotypes. The improvement in P uptake by fungal mycelium is because of several important characteristics like hyphae penetrating the smaller soil pores rather than plant roots. Moreover, because of the size the P-zone depletion around the hyphae is minimal. Mycorrhizal hyphae are very efficient in the soil penetration than the root of the plant and also the per unit carbon cost of hyphal surface is higher than root surface of plants (Gregory, 2006). The dynamics of hyphal network is not studied well. The branched-like and short-lived structures present in AMF are very important in phosphorous uptake from the soil for plants. These structures also have 7–35 days high turnover rates when the condition is axenic than the simple main threads of hyphae which remain in the cytoplasm for 3 months under same axenic environments (de Vries et al., 2009). According to the isotopic signature by carbon estimation, the turnover of AMF hyphae is only 5 to 6 days in *Glomus* species (Staddon et al., 2003). The outcomes are the same with the time span of fine and short-lived hyphae as well as similar with the estimated values of microscopy of soil hyphae which is 5 to 7 days. The hyphal turnover rate of ECM is less than the AMF. According to the carbon flux study, the lifespan of mycelium of ECM is 9 days, but for the rhizomorphs it is for several months (Godbold et al., 2006). The variation of hyphal growth and turnover rates is observed in different fungal taxon as well as within mycorrhizal types and this variation may be because of different strategies of acquisition of phosphorous in different mycorrhizal fungi. The mining of phosphorous is considered longer lived mycelium, but the scavenging of the phosphorous is only carried by mycorrhizal fungi with high rates of turnover, and this hypothesis is tested among the same and different mycorrhizal types (Fig. 2.2).

2.3.3 ACCESS TO RECALCITRANT MINERALIZATION AND WEATHERING FORMS OF P

There are different kinds of information available regarding the impact of various mycorrhizal fungi on the release of phosphorous from recalcitrant forms. The ECM fungi have the ability to discharge Pi from the apatite which is the poorly soluble form of phosphorous in the pot, pure cultures, and also in microcosm studies (Rosling, 2009). Pure culture experiments have been questioned because the high amount of acids is produced when high amount of organic carbon is added in the culture. The pot and microcosm culture also has some issues like the presence of other microorganisms which can affect the sorbed phosphate and can also produce exo-cellular phosphatases that also exist in the soil (Jones and Smith, 2004). The ECM has potential to uptake Pi which is released by microbes and then transfers it to plants. This can increase P uptake from recalcitrant sources even in the situation when mycorrhizal fungi are unable to solubilize the recalcitrant P. When fungi colonized the plant it increased the weathering of apatite up to three folds and also delivered 14°C to the plant. The mechanism behind the dissolution of apatite is enhancement of acidification and chelation of cations of calcium of apatite because of oxalic acid released from fungi. Calcium reacts with oxalic acid and forms calcium oxalate crystals on fungal hyphae which are in contact of apatite. The elevation in the concentration of oxalic acid in the soil, having the mycorrhizal network in the soil of forests, has been observed. It is supported by the fact that apatite grains when applied to the forest soil increased the colonization of mycorrhizal fungi (Turpault et al., 2009). Still the role of ECM in the increase of weathering of apatite is limited. Many studies revealed that under the acidic soil condition preclude

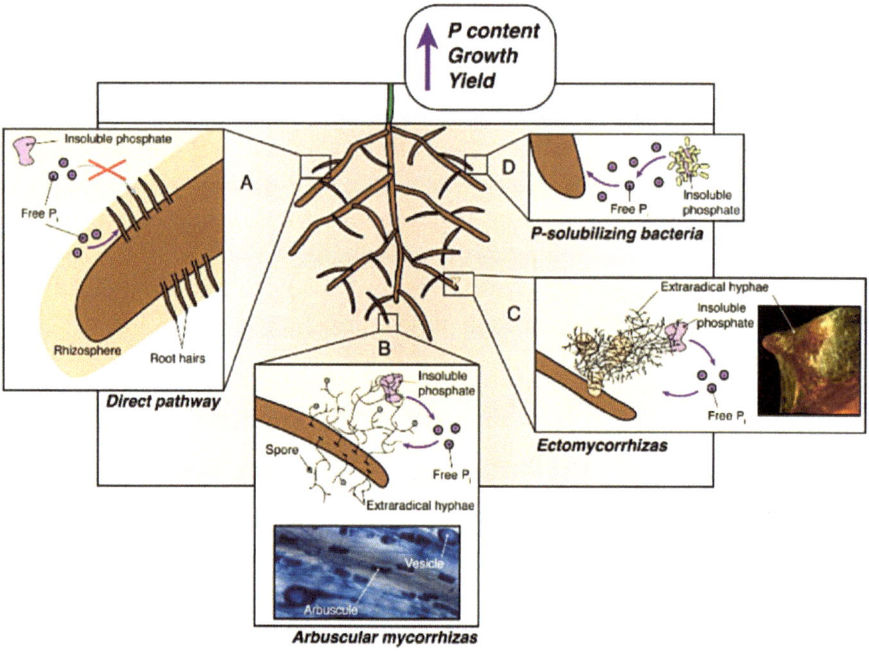

FIGURE 2.2 Schematic model showing the acquisition of phosphorus in plants, facilitated by arbuscular mycorrhizal fungi, ectomycorrhizal fungi, and P-solubilizing bacteria. (A) Root hairs on the roots of plants are able to acquire inorganic phosphate (Pi) from soil that is within close proximity (e.g., within the rhizosphere). This method of Pi acquisition is known as the direct uptake pathway. Insoluble forms of phosphate and free Pi that are outside of the rhizosphere cannot be reached by the plant (red cross). (B) Spores of arbuscular mycorrhizal arbuscules and sometimes vesicles in plant cortical cells. Extraradical hyphae exploring the soil can extract Pi from insoluble forms of phosphate (e.g., minerals and organic matter) and acquire it. Pi is then transported through the hyphae colonizing the roots and delivered to plant cells. (C) Specific plant–fungal structures called ectomycorrhizas can be formed on woody plant roots. As with arbuscular mycorrhizas, extraradical hyphae exploring the soil extract Pi from insoluble forms and acquire it. These hyphae can sometimes form rhizomorphs that are highly differentiated structures (not represented here). Pi is transported toward the roots and then delivered to plant cells. (D) Insoluble forms of phosphate are actively degraded by P-solubilizing bacteria. The resulting free Pi can either be taken up by the roots or other organisms, including mycorrhizal fungi. The improvement of P acquisition through direct or indirect microbial pathways leads to the increase of P content in aboveground tissues (e.g., seeds and fruits), thus increasing plant growth and yield.

Source: Reprinted from Arjun et al. (2019), https://creativecommons.org/licenses/by/4.0/

the role of mycorrhiza in the increased solubilization of apatite by acidification of the mycorhizosphere. No doubt under lab condition ECM plays an important role in releasing P from the minerals having phosphorous in it, but to what extent ECM plays a role in the weathering of apatite in the field is still required further explanation (Rosling et al., 2009).

For the better understanding of ECM fungi in the mobilization of phosphorous from organic substrate different strategies, ranging from axenic system to the microcosm, have been applied. Different kinds of enzymes that are used to degrade the polyphenols are produced by ECM. The experiment performed by applying pollen grains or dead nematodes in the microcosm which includes mycorrhizal as well as nonmycorrhizal seedlings shows more than 96% of the phosphorous present in the pollen applied to the mycorrhizal microcosm was removed and also 25% on average transferred to the seedling of mycorrhiza. On the other hand, only 25% of the phosphorous present in the pollen were removed in the nonmycorrhizal microcosm and only 7% transferred to the nonmycorrhizal plants explained the role of ECM in the uptake of P present in the organic substrate. The overall role of ECM mycelium for mobilization of nutrients has been studied. The enzymatic activity also studied for the mobilization of the nutrient's results in the increase in the activity of phosphor-mono-esterases in the first 28 to 50 days of colonization, but their activity decreased between 50 and 90 days, but it is also difficult to differentiate mycorrhizal and saprotrophic fungi.

The activity of inositol hexaphosphate has increased at the surface of mycorrhizal colonized roots for the uptake of phosphorous, but the role of roots or mycorrhizal roots in the hydrolysis of inositol hexaphosphate is not clear yet. The uptake of phosphorous has been improved when colonized by mycorrhizal fungi because of organic phosphorous recycling between soil and plants through hyphae and secondly because of the effect of ECM on the recycling of P (Tibbett and Sanders, 2002). The AMF hydrolyzed the organic phosphorous and then transferred it to the plant through its mycelium. The process behind the AMF role in mineralization of organic P in unsterile soil condition is not clear (Joner and Johansen, 2000). AMF shows the lower level of phosphatase activity than ECM fungi. The phosphatases of fungi are more efficient in the mobilization of P from lecithin and phytate than plant enzymes (Tarafdar et al., 2001). The saprotrophic nature of ECM to interfere with the microbial mobilization and immobilization cycles for the requisition of N and P from different substrates formed by decomposition of microbial, faunal, and remnants of plants have been clear from different studies.

The action of AMF is not much clear as it is difficult to differentiate between the physiological activities of other fungi and microbes which may be involved with AMF. The AMF enhanced the decomposition and nitrogen uptake from organic matter, but the role of other soil saprotrophs was not clear (Hodge et al., 2001). Further studies are required to understand the potential role of bacteria associated with bacteria. Through several studies on the role of hyphae of mycorrhizal fungi has been confirmed, but how this association helps in the acquisition and transfer of nutrients in the rhizosphere is not sufficient.

2.4 THE ROLE OF MYCORRHIZA AS FILTERS

The role of mycorrhiza for the acquisition of phosphorous under various environmental conditions of different ecosystems with the various host plants has been confirmed through several studies (Jansa et al., 2009). However, mycorrhiza not only play a role in the uptake of phosphorous but also are beneficial for the uptake of other nutrients like nitrogen from all kinds of organic or inorganic substrates (Allen, 2007). Mycorrhiza also helps in the uptake of water from the soil as well as to alleviate the deficiencies of other important micronutrients like zinc, copper, etc., mycorrhizal association also prevents the host plant from the uptake of harmful compounds and pollutants like radiocaesium (Ladeyn et al., 2008). Mycorrhizal fungi also prevent the plants from heavy metals while balancing the supply of P and Zn to the plants (Soares and Siqueira, 2008).

2.5 TRANSPORT OF PHOSPHOROUS IN HYPHAE

To meet the physiological demand of P of fungal part, phosphorous uptake from the soil and the remnants of P may be accumulated in the hyphae or translocated to the plant host. The translocation of the extra phosphorous to the plant part consists of the long-distance pathway via hyphal network and involves polyphosphates (Ezawa et al., 2002). The long-chain polyphosphates are depolymerized and converted into short-chain polyphosphates that are the immediate phosphorous source to the plants (Takanishi et al., 2009). To determine the transfer rate of P to the plant, different techniques like radioisotopy, labeling, and microscopy can be used (Nielsen et al., 2002). Discharge of phosphorous from recalcitrant sources and then translocation to the plant explained the role of mycorrhiza in soil plant cycling of phosphorous by avoiding the release of P from organic substrates by soil microbes (Johnson et al., 2005).

2.6 THE TRANSFER OF P TO THE PLANT

There is no pet knowledge known about the transfer of phosphorous to the plant from mycorrhiza. Also, it is not clear that the transfer of P is via active and passive transport mechanism. Some fungi retain the extra P in their mycelium and thus allow partial immobilization of P during its translocation resulting in a low transfer rate within the hyphae or limited release to the plant ultimately slowing the transfer of P. The death of hyphae caused by disturbance in soil, parasitic activities, and grazing leads to release of phosphorous back to the soil microorganisms from mycorrhizal hyphae. Under the changed environmental condition the partial storage of P is beneficial as it can act as buffer and ensure the continuous supply of P to the plant (Genet et al., 2000). Thus, the rapid transfer of P will not always be beneficial for the plant. The Pi transporters induced by mycorrhiza are expressing themselves in the close vicinity hyphal network and arbuscules already studied and reported in many plant species having AM association (Javot et al., 2007).

2.7 CONSEQUENCES OF PHOSPHOROUS UPTAKE BY MYCORRHIZA AND MAINTENANCE OF SYMBIOTIC ASSOCIATION

The mycorrhizal efficiency for P uptake, partial immobilization of P in the hypha of mycelium, and the controlled release of P to the plant are the main reasons for growth and phosphorous nutrition of the plant. Certain AMF form the mycorrhizal association with the plant and hence improved the phosphorous uptake, but some fungal species shows a negative effect and thus the association may be parasitic (Smith et al., 2004). The controlled phosphoric discharge to the plant from fungal part is because of sustaining the diversity and symbiotic association of mycobiont by the modification of the transfer of C from the plant to fungal partner (Cowden and Peterson, 2009). When roots of plants are colonized by several fungal species, the potential transfer of carbon is reported, which results in more benefit of this mutualistic association (Bever et al., 2009). The transfer of carbon and P is correlated. Under sufficient nitrogen and phosphorous conditions, the AMF obtained the carbon from the plant as the uptake of P in such a condition was limited

(Hoeksema et al., 2010). Also, the presence of myco-heterotrophic plants that obtained carbon as well as mineral nutrition from the fungal partner is difficult to meet the hypothesis about the preferential distribution of carbon to the fungal species in mycorrhizal association. It is also noted that many processes are not regulated by a single-plant species only, but regulated at the level of the ecosystem. So the answer of this hypothesis is depending on the ecology of plant community, the mycorrhizal network, and source–sink relationship (Bever et al., 2009). Limited information is available about the transfer of Phosphorous and C among the plants mediated by mycorrhiza (Philip and Simard, 2008). Experimental data shows the ordinary role of mycorrhizal fungi in phosphorous uptake by plants when growth of plant is not detectable and also the improvement of P uptake is not noted (Smith et al., 2003, 2004). These results are only noticeable when studied through proper radioisotope labeling of nonmycorrhizal mutant genotypes and by analyzing the transporters of both the plant and fungal species (Grace et al., 2009). The result of different studies shows that the mycorrhizal association is functional and beneficial even though the net uptake and plant growth was not noticeable (Facelli et al., 2009).

2.8 MYCORRHIZAL FUNCTIONAL DIVERSITY FOR P UPTAKE

The pattern and efficiency of P uptake is varied among different species and types of mycorrhizal association just like their functional differences (Cavagnaro et al., 2005). To study this most important mechanism is radioisotope labeling. To comprehend the function of mycorrhiza in the field it is compulsory to understand and study the functional diversity in various mycorrhizal fungi. Because the roots of plants can be colonized by many fungi at the same time, it may differ in their types. At the same time when different mycorrhizal fungal species colonize a plant, they complement each other in the uptake of nutrients leads a more beneficial symbiotic association than a plant colonized by a single mycorrhizal fungi (Koide, 2000). The functional complementarity in the uptake of P in mycorrhizal communities is not clear yet through experiments (Jansa et al., 2008). For further understanding, the components of uptake of nutrients via the mycorrhizal pathway structure of functional diversity of different mycorrhizal fungi need to be explored further. This required the proper use of different processes that include molecular biology, isotope labeling, pot experiments, and modeling (Jansa et al., 2005). To understand the whole picture, the interaction between different mycorrhizal fungi, P uptake, costs of carbon, cycling of other important nutrients like N at the community level is required (Smith et al., 2008).

2.9 CONCLUSION

The mutualistic relationship of plant roots with fungi under certain circumstances is beneficial for both partners. Plants when growing in highly weathered and severely phosphorous needed soil are not in the mutualistic association, they fulfill their P requirements by clustering of roots. The phosphorous present in the soil is the component of many important compounds. Instead of so many different forms of phosphorous present in the soil, the only form uptake by hyphae and root is orthophosphate in the form of H_2PO_4 ions. The mycorrhizal association improved the nutrient's uptake, especially the elements having low mobility like P, Zn, Cu, etc., and also helped the plants to cope up the stresses and thus help plants to grow better. Discharge of phosphorous from recalcitrant sources and then translocation to the plant explained the role of mycorrhiza in soil plant cycling of phosphorous by avoiding the release of P from organic substrates by soil microbes. There is no pet knowledge known about the transfer of phosphorous to the plant from mycorrhiza. Also, it is not clear that the transfer of P is via active and passive transport mechanism. The result of different studies shows that the mycorrhizal association is functional and beneficial even though the net uptake and plant growth was not noticeable. For the better understanding the whole mechanism involved in the P acquisition and absorption still required knowledge about the interaction between different mycorrhizal fungi, P uptake, costs of carbon, cycling of other important nutrients like N at the community level.

KEYWORDS

- **AMF**
- **ECM**
- **nutrient uptake**
- **phosphorus acquisition**
- **phosphorus metabolism**
- **types of phosphorus**

REFERENCES

Adams, F.; Reddell, P.; Webb, M. J.; Shipton, W. A. Arbuscular mycorrhizas and ectomycorrhizas on *Eucalyptus grandis* (Myrtaceae) trees and seedlings in native forests of tropical northeastern Australia. *Aust. J. Bot.* **2006**, *54*, 271–281.

Ahemad, M.; Khan, M. S. Phosphate-solubilizing and plant-growth-promoting *Pseudomonas aeruginosa* PS1 improves green gram performance in quizalafop-p-ethyl and clodinafop amended soil. *Arch. Environ. Contam. Toxicol.* **2010**, *58*, 361–372.

Ahemad, M.; Khan, M. S. Toxicological effects of selective herbicides on plant growth promoting activities of phosphate solubilizing *Klebsiella* sp. strain PS19. *Curr. Microbiol.* **2011**, *62*, 532–538.

Allen, M. F. Mycorrhizal fungi: Highways for water and nutrients in arid soils. *Vadose Zone J.* **2007**, *6*, 291–297.

Arjun, K.; Kevin, R. C.; Rachel, R.; Jaya, K. Y. et al. Harnessing soil microbes to improve plant phosphate efficiency in cropping systems. *Agronomy* **2019**, *9*(3), 127. doi: 10.3390/agronomy9030127.

Banerjee, S.; Palit, R.; Sengupta, C.; Standing, D. Stress induced phosphate solubilization by *Arthrobacter* sp. and *Bacillus* sp. isolated from tomato rhizosphere. *Australian Journal of Crop Science* **2010**, *4*, 378–383.

Bever, J. D.; Richardson, S. C.; Lawrence, B. M.; Holmes, J.; Watson, M. Preferential allocation to beneficial symbiont with spatial structure maintains mycorrhizal mutualism. *Ecol. Lett.* **2009**, *12*, 13–21.

Blum, J. D.; Klaue, A.; Nezat, C. A.; Driscoll, C. T.; Johnson, C. E.; Siccama, T. G.; Eagar, C.; Fahey, T. J.; Likens, G. E. Mycorrhizal weathering of apatite as an important calcium source in base-poor forest ecosystems. *Nature* **2002**, *417*, 729–731.

Bolan, N. S. A. Critical review on the role of mycorrhizal fungi in the uptake of phosphorus by plants. *Plant Soil* **1991**, *134*, 189–207.

Burke, D. J.; Martin, K. J.; Rygiewicz, P. T.; Topa, M. A. Ectomycorrhizal fungi identification in single and pooled root samples: Terminal restriction fragment length polymorphism (T-RFLP) and morphotyping compared. *Soil Biol. Biochem.* **2005**, *37*, 1683–1694.

Castillo, C. G.; Borie, F.; Godoy, R.; Rubio, R.; Sieverding, E. Diversity of mycorrhizal plant species and arbuscular mycorrhizal fungi in evergreen forest, deciduous forest and grassland ecosystems of Southern Chile. *J. Appl. Bot. Food Qual.* **2006**, *80*, 40–47.

Cavagnaro, T. R.; Smith, F. A.; Smith, S. E.; Jakobsen, I. Functional diversity in arbuscular mycorrhizas: Exploitation of soil patches with different phosphate enrichment differs among fungal species. *Plant Cell Environ.* **2005**, *28*, 642–650.

Choi, D. S.; Quoreshi, A. M.; Maruyama, Y.; Jin, H. O.; Koike, T. Effect of ectomycorrhizal infection on growth and photosynthetic characteristics of *Pinus densiflora* seedlings grown under elevated CO_2 concentrations. *Photosynthetica* **2005**, *43*, 223–229.

Cowden, C. C.; Peterson, C. J. A. Multi-mutualist simulation: Applying biological market models to diverse mycorrhizal communities. *Ecol. Modell.* **2009**, *220*, 1522–1533.

Dastager, S. G.; Deepa, C. K.; Pandey, A. Potential plant growth-promoting activity of *Serratia nematodiphila* NII-0928 on black pepper (*Piper nigrum* L.). *World J Microbiol Biotechnol.* **2011a**, *27, 259–265.*

Dastager, S. G.; Deepa, C. K.; Pandey, A. Plant growth promoting potential of *Pontibacter niistensis* in cowpea (*Vigna unguiculata* (L.) Walp.). *Appl Soil Ecol.* **2011b**, *49*, 250–255.

de Vries, F. T.; Baath, E.; Kuyper, T. W.; Bloem, J. High turnover of fungal hyphae in incubation experiments. *FEMS Microbiol. Ecol.* **2009**, *67*, 389–396.

Deepa, C. K.; Dastager, S. G.; Pandey, A. Isolation and characterization of plant growth promoting bacteria from non-rhizospheric soil and their effect on cowpea (*Vigna unguiculata* (L.) Walp.) seedling growth. *World J. Microbiol,. Biotechnol.* **2010**, *26*, 1233–1240.

Dou, Z. X.; Ramberg, C. F.; Toth, J. D.; Wang, Y.; Sharpley, A. N.; Boyd, S. E. et al. Phosphorus speciation and sorption-desorption characteristics in heavily manured soils. *Soil Sci. Soc. Am. J.* **2009**, *73*, 93–101.

Ezawa, T.; Smith, S. E.; Smith, F. A. P. Metabolism and transport in AM fungi. *Plant Soil.* **2002**, *244*, 221–230.

Facelli, E.; Smith, S. E.; Smith, F. A. Mycorrhizal symbiosis — overview and new insights into roles of arbuscular mycorrhizas in agro- and natural ecosystems. *Australas. Plant Pathol.* **2009**, *38*, 338–344.

Faeth, S. H.; Fagan, W. F. Fungal endophytes: Common host plant symbionts but uncommon mutualists. *Integr. Comp. Biol.* **2002**, *42*, 360–368.

Finlay, R. D. Ecological aspects of mycorrhizal symbiosis: With special emphasis on the functional diversity of interactions involving the extraradical mycelium. *J. Exp. Bot.* **2008**, *59*, 1115–1126.

Genet, P.; Prevost, A.; Pargney, J. C. Seasonal variations of symbiotic ultrastructure and relationships of two natural ectomycorrhizae of beech (*Fagus sylvatica/Lactarius blennius* var. *viridis* and *Fagus sylvatica/Lactarius subdulcis*). *Trees Struct. Funct.* **2000**, *14*, 465–474.

Godbold, D. L.; Hoosbeek, M. R.; Lukac, M.; Cotrufo, M. F.; Janssens, I. A.; Ceulemans, R. et al. Mycorrhizal hyphal turnover as a dominant process for carbon input into soil organic matter. *Plant Soil* **2006**, *281*, 15–24.

Grace, E. J.; Cotsaftis, O.; Tester, M.; Smith, F. A.; Smith, S. E. Arbuscular mycorrhizal inhibition of growth in barley cannot be attributed to extent of colonization, fungal phosphorus uptake or effects on expression of plant phosphate transporter genes. *New Phytol.* **2009**, *181*, 938–949.

Gregory, P. J. *Plant roots: Growth, activity and interaction with soils*; Blackwell Publishing Ltd: Oxford, **2006**.

Grelet, G. A.; Johnson, D.; Paterson, E.; Anderson, I. C.; Alexander, I. J. Reciprocal carbon and nitrogen transfer between an ericaceous dwarf shrub and fungi isolated from *Piceirhiza bicolorata* ectomycorrhizas. *New Phytol.* **2009**, *182*, 359–366.

Harrison, M. J. Biotrophic interfaces and nutrient transport in plant fungal symbioses. *J. Exp. Bot.* **1999**, *50*, 1013–1022.

Hassen, A. I.; Labuschagne, N. Root colonization and growth enhancement in wheat and tomato by rhizobacteria isolated from the rhizoplane of grasses. *World J. Microbiol. Biotechnol.* **2010**, *26*, 1837–1846.

Heath, R. T. Microbial turnover of organic phosphorus in aquatic systems. In: *Organic phosphorus in the environment*; Turner, B. L., Frossard, E., Baldwin, D. S., Eds.; CABI: Wallingford, **2005**; pp 185–204.

Hodge, A.; Campbell, C. D.; Fitter, A. H. An arbuscular mycorrhizal fungus accelerates decomposition and acquires nitrogen directly from organic material. *Nature* **2001**, *413*, 297–299.

Hoeksema, J. D.; Chaudhary, V. B.; Gehring, C. A.; Johnson, N. C.; Karst, J.; Koide, R. T.; Pringle, A.; Zabinski, C.; Bever, J. D.; Moore, J. C.; Wilson, G. W. T.; Klironomos, J. N.;

Umbanhowar, J. A. Meta-analysis of context-dependency in plant response to inoculation with mycorrhizal fungi. *Ecol Lett.* **2010**, *13*, 394–407.
Indiragandhi, P.; Anandham, R.; Madhaiyan, M.; Sa, T. M. Characterization of plant growth promoting traits of bacteria isolated from larval guts of diamondback moth *Plutella xylostella* (Lepidoptera: Plutellidae). *Curr. Microbiol.* **2008**, *56*, 327–333.
Jansa, J.; Gryndler, M. Biotic environment of the arbuscular mycorrhizal fungi in soil. In: *Arbuscular mycorrhizas: Physiology and function*; Koltai, H., Kapulnik, Y., Eds.; Springer: Heidelberg, **2010**; pp 209–236.
Jansa, J.; Mozafar, A.; Frossard, E. Phosphorus acquisition strategies within arbuscular mycorrhizal fungal community of a single field site. *Plant Soil* **2005**, *276*, 163–176.
Jansa, J.; Oberholzer, H. R.; Egli, S. Environmental determinants of the arbuscular mycorrhizal fungal infectivity of Swiss agricultural soils. *Eur. J. Soil Biol.* **2009**, *45*, 400–408.
Jansa, J.; Smith, F. A.; Smith, S. E. Are there benefits of simultaneous root colonization by different arbuscular mycorrhizal fungi? *New Phytol.* **2008**, *177*, 779–789.
Javot, H.; Penmetsa, R. V.; Terzaghi, N.; Cook, D. R.; Harrison, M. J. A. *Medicago truncatula* phosphate transporter indispensable for the arbuscular mycorrhizal symbiosis. *Proc. Natl. Acad. Sci. USA* **2007**, *104*, 1720–1725.
Jiang, C.; Sheng, X.; Qian, M.; Wang, Q. Isolation and characterization of a heavy metal resistant *Burkholderia* sp. from heavy metal-contaminated paddy field soil and its potential in promoting plant growth and heavy metal accumulation in metal-polluted soil. *Chemosphere* **2008**, *72*, 157–164.
Johnson, D.; Ijdo, M.; Genney, D. R.; Anderson, I. C.; Alexander, I. J. How do plants regulate the function, community structure, and diversity of mycorrhizal fungi? *J. Exp. Bot.* **2005**, *56*, 1751–1760.
Joner, E.; Johansen, A. Phosphatase activity of external hyphae of two arbuscular mycorrhizal fungi. *Mycol. Res.* **2000**, *104*, 81–86.
Jones, M. D.; Smith, S. E. Exploring functional definitions of mycorrhizas: Are mycorrhizas always mutualisms? *Can. J. Bot.* **2004**, *82*, 1089–1109.
Karandashov, V.; Bucher, M. Symbiotic phosphate transport in arbuscular mycorrhizas. *Trends Plant Sci.* **2005**, *10*, 22–29.
Koide, R. T. Functional complementarity in the arbuscular mycorrhizal symbiosis. *New Phytol. 147*, 233–235.
Koide, R. T.; Sharda, J. N.; Herr, J. R.; Malcolm, G. M. Ectomycorrhizal fungi and the biotrophy–saprotrophy continuum. *New Phytol.* **2008**, *178*, 230–233.
Kumar, K. V.; Singh, N.; Behl, H. M.; Srivastava, S. Influence of plant growth promoting bacteria and its mutant on heavy metal toxicity in *Brassica juncea* grown in fly ash amended soil. *Chemosphere* **2008**, *72*, 678–683.
Ladeyn, I.; Plassard, C.; Staunton, S. Mycorrhizal association of maritime pine, *Pinus pinaster*, with *Rhizopogon roseolus* has contrasting effects on the uptake from soil and root-to-shoot transfer of ^{137}Cs, ^{85}Sr and ^{95m}Tc. *J. Environ. Radioactiv.* **2008**, *99*, 853–863.
Lambers, H, Raven, J. A.; Shaver, G. R.; Smith, S. E. Plant nutrient-acquisition strategies change with soil age. *Trends Ecol. Evol.* **2008**, *23*, 95–103.
Li, H. Y.; Zhu, Y. G.; Marschner, P.; Smith, F. A.; Smith, S. E. Wheat responses to arbuscular mycorrhizal fungi in a highly calcareous soil differ from those of clover, and change with plant development and P supply. *Plant Soil* **2005**, *277*, 221–232.

Merryweather, J.; Fitter, A. Phosphorus and carbon budgets—mycorrhizal contribution in *Hyacinthoides non-scripta* (L.) Chouard ex Rothm under natural conditions. *New Phytol.* **1995**, *129*, 619–627.
Mishra, A.; Chauhan, P. S.; Chaudhry, V.; Tripathi, M.; Nautiyal, C. S. Rhizosphere competent *Pantoea agglomerans* enhances maize (*Zea mays*) and chickpea (*Cicer arietinum* L.) growth, without altering the rhizosphere functional diversity. *Antonie van Leeuwenhoek.* **2011**, *100*, 405–413.
Mohammad, S. K.; Ees, A.; Almas, Z.; Mohammad, O. Functional aspect of phosphate-solubilizing bacteria: Importance in crop production. In: *Bacteria in agrobiology: Crop productivity*; Maheshwari, D. K., Meenu, S., Abhinav, A., Eds.; Springer-Verlag: Berlin, Heidelberg, 2013; p 507.
Morgan, J. A. W.; Bending, G. D.; White, P. J. Biological costs and benefits to plant–microbe interactions in the rhizosphere. *J. Exp. Bot.* **2005**, *56*, 1729–1739.
Naz, I.; Bano, A. Biochemical, molecular characterization and growth promoting effects of phosphate solubilizing *Pseudomonas* sp. isolated from weeds grown in salt range of Pakistan. *Plant Soil* **2010**, *334*, 199–207.
Newsham, K. K.; Upson, R.; Read, D. J. Mycorrhizas and dark septate root endophytes in polar regions. *Fungal Ecol.* **2009**, *2*, 10–20.
Nielsen, J. S.; Joner, E. J.; Declerck, S.; Olsson, S.; Jakobsen, I. Phospho-imaging as a tool for visualization and noninvasive measurement of P transport dynamics in arbuscular mycorrhizas. *New Phytol.* **2002**, *154*, 809–819.
Opik, M.; Metsis, M.; Daniell, T. J.; Zobel, M.; Moora, M. Large-scale parallel 454 sequencing reveals host ecological group specificity of arbuscular mycorrhizal fungi in a boreonemoral forest. *New Phytol.* **2009**, *184*, 424–437.
Philip, L. J.; Simard, S. W. Minimum pulses of stable and radioactive carbon isotopes to detect belowground carbon transfer between plants. *Plant Soil* **2008**, *308*, 23–35.
Poonguzhali, S.; Madhaiyan, M.; Sa, T. Isolation and identification of phosphate solubilizing bacteria from chinese cabbage and their effect on growth and phosphorus utilization of plants. *J. Microbiol. Biotechnol.* **2008**, *18*, 773–777.
Rajkumar, M.; Freitas, H. Influence of metal resistant-plant growth-promoting bacteria on the growth of *Ricinus communis* in soil contaminated with heavy metals. *Chemosphere* **2008**, *71*, 834–842.
Redecker, D.; Kodner, R.; Graham, L. E. Glomalean fungi from the Ordovician. *Science* **2000**, *289*, 1920–1921.
Rillig, M. C.; Mummey, D. L. Mycorrhizas and soil structure. *New Phytol.* **2006**, *171*, 41–53.
Rosling, A. Trees, mycorrhiza and minerals—field relevance of in vitro experiments. *Geomicrobiol. J.* **2009**, *26*, 389–401.
Rosling, A.; Roose, T.; Hermann, A. M.; Davidson, F. A.; Finlay, R. D.; Gadd, G. M. Approaches to modelling mineral weathering by fungi. *Fungal Biol. Rev.* **2009**, *23*, 138–144.
Schroeder, M. S.; Janos, D. P. Phosphorus and intraspecific density alter plant responses to arbuscular mycorrhizas. *Plant Soil* **2004**, *264*, 335–348.
Selvakumar, G.; Mohan, M.; Kundu, S.; Gupta, A. D.; Joshi, P.; Nazim, S.; Gupta, H. S. Cold tolerance and plant growth promotion potential of *Serratia marcescens* strain SRM (MTCC 8708) isolated from flowers of summer squash (*Cucurbita pepo*). *Lett. Appl. Microbiol.* **2008**, *46*, 171–175.
Serrasolses, I.; Romanya, J.; Khanna, P. K. Effects of heating and autoclaving on sorption and desorption of phosphorus in some forest soils. *Biol. Fertil. Soils* **2008**, *44*, 1063–1072.

Sikes, B. A.; Cottenie, K.; Klironomos, J. N. Plant and fungal identity determines pathogen protection of plant roots by arbuscular mycorrhizas. *J. Ecol.* **2009**, *97*, 1274–1280.
Smith, F. A.; Grace, E. J.; Smith, S. E. More than a carbon economy: Nutrient trade and ecological sustainability in facultative arbuscular mycorrhizal symbioses. *New Phytol.* **2009**, *182*, 347–358.
Smith, F. W. The phosphate uptake mechanism. *Plant Soil* **2002**, *245*, 105–114.
Smith, S. E.; Read, D. J. *Mycorrhizal symbiosis*. Academic Press: New York, 2008.
Smith, S. E.; Dickson, S.; Smith, F. A. Nutrient transfer in arbuscular mycorrhizas: How are fungal and plant processes integrated? *Aust. J. Plant Physiol.* **2001**, *28*, 683–694.
Smith, S. E.; Smith, F. A.; Jakobsen, I. Mycorrhizal fungi can dominate phosphate supply to plants irrespective of growth responses. *Plant Physiol.* **2003**, *133*, 16–20.
Smith, S. E.; Smith, F. A.; Jakobsen, I. Functional diversity in arbuscular mycorrhizal (AM) symbioses: The contribution of the mycorrhizal P uptake pathway is not correlated with mycorrhizal responses in growth or total P uptake. *New Phytol.* **2004**, *162*, 511–524.
Smits, M. M.; Bonneville, S.; Haward, S.; Leake, J. R. Ectomycorrhizal weathering, a matter of scale. *Mineral Mag.* **2008**, *72*, 135–138.
Soares, C. R. F. S.; Siqueira, J. O. Mycorrhiza and phosphate protection of tropical grass species against heavy metal toxicity in multi-contaminated soil. *Biol. Fertil. Soils.* **2008**, *44*, 833–841.
Sorensen, J. N.; Larsen, J.; Jakobsen, I. Management strategies for capturing the benefits of mycorrhizas in the production of field-grown vegetables. *Acta. Hortic.* **2003**, *627*, 65–71.
Staddon, P. L.; Ramsey, C. B.; Ostle, N.; Ineson, P.; Fitter, A. H. Rapid turnover of hyphae of mycorrhizal fungi determined by AMS microanalysis of 14°C. *Science* **2003**, *300*, 1138–1140.
Takanishi, I.; Ohtomo, R.; Hayatsu, M.; Saito, M. Short-chain polyphosphate in arbuscular mycorrhizal roots colonized by *Glomus* spp.: A possible phosphate pool for host plants. *Soil Biol. Biochem.* **2009**, *41*, 1571–1573.
Tarafdar, J. C.; Yadav, R. S.; Meena, S. C. Comparative efficiency of acid phosphatase originated from plant and fungal sources. *J. Plant Nutr. Soil Sci.* **2001**, *164*, 279–282.
Tatry, M. V.; Kassis, E. E.; Lambilliotte, R.; Corratge, C.; Van A. I.; Amenc, L. K.; Alary, R.; Zimmermann, S.; Sentenac, H.; Plassard, C. Two differentially regulated phosphate transporters from the symbiotic fungus *Hebeloma cylindrosporum* and phosphorus acquisition by ectomycorrhizal *Pinus pinaster*. *Plant J.* **2009**, *57*, 1092–1102.
Taurian, T.; Anzuay, M. S.; Angelini, J. G.; Tonelli, M. L.; Ludueña, L.; Pena, D.; Ibáñez, F.; Fabra, A. Phosphate-solubilizing peanut associated bacteria: Screening for plant growth-promoting activities. *Plant Soil* **2010**, *329*, 421–431.
Tibbett, M.; Sanders, F. E. Ectomycorrhizal symbiosis can enhance plant nutrition through improved access to discrete organic nutrient patches of high resource quality. *Ann. Bot.* **2002**, *89*, 783–789.
Timonen, S.; Finlay, R. D.; Olsson, S.; Soderstrom, B. Dynamics of phosphorus translocation in intact ectomycorrhizal systems: Non-destructive monitoring using a beta scanner. *FEMS Microbiol. Ecol.* **1996**, *19*, 171–180.
Toljander, J. F.; Lindahl, B. D.; Paul, L. R.; Elfstrand, M.; Finlay, R. D. Influence of arbuscular mycorrhizal mycelial exudates on soil bacterial growth and community structure. *FEMS Microbiol. Ecol.* **2007**, *61*, 295–304.
Turpault, M.-P.; Nys, C.; Calvaruso, C. Rhizosphere impact on the dissolution of test minerals in a forest ecosystem. *Geoderma* **2009**, *153*, 147–154.

Wang, B.; Qiu, Y. L. Phylogenetic distribution and evolution of mycorrhizas in land plants. *Mycorrhiza* **2006**, *16*, 299–363.
Yoshihiro, K.; Dynamic phosphate uptake in arbuscular mycorrhizal roots under field conditions. *Front. Environ. Sci.* **2019**, *6*, 156. https://doi.org/10.3389/fenvs.2018.00159.
Zabihi, H. R.; Savaghebi, G. R.; Khavazi, K.; Ganjali, A.; Miransari, M. *Pseudomonas* bacteria and phosphorous fertilization, affecting wheat (*Triticum aestivum* L.) yield and P uptake under greenhouse and field conditions. *Acta. Physiol. Plant.* **2011**, *33*, 145–152.

CHAPTER 3

MYCORRHIZA AS A BIOCONTROL AGENT

SAMIA KHANUM[1], ABDEL RAHMAN MOHAMMAD AL-TAWAHA[2], ABDEL RAZZAQ AL-TAWAHA[3], HIBA ALATRASH[4], ABDUR RAUF[5], ARUN KARNWAL[6], ABHIJIT DEY[7], NUJOUD ALIMAD[8], DEVARAJAN THANGADURAI[9], JEYABALAN SANGEETHA[10], SAHER ISLAM[11], IMRAN[12], AMANULLAH[12], SHAH KHALID[12], PALANI SARANRAJ[13], and ELIF GUNAL[14]

[1]*Department of Botany, University of the Punjab, Lahore, Pakistan*

[2]*Department of Biological Sciences, Al Hussein Bin Talal University, Maan, Jordan*

[3]*Department of Crop Science, Faculty of Agriculture, University Putra Malaysia, Serdang, Selangor, Malaysia*

[4]*General Commission for Scientific Agricultural Research, Syria*

[5]*Department of Chemistry, University of Swabi, Anbar, Khyber Pakhtunkhwa, Pakistan*

[6]*Department of Microbiology, School of Bioengineering and BioSciences, Lovely Professional University, Phagwara, Punjab, India*

[7]*Department of Life Sciences, Presidency University, Kolkata, India*

[8]*Damascus University, Faculty of Agriculture, Damascus, Syria*

[9]*Department of Botany, Karnatak University, Dharwad, Karnataka, India*

Mycorrhizal Technology: Managing Plant Stress and Mitigating Climate Change using Mycorrhizae.
Jeyabalan Sangeetha, Abdel Rahman Mohammad Al-Tawaha, & Devarajan Thangadurai (Eds.)

[10]*Department of Environmental Science, Central University of Kerala, Kasaragod, Kerala, India*

[11]*Institute of Biochemistry and Biotechnology, University of Veterinary and Animal Sciences, Lahore, Pakistan*

[12]*Department of Agronomy, The University of Agriculture, Peshawar, Pakistan*

[13]*Department of Microbiology, Sacred Heart College (Autonomous), Tirupattur, Tamil Nadu, India*

[14]*Department of Soil Science and Plant Nutrition, Faculty of Agriculture, Harran University, Sanliurfa, Turkey*

ABSTRACT

Biological control is the way to protect plants against pathogen by controlling factors that influence growth of plants. Plant roots interact with other microorganisms present in the soil or present on any substrate and hence create rhizosphere. The interaction between roots, soil, and microorganism is by chemical, biological, and sometimes physical. Commonly plants have an association with their roots known as mycorrhizae. Mycorrhiza is defined as a mutualistic association of plants with soil-borne fungi. Arbuscular mycorrhiza is considered more common type of such an association. This association saves plants from different pathogens by acting as biocontrol agents. Many studies revealed that many factors affect the efficiency of arbuscular mycorrhizal fungi as a biocontrol agent. These factors are soil moisture, soil content, genotype of host, inoculum of mycorrhiza, inoculation time of mycorrhiza, virulence of mycorrhizal species, inoculation potential of pathogen, and microflora of soil. Mycorrhizal association helps plants in uptake of mineral nutrients, promotes growth, inhibits soil-borne pathogen, competes with pathogen for space, photosynthates and nutrient's uptake, and increases resistance of the plant against such pathogens.

3.1 INTRODUCTION

It is studied that microorganisms in soil and in other substrate grow around the plant roots and hence create rhizosphere (Azcon-Aguilar and Barea, 1992). Rhizosphere develops at the interface of root and soil so microorganism interacts with both plants and soil. The interaction between roots, soil, and microorganism is by chemical, biological, and sometimes physical. So, the interaction between these organisms can be categorized into three major categories. There is an interaction of a plant with the other plant due to rhizosphere's interaction. Due to this interaction competition for nutrients increases among plants. The second interaction is between plant roots and microorganisms. It is due to activities of plants that enhance the activity of the microorganism to grow in rhizosphere and to infect plants for beneficial or to cause disease. The third interaction is between different microbes (Bowen, 1980; Lynch, 1990).

Biological control is defined so as to protect plants against pathogen by controlling factors that influence growth of plants (Kennedy and Smith, 1995). Commonly, plants have an association with their roots known as mycorrhizae (Harley and Smith, 1983). Mycorrhiza is defined as a mutualistic association of plants with soil-borne fungi. Arbuscular mycorrhiza is considered to be the more common type of such an association. This association saves plants from different pathogens by acting as biocontrol agents (Harley and Smith, 1983; Gianinazzi and Schuepp, 1994).

3.2 TYPES OF MYCORRHIZAL FUNGI

Basically mycorrhizal fungi are of two types: (1) Ectomycorrhiza and (2) Endomycorrhiza.

3.2.1 ECTOMYCORRHIZA

Ectomycorrhizal fungi are those that form an association with plant roots result in the formation of mantle around the tip of the root. These hyphae enter the root cells and then form a net-like structure known as the hartig net. Sheath and hartig net act as interface for exchange of nutrients between plants and fungus (Kumar and Satyanaryana, 2002). In ectomycorrhiza, fungal hyphae cannot penetrate inside the root cells, but only surround them (Suverch et al., 1991). It is studied that the ectomycorrhizal association is most commonly found in ornamental and forest trees of family Salicaceae, Dipterocarpaceae, etc. (Shalini et al., 2002). So the distinguished features of ecto from endomycorrhiza are the presence of the mantle and hartig net. These structures also enhance the surface area for improved absorption of nutrients and water from the soil (Peterson et al., 2004).

3.2.2 ENDOMYCORRHIZA

It is studied that endomycorrhizal fungi form association with crop plants mostly. They provide protection of plants from pathogens (Smith and Read, 2008) and are found in almost all types of ecological systems in different crops and horticultural species like Wheat, Maize, Rice, Grapes, Soybean, Cotton, etc. (Peterson et al., 2004). Endomycorrhizal fungi are obligatory in their mode of nutrition and obtain food in the form of photosynthates and carbon compounds from their host plant and in the return plant get mineral nutrients from the fungal partner which absorbs these nutrients from the root depletion zone with the help of extra radical mycelium (Smith and Read, 1997). Endomycorrhizal fungi form vesicles and arbuscules in the cells of root of plant while penetrating these cells. These vesicles are filled with lipids and turned into resting spores after passing of some time.

3.3 MYCORRHIZA AS BIOCONTROL AGENT

3.3.1 PROTECTION AGAINST PATHOGENS

Many agricultural practices are used to control soil-borne pathogens. These techniques include resistant cultivars, seed certification, crop rotation, fumigation of soil, and many chemicals like fungicides. Due to less resistance in plants against pathogen and less chances of control of pathogen inoculum, it is difficult to control plant pathogens (Azcon-Aguilar and Barea, 1997). To overcome the drawbacks of the different techniques in controlling plant pathogens, researchers introduced beneficial microorganisms into the soil that compete with soil-borne plant pathogens for nutrients and space by secreting antibiotics and acting like parasites for those pathogens and ultimately supporting the plant against plant pathogens (Grosch et al., 2005; Berg et al., 2007).

It is studied that using chemicals against soil-borne pathogens also damages plants' growth as well so creating an alarming situation; that is why scientists suggest and use alternative agents for controlling pathogens like fungi and other microbes like bacteria. Many researchers studied that arbuscular mycorrhizal association is very beneficial against soil-borne pathogens. There is a complex interaction between soil pathogen, the plant, and fungal partner which reduced the disease effect caused by pathogens in the host plant that is colonized with arbuscular mycorrhizal fungi (Dehne, 1982; Mukerji et al., 2002).

It is reported that when tomato roots were colonized by *Glomus mosseae* it reduced the disease effect of pathogen *Phytophthora parasitica*. It is also reported that when the oil palm plant was colonized by AMF it reduced the disease effect of *Ganoderma boninense* and seedlings of oil palm plant are more resistant to *Ganoderma* (Tahat et al., 2010).

AMF interacts with soil-borne pathogen and reduces their effect on plant by various mechanisms (Harrier and Watson, 2004) (Fig. 3.1). During this association arbuscular mycorrhizal association causes certain changes like biochemical changes in the host plant, microbial changes in rhizosphere, status of nutrients in soil, change the anatomy of plant cells, changes morphology of roots, and reduced stress (Hooker et al., 1994) (Table 3.1).

3.3.2 DAMAGE COMPENSATION

Arbuscular mycorrhizal fungi also increase resistance in plants against pathogen by compensating the damage caused by pathogen by increasing root growth and development. The root biomass loss and damages in roots by fungi and nematodes reduce this way (Linderman, 1994; Cordier et al., 1996). Arbuscular mycorrhizal fungi when colonizing the plants reduce these damages by increasing root surface area for better uptake of mineral nutrition for better plant growth by fungal hyphae that grow outside in the soil as well as inside the root cell by the formation of arbuscules (Cordier et al., 1996).

3.3.3 COMPETITION WITH HOST PHOTOSYNTHATES

It is revealed by many studies that the plant photosynthates are the major reason for the growth of soil microbes and arbuscular mycorrhizal fungi; that is why both organisms compete for these photosynthates, like carbon compounds. In this competition the arbuscular mycorrhizal fungus is better than pathogens for taking carbon compounds from the plant; that is why it inhibits the growth of pathogens. It is studied that 4–20% of carbon compounds are transferred to AMF rather than pathogens, while there is not considerable data available in support of this concept (Tahat et al., 2010).

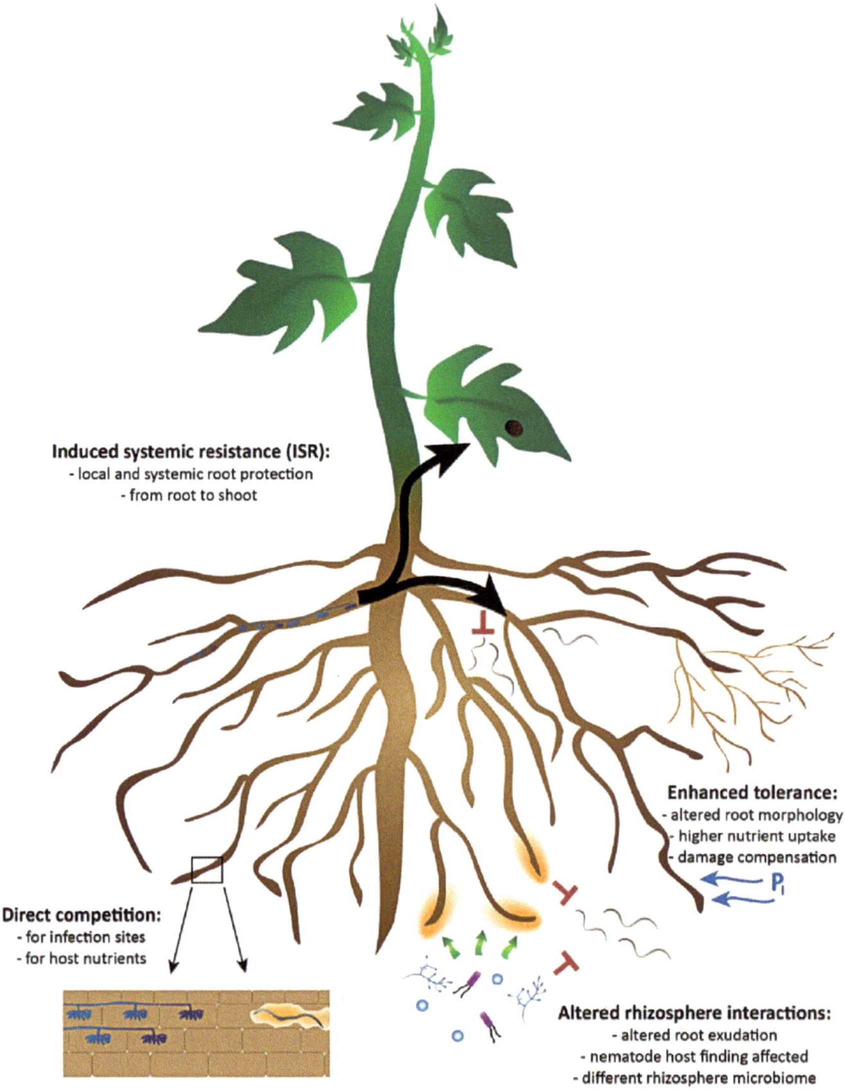

FIGURE 3.1 Overview of the possible mechanisms by which arbuscular mycorrhizal fungi can exert biocontrol against plant-parasitic nematodes. They consist of direct effects of AMF on the pathogen, involving competition for space and nutrients (bottom left) and indirect plant-mediated effects, involving damage compensation and enhanced tolerance (top right). The latter can be further divided into the effects of AMF on plant defense induction (ISR; top left) and altered rhizosphere interactions through changes in plant root exudation (bottom right). The different mechanisms cannot be considered completely independent of each other and biocontrol probably results from a combination of different mechanisms.

Source: Reprinted from Schouteden et al. (2015), https://creativecommons.org/licenses/by/4.0/

TABLE 3.1 Elaborates on the Diversity of Symbiotic Association of Fungi to Mitigate Environmental Stresses Subjected to Higher Plants.

Fungal endophytes	Environmental stress	Host plant species	References
Neotyphodium sp.	Drought	*Festuca arizonica*	Morse et al. (2002)
Neotyphodium lolii	Water scarcity	*P. ryegrass*	Ravel et al. (1997)
Neotyphodium coenophialum	Drought stress	*T. fescue*	De Battista et al. (1990)
Neotyphodium uncinatum	Water stress	*M. fescue*	Malinowski (1995)
Curvularia sp.	High temperature and water scarcity	*L. esculentum*	Rodriguez et al. (2009)
C. protuberata	Famine stress	*T. aestivum, Citrullus lanatus*	Rodriguez et al. (2009)
C. protuberata	High temperature	*L. esculentum*	Rodriguez et al. (2009)
Rhizophagus intraradices	Famine stress	*Zea mays*	Zhao et al. (2015)

Source: Reproduced with permission from Hussain et al. (2021), Copyright © Springer Nature (2021).

3.3.4 INTERACTIONS OF SOIL MICROBIAL POPULATION

The interaction of AMF with pathogens and their ability to increase plant nutrition is studied in relation with plant growth. There is little information about how these interactions have any effect on plant (Schreiner and Bethlenfalvay, 1995). It is studied that plants with mycorrhizal association differ from the plant having no such association in the rhizosphere, due to these interactions, the quality of respiration of root and the quantity of the chemicals secreted from roots altered (Marschner et al., 2001). These interactions between pathogens and fungi also have an impact on fungal growth. In the presence of bacteria in soil, the hyphae emerging from the germination of fungal spore develop smoothly and form small vesicles and more branched and longer than the hyphae growth in the absence of such bacteria.

AMF changes the mycorrhizosphere by changing the number of such pathogens as well as their activity as a pathogen (Burke et al., 2002). Plants with arbuscular mycorrhizal association with their roots effect the composition of bacterial community differently within the rhizosphere. It is studied that many factors affect the efficiency of arbuscular mycorrhizal fungi as a biocontrol agent. These factors are soil moisture and content, genotype of host, inoculum of mycorrhiza, inoculation time of mycorrhiza, virulence of mycorrhizal species, inoculation potential of microbes, and microflora of soil (Singh et al., 2000).

It is studied that the protection of plants against all such types of pathogens is totally dependent upon the rate of AMF colonization of plant roots. There are two zones of interaction between pathogens and AMF and these two are mycosphere and rhizosphere. There are some bacteria present in the rhizosphere known as mycorrhization helper bacteria which increase the colonization ability of AMF (Tahat et al., 2010).

3.3.5 COMPETITION FOR COLONIZATION AND INFECTION SITE

There is a competition between pathogens and arbuscular mycorrhizal fungi in rhizosphere for occupying more space, and it is important for explaining the communication among pathogens and AMF (Bansal and Mukerji, 1996). One phenomenon as already mentioned is that mycorrhizal fungi rely on the host photosynthetic compounds like carbon compounds which are the major cause of suppression of pathogen growth in rhizosphere (Cordier et al., 1998). It is also noticed through several studies that the presence of arbuscules in root cells causes hindrance of the penetration of pathogens in the cells. So there is a competition for space between these organisms to penetrate inside the cells of cortex and epidermis.

3.4 MORPHOLOGICAL AND ANATOMICAL CHANGES

The roots of plants which are in mycorrhizal association with AMF have some morphological and anatomical modification in their root systems (Tahat et al., 2008). Colonized roots are having high branching system than non-mycorrhizal plant roots and adventitious roots also have large diameter in mycorrhizal-associated plants. Because of such modification in root system morphology certain pathogenic attacks are reduced; these changes also include highly lignified epidermal cells of roots. The high rate of lignification in cells prevents the penetration of other pathogens and it also increases the phenolic metabolites inside the host plant. Other changes include leading to a bigger size of roots, dense branching, and increase the number of root tips, etc. (Tahat et al., 2010).

3.4.1 CHANGES IN ROOT EXUDATES

It is studied that some toxic chemicals like phytoalexins are released and produced by AMF. These chemicals are not observed during the early establishment stages of AMF, but are observed in later stages of this association (Morandi, 1996). But activity of peroxidases is observed in the early stages (Azcon-Aguilar et al., 2002). For example, development of *Phytophthora parasitica* decreases in tomato roots when colonized by *Glomus mosseae* but shows defense responses and accumulates phenolic compounds in noncolonized parts. Cortical cells of the same plant also show the localized immune response so they are immune to pathogens (Cordier et al., 1998).

During this study it is noticed that some corresponding proteins are also involved in the plant defense mechanism when associated with AMF. These include glycoprotein rich with hydroxyproline, chitinases, phenolics, peroxidases, PR-pathogenesis proteins, and β-1-3-glucanases-callose proteins. In the establishment of arbuscular mycorrhizal association root exudates are playing an important role. Growth and germination of many pathogens inhibited in the presence of such exudates. The spore production of *Phytophthora fragariae* is suppressed by exudates released from the strawberry plant which is associated with AMF that is the best example to understand the effect of chemicals discharged from the roots. It is reported that when phenolic compound interacts and is linked with the cell wall, the resistance against digestive enzymes and pathogenic fungi increased (Tahat et al., 2010).

3.4.2 NUTRIENTS UPTAKE

The actual role of mycorrhizal association is the increase in the yield of production of plants. AMF is important in increasing the rate of uptake of mineral nutrients of plants. Due to the increased rate of uptake of mineral nutrients plants have the ability to stand against the pathogen and reduce uptake of poisonous compounds (Clark, 1997). Arbuscular mycorrhizal fungi provides the large root surface area which covers the large volume of soil and these roots penetrate in the depletion zone for better uptake of nutrients and water (Smith and Read, 2008).

All the parameters of roots of a mycorrhizal plant are increased overall (Hetrick et al., 1988). The width of fungal hyphae is less than the root hair; that is why they can breach soil pores where root hairs are unable to reach. AM roots increased the nutrient uptake from the soil, especially the phosphorous uptake than other nutrients (Jakobsen, 1995). Absorption of phosphorous from the soil is more rapid in mycorrhizal associated plants due to small diffusion distance of phosphate ions for fungal hyphae than the roots of plant. Due to increase in the rate of mineral nutrition uptake, there are many physiological changes in root (Akthar and Siddiqui, 2008).

It can consider that increase in phosphorous uptake is one of the most important benefits of mycorrhiza. Phosphate present in the soil is changed into polyphosphate by enzyme in the vacuole. These polyphosphates were transferred to the sink of symbiotic interface from fungal hyphae tips. The rate of translocation is based on the net efflux of phosphorous at hyphal tips (Pearson and Tinker, 1975). AMF is also increasing efficiency of nitrogen uptake and growth in legumes. Due to high uptake of phosphorous, nitrogen uptake also increases to accomplish high demand of nitrogen. Hyphae of AMF have the ability to uptake nitrogen and are transported to the plants. The nitrogen uptake by fungal hyphae is regulated by the nitrogen demand of plants. When nitrogen present in the soil in the form of ammonium and nitrate ions, then AMF uptake these minerals and transport them to the plants. When nitrogen is applied to as $(NH_4)_2SO_4$ then uptake of nitrogen by hyphae is about 40% and some nitrogen is transported over a distance of 5 cm. Other than phosphorous and nitrogen AMF also increases the uptake of many other minerals like Zn, S, Fe, and Ca. When plants associated with AM are grown in low pH, then the Fe uptake increases than when grown in high pH. But, it was noticed that Manganese uptake is relatively lower in AM as compared to non-AMF (Tahat et al., 2010).

3.4.3 BIOREMEDIATION

AMF plants also show an effect on uptake of trace elements (Clark and Zeto, 2000). When AMF associated with plants is grown in a soil having the low concentration of Cu and Zn, their uptake is increased by AMF. The accumulation of Cu and Zn is increased in shoot and leaves in mycorrhizal plants. Uptake of sulfur also increases in plants colonized with AMF (Raju et al., 1990). In acidic and alkaline soil uptake of Boron also increases; on the other hand, uptake of potassium, calcium, and magnesium is also enhanced by AMF as compared to nonmycorrhizal. In acidic soil uptake of toxic aluminum is lower in AM than in non-AM (Clark, 1997). AMF also increases the acquisition of trace elements from the soil like Zn, Mg, Mn, Cd, Cu, Co, Cs, Ni, and Br.

3.4.4 MYCORRHIZOSPHERE

Mycorrhizosphere is the zone of soil associated or in interaction with mycorrhizal fungi (Oswald and Ferchau, 1968). It is observed that the rhizosphere soil extract of mycorrhizal plants reduced the formation of sporangia of pathogenic fungus than the rhizosphere soil extract of nonmycorrhizal. The reason behind this is, maybe, the absence of sporulation. Microbial communities around the roots are affected by changes in exudates of roots (Varma et al., 1999). In mycorrhizosphere, the mycorrhiza influences the microbes of rhizosphere. The microflora in mycorrhizosphere differ quantitatively as well as qualitatively as compared with nonmycorrhizal. Mycorrhizosphere has two components: a layer of soil surrounded the mycorrhizal roots and a layer of soil surrounded the arbuscular mycorrhizal hyphae in the soil known as hyphosphere. There are two possible interactions between soil-borne pathogens and AMF; these interactions may be positive (synergistic) or negative (antagonism) (Mukerji et al., 2002). AMF interacts positively with plant growth-promoting, nitrogen-fixing, and phosphorous-solubilizing bacteria that enhance the germination of AMF spores (Mayo et al., 1986). In negative interaction is the role of AMF to reduce the presence of soil-borne pathogens (Dehne, 1982).

3.5 CONCLUSION

Microorganisms in soil and other substrates grow around the plant roots and hence create rhizosphere. Rhizosphere develops at the interface of root and soil so microorganism interacts with both plants and soil. The interaction between roots, soil, and microorganism are by chemical, biological, and sometimes physical. So, the interaction between these organisms can be categorized into three major categories. There is an interaction of a plant with the other plant due to rhizosphere's interaction. Due to this interaction competition for nutrients increases among plants. The second interaction is between plant roots and microorganisms. It is due to activities of plants that enhance the activity of the microorganism to grow in rhizosphere and to infect plants for beneficial or to cause disease. The third interaction is between microbes. Biological control is defined so as to protect plants against pathogen by controlling factors that influence growth of plants. Mycorrhiza is defined as a mutualistic association of plants with soil-borne fungi. Arbuscular mycorrhiza is considered the most common type of such an association. This association saves plants from different pathogens by acting as biocontrol agents. Mycorrhizal association helps plants in uptake of mineral nutrients, promotes growth, inhibits soil-borne pathogen, competes with pathogen for space, photosynthates and nutrient uptakes, and increases resistance of the plant against such pathogens. Hence, it proved that mycorrhizal association acts as a biocontrol agent.

KEYWORDS

- **AMF**
- **biocontrol agent**
- **bioremediation**
- **ECM**
- **microbial competition**
- **mutualistic association**
- **soil pathogens**

REFERENCES

Akthar, M. S.; Siddiqui, Z. A. Arbuscular mycorrhizal fungi as potential bioprotectants against plant pathogens. In: *Mycorrhizae: Sustainable agriculture and forestry*; Siddiqui, Z. A., Akhtar, M. S., Futai, K., Eds.; Springer: The Netherlands, 2008; pp 61–97.

Azcon-Aguilar, C.; Barea, J. M. Interactions between mycorrhizal fungi and other rhizosphere microorganisms. In: *Mycorrhizal functioning: An integrative plant-fungal process*, Allen, M. J., Ed.; Chapman and Hall: New York, 1992; pp 163–198.

Azcon-Aguilar, C.; Barea, J. M. Applying mycorrhiza biotechnology to horticulture: Significance and potentials. *Sci. Hortic.* **1997**, 68, 1–24.

Azcon-Aguilar, C.; Jaizme-Vega, M. C.; Calvet, C. The contribution of arbuscular mycorrhizal fungi for bioremediation. In: *Mycorrhizal technology in agriculture, from genes to bioproducts*; Gianinazzi, S., Schuepp, H., Barea, J. M., Haselwandter. K., Eds.; Birkhauser Verlag: Berlin, 2002; pp 187–197.

Bansal, M.; Mukerji, K. G. Root exudates and its rhizosphere biology. In: *Concepts in applied microbiology and biotechnology*; Mukerji, K. G., Singh, V. P., Dwivedi, S., Eds.; Adita Books Private Ltd, New Delhi, 1996; pp 79–119.

Berg, G.; Grosch, R.; Scherwinski, K. Risk assessment for microbial antagonists: Are there effects on non-target organisms? *Gesunde Pflanzen.* **2007**, *59*, 107–117.

Bowen, G. D. Misconceptions, concepts and approaches in rhizosphere biology. In: *Contemporary microbial ecology*; Ellwood, D. C., Latham, M. J., Hedger, J. N., Lynch, J. N., Slater, J. M., Eds.; Academic Press, London, 1980.

Burke, D. J.; Hamerlynck, E. P.; Hahn, D. Interactions among plant species and microorganisms in salt march sediments. *Appl. Environ. Microbiol.* **2002**, *68*, 1157–1164.
Clark, R. B. Arbuscular mycorrhizal adaptation, spore germination, root colonization and host plant growth and mineral acquisition at low pH. *Plant Soil* **1997**, *192*, 15–22.
Clark, R. B.; Zeto, S. K. Mineral acquisition by arbuscular mycorrhizal plants. *J. Plant Nutr.* **2000**, *23*, 867–902.
Cordier, C.; Gianinazzi, S.; Gianinazzi-Pearson, V. Colonization patterns of root tissues by *Phytophthora nicotianae* var. *parasitica* related to reduced disease in mycorrhizal tomato. *Plant Soil* **1996**, *185*, 223–232.
Cordier, C.; Pozo, M. J.; Barea, J. M.; Gianinazzi, S.; Gianinazzi-Pearson, V. Cell defense responses associated with localized and systemic resistance to *Phytophthora parasitica* induced in tomato by an arbuscular mycorrhizal fungus. *Mol. Plant Microbe Interact.* **1998**, *11*, 1017–1028.
De Battista, J. P.; Bacon, C. W.; Severson, R.; Plattner, R. D.; Bouton, J. H. Indole acetic acid production by the fungal endophyte of tall fescue. *Agronom. J.* **1990**, *82*, 878–880.
Dehne, H. W. Interaction between vesicular-arbuscular mycorrhizal fungi and plant pathogens. *Phytopathology* **1982**, *72*, 1115–1119.
Gianinazzi, S.; Schüepp, H. *Impact of arbuscular mycorrhizas on sustainable agriculture and natural ecosystems*; Springer Science + Business Media: Berlin, 1994.
Grosch, R.; Lottmann, J.; Faltin, F.; Berg, G. Use of bacterial antagonists to control diseases caused by *Rhizoctonia solani. Gesunde Pflanzen.* **2005**, *57*, 199–205.
Harley, J. L.; Smith, S. E. *Mycorrhizal symbiosis*. Academic Press: New York, 1983.
Harrier, L. A.; Watson, C. A. The potential role of arbuscular mycorrhizal (AM) fungi in the bioprotection of plants against soil-borne pathogens in organic and/or other sustainable farming systems. *Pest Manage. Sci.* **2004**, *60*, 149–157.
Hetrick, B. A. D.; Kitt, D. G.; Wilson, G. W. T. Mycorrhizal dependency and growth habit of warm season and cool season tall grass prairie plants. *Can. J. Bot.* **1988**, *66*, 1376–1380.
Hooker, J. E.; Jaizme-Vega, M.; Alkinson, D. Biocontrol of plant pathogen using arbuscular mycorrhizal fungi. In: *Impact of arbuscular mycorrhizas on sustainable agriculture and natural ecosystems* Gianinazzi, S., Schhepp, H., Eds.; Birkhauser Verlag: Basel, Switzerland, 1994; pp 191–209.
Hussain, T.; Usmaan, M.; Numan, M.; Khan, A. A.; Abbas, F.; Gul, A. Mycorrhiza: Plant growth-promoting and biocontrol agent ability under the abiotic stress conditions. In: *Soil microbiomes for sustainable agriculture: Sustainable development and biodiversity*; Yadav, A. N., Ed., Vol. 27; Springer: Cham, 2021. https://doi.org/10.1007/978-3-030-73507-4_16.
Jakobsen, I. Transport of phosphorus and carbon in VA mycorrhizas. In: *Mycorrhiza: Structure, function, molecular biology and biotechnology*; Varma, A., Hock, B., Eds.; Springer, USA, 1995; pp 297–324.
Kennedy, A. C.; Smith, K. L. Soil microbial diversity and the sustainability of agricultural soils. *Plant Soil* **1995**, *170*, 75–86.
Kumar, S.; Satyanarayana, T. Isolation of ectomycorrhizal fungi: Methods and techniques. In: *Techniques in mycorrhizal studies*; Mukerji, K. G., Manoharachary, C., Chamola, B. P., Eds.; Kluwer Academic Publishers: London, 2002; pp 143–166.
Linderman, R. G. Role of VAM fungi in biocontrol. In: *Mycorrhizae and plant health*; Pfleger, F. L., Linderman, R. G., Eds.; The American Phytopathological Society: St. Paul, MN, 1994; pp 1–27.

Lynch, J. M. The potential for gene exchange between rhizosphere bacteria. In *Bacterial genetics in natural environments*; Springer: Dordrecht, 1990; pp 172–181.

Malinowski, D. Rhizomatous ecotypes and symbiosis with endophytes as new possibilities of improvement in competitive ability of meadow fescue (*Festuca pratensis* Huds.); ETH, Zurich, 1995.

Marschner, P.; Crowley, D.; Lieberei, R. Arbuscular mycorrhizal infection changes bacterial 16S DNA community composition in the rhizosphere of maize. *Mycorrhiza* **2001**, *11*, 297–302.

Mayo, K.; Davis, R. E.; Motta, J. Stimulation of germination of spores of *Glomus versiforme* by spore-associated bacteria. *Mycologia* **1986**, *78*, 426–431.

Morandi, D. Occurrence of phytoalexins and phenolic compounds in endomycorrhizal interactions and their potential role in biological control. *Plant Soil* **1996**, *185*, 241–251.

Morse, L. J.; Day, T. A.; Faeth, S. H. Effect of *Neotyphodium* endophyte infection on growth and leaf gas exchange of *Arizona fescue* under contrasting water availability regimes. *Environ. Exp. Bot.* **2002**, *48*, 257–268

Mukerji, K. G.; Manoharachary, C.; Chamola, B. P. *Techniques in mycorrhizal studies*; Kluwer Academic Publishers: Netherlands, 2002; pp 285–296.

Oswald, E. T.; Ferchau, H. A. Bacterial associations of coniferous mycorrhizae. *Plant Soil* **1968**, *28*, 187–192.

Pearson, V.; Tinker, P. B. Measurement of phosphorus fluxes in the external hyphae of endomycorrhizas. In: *Endomycorrhizas*; Sanders, F. E., Mosse, B., Tinker, P. B., Eds.; Academic Press Inc.: London, 1975; pp 277–287.

Peterson, R. L.; Massicotte, H. B.; Melville, L. H. *Mycorrhizas: Anatomy and cell biology*; NCR Research Press: Ottawa, 2004; p 173.

Raju, P. S.; Clark, R. B.; Ellis, J. R.; Maranville, J. W. Effects of species of VA-Mycorrhizal fungi on growth and mineral uptake of sorghum at different temperatures. *Plant Soil* **1990**, *121*, 165–170.

Ravel, C.; Courty, C.; Coudret, A.; Charmet, G. Beneficial effects of *Neotyphodium lolii* on the growth and the water status in perennial ryegrass cultivated under nitrogen deficiency or drought stress, *Agronomie* **1997**, *3*(17), 173–181.

Rodriguex, R. Z.; White Jr, J. F.; Arnold, A. E.; Redmann, R. S. Fungal endophytes: diversity and functional roles. *New Phytologist* **2009**, *182*(2), 314–330.

Schouteden, N.; De Waele, D.; Panis, B.; Vos, C. M. Arbuscular mycorrhizal fungi for the biocontrol of plant-parasitic nematodes: A review of the mechanisms involved. *Front. Microbiol.* **2015**, *6*, 1280. https://doi.org/10.3389/fmicb.2015.01280.

Schreiner, R. P.; Bethlenfalvy, G. J. Mycorrhizal interactions in sustainable agriculture. *Crit. Rev. Biotechnol.* **1995**, *15*, 271–287.

Shalini, R.; Chamola, B. P.; Mukerji, K. G. Evolution of mycorrhiza. In: *Mycorrhizal biology*; Mukerji, K. G., Chamola, B. P., Singh, J., Eds.; Kluwer Academic/Plenum Publishers: New York, 2000; pp 1–25.

Singh, R.; Adholega, A.; Mukerji, K. G. Mycorrhiza in control of soil borne pathogens. In: *Mycorrhizal biology*; Mukerji, K. G., Chamola, B. P., Singh, J., Eds.; Kluwer Academic/ Plenum Publishers: New York, 2000; pp 173–196.

Smith, S. E.; Read, D. J. *Mycorrhizal symbiosis*; Academic Press: London, 1997; p 605.

Smith, S. E.; Read, D. J. Mineral nutrition, toxic element accumulation and water relations of arbuscular mycorrhizal plants. In: *Mycorrhizal symbiosis*; Academic Press: London, 2008; pp 145–148.

Suverch, K. G.; Mukerji, K.; Arora, D. K. Ectomycorrhizal. In: *Soil and Plants*; Arora, D. K., Rai, B., Mukerji, K. G., Kndson, G. R., Eds.; Marcel Dekker Inc.: New York, 1991; pp 187–215.

Tahat, M. M.; Kamaruzaman, S.; Radziah, O.; Kadir, J.; Masdek, H. N. Response of (*Lycopersicum esculentum* Mill.) to different arbuscular mycorrhizal fungi species. *Asian J. Plant Sci.* **2008**, *7*, 479–484.

Tahat, M. M.; Kamaruzaman, S., Othman, R. Mycorrhizal fungi as a biocontrol agent. *Plant Pathol. J.* **2010**, *9*(4), 198–207.

Varma, A. Function and applications of arbuscular mycorrhizal fungi in arid and semi-arid soils. In: *Mycorrhiza: Structure, function, molecular biology and biotechnology*; Varma, A., Hock, B., Eds.; Springer-Verlag: Germany, 1999; pp 521–556.

Zhao, R.; Guo, W.; Bi, N.; Guo, J.; Wang, L.; Zhao, J.; Zhang, J. Arbuscular mycorrhizal fungi affect the growth, nutrient uptake and water status of maize (*Zea mays* L.) grown in two types of coal mine spoils under drought stress. *Appl. Soil Ecol.* **2015**, *88*, 41–49.

PART II

MYCORRHIZAE IN AGRICULTURAL, HORTICULTURAL, AND BIOTECHNOLOGICAL SECTORS

CHAPTER 4

APPLICATION OF ARBUSCULAR MYCORRHIZA FUNGI IN AGRICULTURAL AND HORTICULTURAL CROPS

DEBASIS MITRA[1], PAULAMI DAM[2], RITTICK MONDAL[2], BHASWATIMAYEE MAHAKUR[3], ABDEL RAHMAN MOHAMMAD AL-TAWAHA[4], JEYABALAN SANGEETHA[5], DEVARAJAN THANGADURAI[6], and PAVITRA CHIPPALAKATTI[6]

[1]*Department of Microbiology, Raiganj University, Raiganj, Uttar Dinajpur, West Bengal, India*

[2]*Department of Sericulture, Raiganj University, Raiganj, Uttar Dinajpur, West Bengal, India*

[3]*Department of Botany and Biotechnology, Ravenshaw University, Cuttack, Odisha, India*

[4]*Department of Biological Sciences, Al Hussein Bin Talal University, Maan, Jordan*

[5]*Department of Environmental Science, Central University of Kerala, Kasaragod, Kerala, India*

[6]*Department of Botany, Karnatak University, Dharwad, Karnataka, India*

Mycorrhizal Technology: Managing Plant Stress and Mitigating Climate Change using Mycorrhizae.
Jeyabalan Sangeetha, Abdel Rahman Mohammad Al-Tawaha, & Devarajan Thangadurai (Eds.)

ABSTRACT

Climate change and agricultural malpractices like excessive use of fertilizers and pesticides have aggravated the effects of abiotic stresses on crop productivity and degraded the ecosystem. On the other hand, eco-friendly techniques such as the use of arbuscular mycorrhizal fungi to increase crop productivity are urgently required. Arbuscular mycorrhizal fungi are beneficial microbes that colonize the majority of agricultural plants. They are one of the most widespread symbiotic fungi that colonize the majority of agricultural plants. AMF has also been shown to boost plant performance under water shortages, in addition to improving plant nutrition. In addition, AMF inoculation is generally considered to provide host plants with tolerances to various stressful situations such as heat, salinity, drought, metals, and temperature conditions. On the other hand, AMF can colonize in more than 80% families of plants by forming filamentous hyphae. AMF can potentially improve the adaptation of plants to changing environment as a bio-fertilizer. This chapter provides comprehensive, up-to-date information on AMF and their impact on horticultural crops.

4.1 INTRODUCTION

Arbuscular mycorrhizal fungi (AMF) are one of the most important beneficial relations between plant roots (mainly vascular plants) and soil fungus. Majority of the AMF present in the normal soil environment and plays a major role in plant growth promoting attributes like increasing the plant nutrition uptake, stress resistance and tolerance, and also soil fertility and structure. AMF can colonize in more than 80% families of plants by forming filamentous hyphae (Gianinazzi et al., 2010). Fungal–host interactions have been shown to be capable of producing micro-associated molecular patterns (MAMPs), exopolysaccharides, volatile organic compounds (VOCs), Myc factors, and nodes (Goh et al., 2014; Basu et al., 2018). In addition, VOCs demonstrated the ability to modulate root system architecture that favors symbiotic interactions and also modulate the signaling pathway for AMF colonization through Nod factor (Maillet et al., 2011). The filamentous network attached to host roots helps in uptake of phosphorus and other essential minerals even from distant soil. AMF exchange absorbed minerals and water with host plant for 20% of fixed photosynthetic carbon (Smith et al., 2008). Nutrient translocation happens in the arbuscules located at root cortical cells (Balestrini et al., 2015). Soil quality and different functional strategies of plants depend on these symbiotic fungi. AMF network along with glomalin-related soil protein support soil aggregate formation and stabilize soil structure (Bedini et al., 2009). Plants allow AMF to live within their roots and provide the fungi with sugars that enables them to grow. To accomplish this, the fungus developed long hyphae or threads, these hyphae like structure look like an extended root. Most importantly, these symbiotic associations help the plant growth under stress condition (limited nutrient, drought, etc.) and show resistant properties against pathogen.

4.2 IMPACT OF AMF IN SUSTAINABLE MANAGEMENT OF AGRICULTURAL ECOSYSTEMS

AMF greatly contribute in sustainable management of agricultural ecosystems as a potential alternative of some pesticides and chemical fertilizers. Nowadays agriculture industries can hardly keep balance with "ecological services" due to business-focused agriculture. Use of nonscientific techniques and machines, over-use of insecticides, pesticides, and fertilizers, mono-culture, lack of crop rotation, etc., are destroying the natural resources, polluting water, degrading soil quality, shrinking arable fields, and thus destroying the balance of ecosystem (Foley et al., 2005). So, effective and beneficial methods are needed for sustainable food production for world population. Approximately 50.9 million hectares of lands are now being used in organic agriculture all over the world, especially in Europe, USA, Australia, etc. (Willer et al., 2012; Willer et al., 2017). New agricultural management strategies are now using AMF in controlling development and growth, prevention of disease, nutrient absorption, soil strength, stress tolerance, and quality of surrounding ecosystem (Gianinazzi et al., 2010). Application-based research showed that inoculation of AMF not only supports plant health but also increases crop production by both nutritional and non-nutritional levels. The inoculation process is tough, so it is applied on valuable crops and horticulture (Chen et al., 2018). The inoculation success depends on several factors such as compatibility with host plant, competition with neighboring biota, environmental factors, etc. (Verbruggen et al., 2013) (Fig. 4.1).

In a recent study carried out in 45 rice fields in Bangladesh, 77 OTUs from eight AMF families were found to be present. Even where the salinity of soil was high and the soil was contaminated with heavy metals (e.g., arsenic)

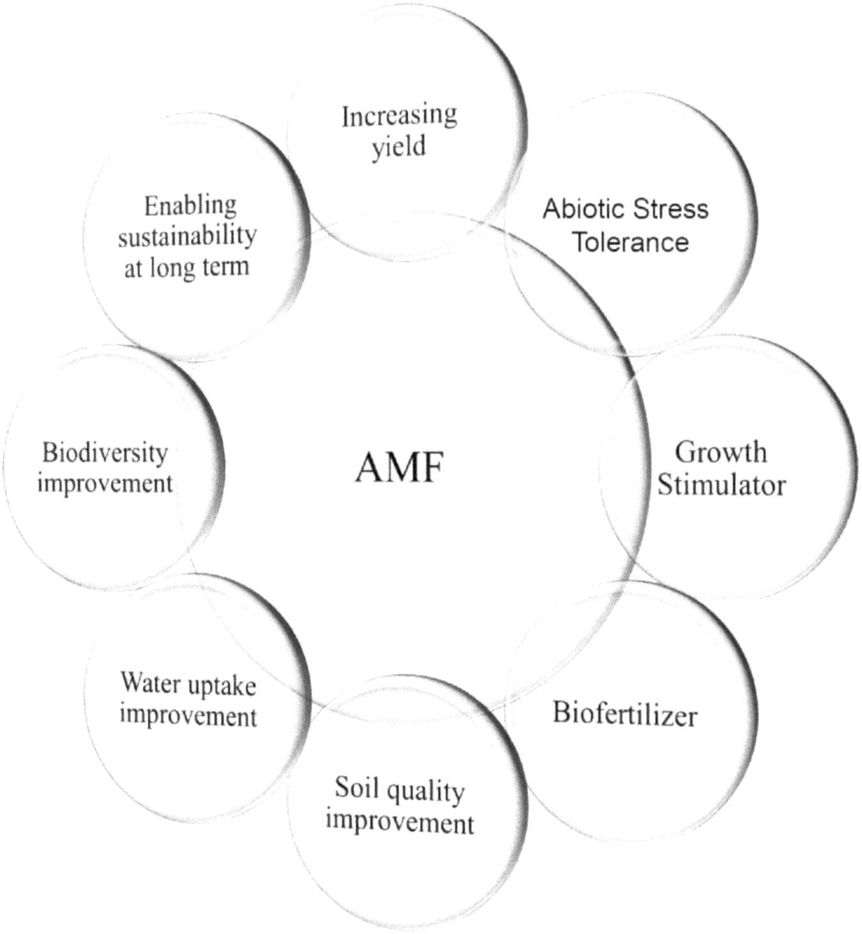

FIGURE 4.1 AMF benefits in agriculture and ecosystem sustainability.

Source: Reproduced and modified with permission from El Omari and El Ghachtouli (2021), Copyright © Springer Nature (2021).

some of the AM families were still present (Parvin et al., 2019). Another study revealed that carbon and phosphorus were exchanged between AMF and the host plants in flooded land depending on water abundance (Bao et al., 2019). First-time evidence from USA showed that AMF-dependent rice growth and pest defense rely on soil types (Bernaola et al., 2019). Zhang et al. (2019) found that after AM inoculation 77% plants had 16% more yield of cereal crops than non-AMF plants. In an experiment by Aziez et al. (2019) in Indonesia, eight rice plant varieties showed positive correlation between plant growth, net assimilation, and plant area index after AMF application. Plant waste products like rice straw can be used as a compost to increase AMF colonization. An experiment where rice straw compost was applied in soil resulted in more AMF symbionts in cereal crops such as sorghum and rice (Kadian et al., 2019). According to Perez et al. (2016), *G. cubense*, a strain of AMF when inoculated with several spring wheat line varieties, resulted in better productivity in contrast to other AMF strains. AMF inoculation in wheat plants increases their chance of survival during persistent stress such as soil aluminum and nickel toxicity (Seguel et al., 2016; Toidi et al., 2018). According to Seguel et al. (2016) and Aguilera et al. (2018), AMF from high Al-concentrated soil of Chile decreases aluminum transfer to plant parts and increases aluminum elimination in wheat. AMF helps to reduce Al toxicity and promotes Al tolerance of Chilean wheat in acidic soil. In modern agricultural biotechnology AMF is essential for producing better quality crops to meet the nutritional needs. AMF colonization not only increases nitrogen transfer but also enriches the durum wheat grains with enough zinc and iron (Ercoli et al., 2017). According to Li et al. (2016), AMF inoculation decreased organic and inorganic arsenic concentration in rice grains. AMF colonization enhances carbon accumulation, nitrogen absorption, and also nitrogen use efficiency into wheat plants cultivated at increased CO_2 (Zhu et al., 2016). AMF symbiosis under heat stress of wheat plants results in increasing the number of grains, number of root arbuscules, electron transfer by PS-II, and in alteration of nutrient distribution (Cabral et al., 2016). El-sharkawy and team found that in greenhouse conditions, co-inoculation of AMF-*Trichoderma* controls stem rust disease of wheat (El-sharkawy et al., 2018). Combined inoculation of AMF, bacteria (*Pseudomonas* sp.), and nematodes helps in biotic stress such as pest infestation in wheat and maize plants (Imperiali et al., 2017; Jaffuel et al., 2019).

4.3 IMPACT OF AMF IN HORTICULTURAL CROPS

Arbuscular mycorrhizal inoculation is easier in horticulture because of the transplantation stage. AMF supports plant growth by absorbing the minerals through symbiotic layer and induces primary and secondary metabolism of host plants. Higher metabolism leads to the extended synthesis of phytochemicals (e.g., carotenoids, anthocyanin, flavonoids, phenolics, etc.). Carotinoids are oxidized to apocarotenoids. Strigolactone, an example of apocarotenoids, is a type of secondary metabolite of plants and is responsible for colonization of AMF in the host plant roots (Fiorilli et al., 2019). Phytochemicals present in commercial food products help people to fight several chronic diseases and boost the immunity up. As Strigolactone is the symbiotic signal from plant, AMF can also release "myc factors" to establish host-pathogen signaling cascade (Basu et al., 2018). AMF as a "bioregulator" stimulates the production of plant hormones and in turn plant development. AM induces biotic and abiotic stress tolerance and plant defense acting as a "bioprotector" (Chen et al., 2018). AMFs act as "biofertilizer" in production of vegetables such as tomato, onion, bell pepper, cucumber, carrot, beans, peas, lettuce, etc. (Baum et al., 2015). Sustainable horticultural management approaches are proper inoculation of AM fungi, detection of proper host stage, and selection of compatible host-AMF. The better horticultural management reduces the application of chemical fertilizers and pesticides. AMF inoculation during micro-propagation and transplantation of flower and fruit trees supports survival rate, growth, and quality. Mycorrhizal establishment greatly enhances commercial characteristics of plants when out-planted (Azcón-Aguilar et al., 1997). Large woody plants that difficultly survive after transplantation have shown increased survival rate after AMF inoculation (Chen et al., 2018). From the last two decades innovative tools are being applied in the field of experimental horticulture. Some elements or microbiota increase absorption of minerals, nutrient transfer, stress tolerance, and stimulate plant functions are defined as a "biostimulant." It is a sustainable approach as it reduces the use of chemicals. Biostimulants are not in any group of fertilizers or pesticides but effectively boost up plant metabolism and performance (Rouphael et al., 2015). AMF is an obligate symbiont and effective biostimulant. AMF (e.g., *Rhizophagus intraradices*, *Funneliformis mosseae*, etc.) is broadly inoculated in horticulture (Krüger et al., 2012). As AMF significantly help in phosphorus translocation, they are more demanding than nitrogen-fixing microbes available in the market. Co-inoculation of AMF with bacteria (AMF-bacteria) or fungi (AMF-fungi) is thought to be a good strategy toward sustainable horticultural management. The triple response from AMF-bacteria-host group or AM-fungi-host group is being deeply researched (Rouphael et al., 2015) (Table 4.1).

A specific case of analysis of potato yield in 231 land trials in North America and Europe resulted in a rise in tuber product into 3.9 tons/ha (9.5% of gross production) after application of *R. irregularis*, a commercial strain (Hijri, 2016). During low phosphorus concentration or low water availability in soil AMF colonization helps to overcome abiotic

TABLE 4.1 Impact of AMF on Agricultural and Horticultural Crops.

Agricultural and horticultural crops	Arbuscular mycorrhizal fungi	Impact
Capsicum annuum	*Glomus fasciculatum*, *G. albidum*, *G. macrocarpum*	Shoot biomass increased, numbers of flowers, fruit yield increased
Abelmoschus esculentus	*Glomus mosseae*	Nematodes stress resistance, increased uptake of minerals and nutrients
Musa spp.	*Glomus intraradices*	Nematodes stress resistance, increased uptake of minerals and nutrients
Allium cepa	*Glomus intraradices*	Shoot biomass increased
Solanum lycopersicum	*Glomus mosseae*, *G. clarum*, *G. etunicatum*, *G. intraradices*, *G. caledonium*, *G. mosseae*, *Rhizophagus irregularis*, *Funneliformis mosseae*, and *Rhizoglomus irregulare*	Shoot biomass increased, root biomass and fruit biomass increased, early flowering, enhancement of secondary metabolites
Ocimum basilicum	*Glomus fasciculatum*, *Rhizophagus intraradices*	Enhancement of secondary metabolites like essential oil, phenolics, anthocyanins, and organic acids
Daucus carota	*Glomus mosseae*	Nematodes stress resistance, increased uptake of minerals and nutrients
Pyrethrum spp.	*Glomus etunicatum*	Nematodes stress resistance, increased uptake of minerals and nutrients
Ipomoea aquatica	*Glomus caledonium*, *G. versiforme*	Shoot biomass increased
Prunus dulcis	*Glomus intraradices*	Nematodes stress resistance, increased uptake of minerals and nutrients
Allium sativum	*Glomus fasciculatum*	Enhancement of production of enzyme alliinase
Vigna radiata	*Glomus mosseae*	Nematodes stress resistance, increased uptake of minerals and nutrients
Lactuca sativa	*Glomus mosseae*, *G. fasciculatum*, *G. coronatum*, *G. intraradices*, *G. claroideum*, *G. mosseae*	Shoot biomass increased, enhancement of secondary metabolites
Spinacia oleracea	*G. fasciculatum*, *G. mosseae*	Enhanced chlorophyll, flavonoid, and total phenolic compounds
Punica granatum	*Glomus mosseae*, *Acaulospora laevis*, and *G. manihotis*	Increased secondary metabolite production
Vigna unguiculata	*Glomus fasciculatum*	Nematodes stress resistance, increased uptake of minerals and nutrients
Solanum melongena	*Glomus fasciculatum*	Nematodes stress resistance, increased uptake of minerals and nutrients
Citrullus lanatus	*Glomus* spp.	Nematodes stress resistance

TABLE 4.1 *(Continued)*

Agricultural and horticultural crops	Arbuscular mycorrhizal fungi	Impact
Cucumis sativus	*Glomus intraradices, G. mosseae, G. versiforme*	Nematodes stress resistance
Vigna mungo	*Glomus mosseae*	Nematodes stress resistance, increased uptake of minerals and nutrients
Fragaria ananassa	*Glomus intraradices*	Enhancement of increased production of flavonoid and phenol
Zingiber officinale	*Glomus fasciculatum*	Nematodes stress resistance, increased uptake of minerals and nutrients
Helianthus annuus L.	*Haematophilum, R. irregularis*	Increased biomass, increased uptake of nutrients
Carica papaya	*Glomus mosseae, G. manihotis*	Nematodes stress resistance, increased uptake of minerals and nutrients

Source: Reproduced and modified with permission from Baum et al. (2015), Copyright © Elsevier and Emmanuel and Babalola (2020), Copyright © Elsevier; Adapted with slight modification from Abbasi et al. (2015).

stress in potato plants (Liu et al., 2018). In an experiment conducted by Issa et al. (2017) in sweet potato plants, AMF acted as a bioregulator and bioprotector against plant pest. In a recent study in tomato plants AMF inoculation resulted in promoting the nutritional value by increasing anti-oxidants; minerals such as phosphorus, nitrogen, and copper; carotenoids; sugars, etc., in tomato fruits (Hart et al., 2015). An experiment revealed that co-inoculation of AMF with bacteria (*Pseudomonas* spp.) increased quality of fruits by increasing the micro-nutrients as well as vitamins. Co-inoculation increased the amount of citric acid, sugars, malic acid, and beta carotene (Bona et al., 2017, 2018). Another experiment in strawberry plants where co-inoculation of AMF was with plant growth promoting bacteria showed better fruit quality and larger fruits enriched with folic acid as well as ascorbic acid (Bona et al., 2015). AMF modulates kernel proteome of *Zea mays* (Bona et al., 2016). Srivastava et al. (2015) showed application of *R. intraradices* in tomato plants promoted plant defense by increasing terpinoid synthesis. Another experiment in Bangladesh resulted in increased mineral uptake and protein concentration after inoculation of AMF, cow dung, and phosphorus in red amaranth and Indian spinach

(Tahjib et al., 2018). Co-inoculation of AMF with cow (fed with rock phosphate enriched diet) dung compost increases phosphorus solubility and absorption by mung bean plants. The organic method can be a suitable alternative of phosphate fertilizers available in the local market (Wahid et al., 2019). According to Hu et al. (2016), AMF association was responsible for lowering cadmium accumulation in cucumber plant parts cultivated in cadmium-contaminated soil for 8 weeks. AMF inoculation increases the synthesis of polyphenols and also antioxidants in selectively breeded oak leaf lettuce (Avio et al., 2017). After an extensive study Pedone and team revealed that AMF application enhanced synthesis of plant secondary metabolites of medicinal value (Pedone et al., 2015). Castellanos-Morales et al. (2010) reported that the beneficial rhizosphere microorganisms not only can enhance the nutrient level of crops but also can improve the crops quality, for example, results showed that AMF-colonized *Fragaria* × *Ananassa* plant exhibited increased levels of secondary metabolites resulting in improved antioxidant property. Through influencing and developing carotenoids and certain volatile compounds, AMF may improve the dietary quality of crops (Hart et al., 2015). Bona et al. (2017) showed beneficial effects of AMF on tomato production. Zeng et al. (2014) have reported increased contents of organic acids, sugars, flavonoids, vitamin C, and minerals due to *G. versiforme* treatment in increased citrus fruit quality. AMF has been used in large-scale field maize production (Sabia et al., 2015), yam (Lu et al., 2015), and potato production (Hijri, 2016), suggesting that AMF has considerable potential to increase crop yields. AMF can also improve the biosynthesis of valuable phytochemicals in edible plants and make them suitable for the healthy food chain (Sbrana et al., 2014; Rouphael et al., 2015). AMF inoculation resulted in positive effects even in floriculture. Different experiments showed tolerance to heavy metal toxicity in marigold, drought tolerance in pelargonium, and increased floral quality in some other plants (Bianciotto et al., 2016).

4.4 CONCLUSIONS

The application of AM in horticulture or agriculture is highly beneficial, but in large aspect the external and environmental factors are being considered for a better result. In conclusion, AMF inoculation in agro-ecosystems to manage scientific agricultural cropping system has become a great support of ecological services and the medium of changing the status of global food insecurity to future food security.

KEYWORDS

- **arbuscular mycorrhizal fungi**
- **bio-fertilizer**
- **drought**
- **heat**
- **microbes**
- **salinity**

REFERENCES

Aguilera, P.; Larsen, J.; Borie, F.; Berríos, D.; Tapia, C.; Cornejo, P. New evidences on the contribution of arbuscular mycorrhizal fungi inducing Al tolerance in wheat. *Rhizosphere* **2018**, *5*, 43–50.

Avio, L.; Sbrana, C.; Giovannetti, M.; Frassinetti, S. Arbuscular mycorrhizal fungi affect total phenolics content and antioxidant activity in leaves of oak leaf lettuce varieties. *Sci. Hort.* **2017**, *224*, 265–271.

Azcón-Aguilar, C.; Barea, J. M. Applying mycorrhiza biotechnology to horticulture: Significance and potentials. *Sci. Hort.* **1997**, *68*(1–4), 1–24.

Aziez, A. F.; Indradewa, D.; Cahyono, O.; Priyadi, S. Effect of vesicular arbuscular mycorrhiza on the growth and the characteristics of rice varieties in rainfed lowland rice cultivation. *J. Agron.* **2019**, *18*(1), 27–32.

Balestrini, R.; Lumini, E.; Borriello, R.; Bianciotto, V. Plant-soil biota interactions. In *Soil Microbiology, Ecology and Biochemistry,* Paul, E. A. (ed.), **2015**, pp 311–338.

Bao, X.; Wang, Y.; Olsson, P. A. Arbuscular mycorrhiza under water-carbon–phosphorus exchange between rice and arbuscular mycorrhizal fungi under different flooding regimes. *Soil Biol. Biochem.* **2019**, *129*, 169–177.

Basu, S.; Rabara, R. C.; Negi, S. AMF: The future prospect for sustainable agriculture. *Physiol. Mol. Plant Pathol.* **2018**, 102, 36–45.

Baum, C.; El-Tohamy, W.; Gruda, N. Increasing the productivity and product quality of vegetable crops using arbuscular mycorrhizal fungi: A review. *Sci. Hort.* **2015**, *187*, 131–141.

Baum, C.; El-Tohamy, W.; Gruda, N. Increasing the productivity and product quality of vegetable crops using arbuscular mycorrhizal fungi: A review. *Sci. Hort.* **2015**, *187*, 131–141.

Bedini, S.; Pellegrino, E.; Avio, L.; Pellegrini, S.; Bazzoffi, P.; Argese, E.; Giovannetti, M. Changes in soil aggregation and glomalin-related soil protein content as affected by the arbuscular mycorrhizal fungal species *Glomus mosseae* and *Glomus intraradices*. *Soil Biol. Biochem.* **2009**, *41*(7), 1491–1496.

Bernaola, L.; Stout, M. J. Effects of arbuscular mycorrhizal fungi on rice-herbivore interactions are soil-dependent. *Sci. Rep.* **2019**, *9*(1), 1–12.

Bianciotto, V.; Victorino, I.; Scariot, V.; Berruti, A. Arbuscular mycorrhizal fungi as natural biofertilizers: Current role and potential for the horticulture industry. *III Int. Symp. Woody Ornamentals Temperate Zone* **2016**, *1191*, 207–216.

Bona, E.; Cantamessa, S.; Massa, N.; Manassero, P.; Marsano, F.; Copetta, A.; Lingua, G.; D'Agostino, G.; Gamalero, E.; Berta, G. Arbuscular mycorrhizal fungi and plant growth-promoting pseudomonads improve yield, quality and nutritional value of tomato: A field study. *Mycorrhiza* **2017**, *27*(1), 1–11.

Bona, E.; Lingua, G.; Manassero, P.; Cantamessa, S.; Marsano, F.; Todeschini, V.; Copetta, A.; D'Agostino, G.; Massa, N.; Avidano, L.; Gamalero, E.; Berta, G. AM fungi and PGP pseudomonads increase flowering, fruit production, and vitamin content in strawberry grown at low nitrogen and phosphorus levels. *Mycorrhiza* **2015**, *25*(3), 181–193.

Bona, E.; Scarafoni, A.; Marsano, F.; Boatti, L.; Copetta, A.; Massa, N.; Gamalero, E.; D'Agostino, G.; Cesaro, P.; Cavaletto, M.; Berta, G. Arbuscular mycorrhizal symbiosis affects the grain proteome of *Zea mays*: a field study. *Sci. Rep.* **2016**, *6*(1), 1–14.

Bona, E.; Todeschini, V.; Cantamessa, S.; Cesaro, P.; Copetta, A.; Lingua, G.; Gamalero, E.; Berta, G.; Massa, N. Combined bacterial and mycorrhizal inocula improve tomato quality at reduced fertilization. *Sci. Hort.* **2018**, *234*, 160–165.

Cabral, C.; Ravnskov, S.; Tringovska, I.; Wollenweber, B. Arbuscular mycorrhizal fungi modify nutrient allocation and composition in wheat (*Triticum aestivum* L.) subjected to heat-stress. *Plant Soil* **2016**, *408*(1–2), 385–399.

Chen, M.; Arato, M.; Borghi, L.; Nouri, E.; Reinhardt, D. Beneficial services of arbuscular mycorrhizal fungi–from ecology to application. *Front. Plant Sci.* **2018**, *9*, 1–14.

Costanza, R.; d'Arge, R.; de Groot, R.; Farber, S.; Grasso, M.; Hannon, B.; Limburg, K.; Naeem, S.; O'Neill, R. V.; Paruelo, J.; Raskin, R. G.; Sutton, P.; Van den Belt, M. The value of the world's ecosystem services and natural capital. *Nature* **1997**, *387*, 253–260.

Duhamel, M.; Vandenkoornhuyse, P. Sustainable agriculture: possible trajectories from mutualistic symbiosis and plant neodomestication. *Trends Plant Sci.* **2013**, *18*(11), 597–600.

El Omari, B.; El Ghachtouli, N. Arbuscular mycorrhizal fungi-weeds interaction in cropping and unmanaged ecosystems: A review. *Symbiosis* **2021**, *83*, 279–292. https://doi.org/10.1007/s13199–021–00753–9.

El-Sharkawy, H. H.; Rashad, Y. M.; Ibrahim, S. A. Biocontrol of stem rust disease of wheat using arbuscular mycorrhizal fungi and *Trichoderma* spp. *Physiol. Mol. Plant Pathol.* **2018**, *103*, 84–91.

Emmanuel, O. C.; Babalola, O. O. Productivity and quality of horticultural crops through co-inoculation of arbuscular mycorrhizal fungi and plant growth promoting bacteria. *Microbiol. Res.* **2020**, *239*, 126569. doi: 10.1016/j.micres.2020.126569

Ercoli, L.; Schüßler, A.; Arduini, I.; Pellegrino, E. Strong increase of durum wheat iron and zinc content by field-inoculation with arbuscular mycorrhizal fungi at different soil nitrogen availabilities. *Plant Soil* **2017**, *419*(1–2), 153–167.

Fiorilli, V.; Wang, J. Y.; Bonfante, P.; Lanfranco, L.; Al-Babili, S. Apocarotenoids: Old and new mediators of the arbuscular mycorrhizal symbiosis. *Front. Plant Sci.* **2019**, *10*(1186), 1–9.

Foley, J. A.; DeFries, R.; Asner, G. P.; Barford, C.; Bonan, G.; Carpenter, S. R.; Chapin, F. S.; Coe, M. T.; Daily, G. C.; Gibbs, H. K.; Helkowski, J. H.; Holloway, T.; Howard, E. A.; Kucharik, C. J.; Monfreda, C.; Patz, J. A.; Prentice, I. C.; Ramankutty, N.; Snyder, P. K. Global consequences of land use. *Science* **2005**, *309*, 570–574.

Gianinazzi, S.; Gollotte, A.; Binet, M. N.; van Tuinen, D.; Redecker, D.; Wipf, D. Agroecology: The key role of arbuscular mycorrhizas in ecosystem services. *Mycorrhiza* **2010**, *20*(8), 519–530.

Goh, H. H.; Sloan, J.; Malinowski, R.; Fleming, A. Variable expansion expression in *Arabidopsis* leads to different growth responses. *J. Plant Physiol.* **2014**, *171*(3–4), 329–339.

Hart, M.; Ehret, D. L.; Krumbein, A.; Leung, C.; Murch, S.; Turi, C.; Franken, P. Inoculation with arbuscular mycorrhizal fungi improves the nutritional value of tomatoes. *Mycorrhiza* **2015**, *25*(5), 359–376.

Hijri, M. Analysis of a large dataset of mycorrhiza inoculation field trials on potato shows highly significant increases in yield. *Mycorrhiza* **2016**, *26*(3), 209–214.

Hu, J.; Tsang, W.; Wu, F.; Wu, S.; Wang, J.; Lin, X.; Wong, M. H. Arbuscular mycorrhizal fungi optimize the acquisition and translocation of Cd and P by cucumber (*Cucumis sativus* L.) plant cultivated on a Cd-contaminated soil. *J. Soils Sediments* **2016**, *16*(9), 2195–2202.

Imperiali, N.; Morales, X. C.; Schlaeppi, K.; Fesselet, M.; Villacrés, D.; Jaffuel, G.; Bender S. F.; Dennert, F. G.; Blanco-Perez, R.; Van Der Heijden, M. G. A.; Maurhofer, M.; Mascher, F.; Turlings, T.; Keel, C. J.; Campos-Herrera, R. Combined field inoculations of *Pseudomonas* bacteria, arbuscular mycorrhizal fungi and entomopathogenic nematodes and their effects on wheat performance. *Front. Plant Sci.* **2017**, *8*(1809), 1–17.

Issa, M.; Ngakou, A.; Haouvang, C. L.; Nukenine, E. N. Potentials of arbuscular mycorrhiza fungi (AMF) and neem (*Azadirachta indica* A. Juss.) leaves extract as biological control agents against the sweet potato weevil (*Cylas puncticollis* Boh.) in two agro-ecological zones of Cameroon. *J. Exp. Agric. Int.* **2017**, *17*(1), 1–13.

Jaffuel, G.; Imperiali, N.; Shelby, K.; Campos-Herrera, R.; Geisert, R.; Maurhofer, M.; Loper, J. E.; Keel, C. H.; Turlings, T.; Hibbard, B. E. Protecting maize from rootworm damage with the combined application of arbuscular mycorrhizal fungi, *Pseudomonas* bacteria and entomopathogenic nematodes. *Sci. Rep.* **2019**, *9*(1), 1–12.

Kadian, N.; Yadav, K.; Jangra, E.; Aggarwal, A. Influence of host plant and rice straw as substrate on mass multiplication of arbuscular mycorrhizal fungi for large-scale agricultural application. *Int. J. Recycl. Org. Waste Agric.* **2019**, 8, 21–26.

Krüger, M.; Krüger, C.; Walker, C.; Stockinger, H.; Schüßler, A. Phylogenetic reference data for systematics and phylotaxonomy of arbuscular mycorrhizal fungi from phylum to species level. *New Phytol.* **2012**, *193*(4), 970–984.

Li, H.; Chen, X. W.; Wong, M. H. Arbuscular mycorrhizal fungi reduced the ratios of inorganic/organic arsenic in rice grains. *Chemosphere* **2016**, *145*, 224–230.

Liu, C.; Ravnskov, S.; Liu, F.; Rubæk, G. H.; Andersen, M. N. Arbuscular mycorrhizal fungi alleviate abiotic stresses in potato plants caused by low phosphorus and deficit irrigation/partial root-zone drying. *J. Agric. Sci.* **2018**, *156*(1), 46–58.

Lu, F.; Lee, C.; Wang, C. The influence of arbuscular mycorrhizal fungi inoculation on yam (*Dioscorea* spp.) tuber weights and secondary metabolite content. *Peer J.* **2015**, 3, 12–66.

Maillet, F.; Poinsot, V.; André, O.; Puech-Pagès, V.; Haouy, A.; Gueunier, M.; Cromer, L.; Giraudet, D.; Formey, D.; Niebel, A.; Martinez, E. A. Fungal lipochitooligosaccharide symbiotic signals in arbuscular mycorrhiza. *Nature* **2011**, *469*(7328), 58–63.

Owen, D.; Williams, A. P.; Griffith, G. W.; Withers, P. J. Use of commercial bio-inoculants to increase agricultural production through improved phosphrous acquisition. *Appl. Soil Ecol.* **2015**, *86*, 41–54.

Parvin, S.; Van Geel, M.; Yeasmin, T.; Lievens, B.; Honnay, O. Variation in arbuscular mycorrhizal fungal communities associated with lowland rice (*Oryza sativa* L.) along a gradient of soil salinity and arsenic contamination in Bangladesh. *Sci. Total Environ.* **2019**, *686*, 546–554.

Pedone-Bonfim, M. V. L.; da Silva, F. S. B.; Maia, L. C. Production of secondary metabolites by mycorrhizal plants with medicinal or nutritional potential. *Acta Physiologiae Plantarum* **2015**, *37*(2), 27.

Pérez, Y. M.; Charest, C.; Dalpé, Y.; Séguin, S.; Wang, X.; Khanizadeh, S. Effect of inoculation with arbuscular mycorrhizal fungi on selected spring wheat lines. *Sustain. Agric. Res.* **2016**, *5*(4), 1–29.

Pimm, S. L. The value of everything. *Nature* **1997**, *387*, 231–232.

Robertson, G. P.; Swinton, S. M. Reconciling agricultural productivity and environmental integrity is a grand challenge for agriculture. *Ecol. Soc. Am.* **2005**, *3*, 39–46

Rouphael, Y.; Franken, P.; Schneider, C.; Schwarz, D.; Giovannetti, M.; Agnolucci, M. Arbuscular mycorrhizal fungi act as bio-stimulants in horticultural crops. *Sci. Hort.* **2015**, *196*, 91–108.

Rouphael, Y.; Franken, P.; Schneider, C.; Schwarz, D.; Giovannetti, M.; Agnolucci, M.; Colla, G. Arbuscular mycorrhizal fungi act as biostimulants in horticultural crops. *Sci. Hort.* **2015**, *196*, 91–108.

Sabia, E.; Claps, S.; Morone, G.; Bruno, A.; Sepe, L.; Aleandri, R. Field inoculation of arbuscular mycorrhiza on maize (*Zea mays* L.) under low inputs: Preliminary study on quantitative and qualitative aspects. *Italian J. Agron.* **2015**, *10*, 30–33.

Sbrana, C.; Avio, L.; Giovannetti, M. Beneficial mycorrhizal symbionts affecting the production of health-promoting phytochemicals. *Electrophoresis* **2014**, *35*, 1535–1546.

Seguel, A.; Castillo, C. G.; Morales, A.; Campos, P.; Cornejo, P.; Borie, F. Arbuscular mycorrhizal symbiosis in four Al-tolerant wheat genotypes grown in an acidic Andisol. *J. Soil Sci. Plant Nutr.* **2016**, *16*(1), 164–173.

Seguel, A.; Cumming, J.; Cornejo, P.; Borie, F. Aluminum tolerance of wheat cultivars and relation to arbuscular mycorrhizal colonization in a non-limed and limed Andisol. *Appl. Soil Ecol.* **2016,** *108*, 228–237.

Shrivastava, G.; Ownley, B. H.; Augé, R. M.; Toler, H.; Dee, M.; Vu, A.; Kőllner, T.; Chen, F. Colonization by arbuscular mycorrhizal and endophytic fungi enhanced terpene production in tomato plants and their defense against a herbivorous insect. *Symbiosis* **2015**, *65*(2), 65–74.

Smith, S. E.; Read, D. J. *Mycorrhizal symbiosis*. Academic Press: Cambridge, 2008; p 787.

Tahjib-Ul-Arif, M.; Ghosh, A.; Chamely, S. G.; Haque, M. R.; Rahman, M. M. Arbuscular mycorrhizal fungi inoculation with organic matter and phosphorus supplementation enhance nutrient contents of *Amaranthus tricolor* L. and *Basella alba* L. by improving nutrients uptake. *Trop. Plant Res.* **2018**, *5*(3), 375–384.

Tohidi Moghadam, H. R.; Heydarian, A.; Donath, T. W.; Sohrabi, M. Study of effect of arbuscular mycorrhiza (*Glomus intraradices*) fungus on wheat under nickel stress. *Agron. Res.* **2018**, *16*(4), 1660–1667.

Verbruggen, E.; Röling, W. F.; Gamper, H. A.; Kowalchuk, G. A.; Verhoef, H. A.; van der Heijden, M. G. Positive effects of organic farming on below-ground mutualists: Large-scale comparison of mycorrhizal fungal communities in agricultural soils. *New Phytol.* **2010**, *186*(4), 968–979.

Wahid, F.; Sharif, M.; Fahad, S.; Adnan, M.; Khan, I. A.; Aksoy, E.; Ali, A.; Sultan, T.; Alam, M.; Saeed, M.; Ullah, H.; Basir, A.; Noor, M. A.; Khan, N. A. Arbuscular mycorrhizal fungi improve the growth and phosphorus uptake of mung bean plants fertilized with composted rock phosphate fed dung in alkaline soil environment. *J. Plant Nutr.* **2019**, *42*(15), 1760–1769.

Willer, H.; Kilcher, L. *The world of organic agriculture-statistics and emerging trends 2012*. Research Institute of Organic Agriculture (FiBL), Frick, International Federation of Organic Agriculture Movements (IFOAM), Bonn, 2012.

Willer, H.; Lernoud, J. *The world of organic agriculture, statistics and emerging trends 2017*; Research Institute of Organic Agriculture FiBL and IFOAM-Organics International, 2017; pp 1–336.

Zhang, S.; Lehmann, A.; Zheng, W.; You, Z.; Rillig, M. C. Arbuscular mycorrhizal fungi increase grain yields: A meta-analysis. *New Phytol.* **2019**, *222*(1), 543–555.

Zhu, X.; Song, F.; Liu, S.; Liu, F. Arbuscular mycorrhiza improve growth, nitrogen uptake, and nitrogen use efficiency in wheat grown under elevated CO_2. *Mycorrhiza* **2016**, *26*(2), 133–140.

CHAPTER 5

ARBUSCULAR MYCORRHIZA IN CITRUS

SAMIA KHANUM[1], ABDEL RAHMAN MOHAMMAD AL-TAWAHA[2], ABDEL RAZZAQ AL-TAWAHA[3], HIBA ALATRASH[4], ABDUR RAUF[5], ARUN KARNWAL[6], ABHIJIT DEY[7], NUJOUD ALIMAD[8], DEVARAJAN THANGADURAI[9], JEYABALAN SANGEETHA[10], SAHER ISLAM[11], IMRAN[12], AMANULLAH[12], SHAH KHALID[12], PALANI SARANRAJ[13], and ELIF GUNAL[14]

[1]*Department of Botany, University of the Punjab, Lahore, Pakistan*

[2]*Department of Biological Sciences, Al Hussein Bin Talal University, Maan, Jordan*

[3]*Department of Crop Science, Faculty of Agriculture, University Putra Malaysia, Serdang, Selangor, Malaysia*

[4]*General Commission for Scientific Agricultural Research, Syria*

[5]*Department of Chemistry, University of Swabi, Anbar, Khyber Pakhtunkhwa, Pakistan*

[6]*Department of Microbiology, School of Bioengineering and BioSciences, Lovely Professional University, Phagwara, Punjab, India*

[7]*Department of Life Sciences, Presidency University, Kolkata, India*

[8]*Faculty of Agriculture, Damascus University, Damascus, Syria*

[9]*Department of Botany, Karnatak University, Dharwad, Karnataka, India*

Mycorrhizal Technology: Managing Plant Stress and Mitigating Climate Change using Mycorrhizae.
Jeyabalan Sangeetha, Abdel Rahman Mohammad Al-Tawaha, & Devarajan Thangadurai (Eds.)

[10]Department of Environmental Science, Central University of Kerala, Kasaragod, Kerala, India

[11]Institute of Biochemistry and Biotechnology, University of Veterinary and Animal Sciences, Lahore, Pakistan

[12]Department of Agronomy, The University of Agriculture, Peshawar, Pakistan

[13]Department of Microbiology, Sacred Heart College (Autonomous), Tirupattur 635601, Tamil Nadu, India

[14]Department of Soil Science and Plant Nutrition, Faculty of Agriculture, Harran University, Sanliurfa, Turkey

ABSTRACT

Plant root and fungus combined, results in a single structure through which exchange of nutrients takes place. There are many other mycorrhizal species found in the rhizosphere of citrus plants, but the major AMF found is *Glomus* species. The mycorrhizal association is also helpful in maintaining yield of the citrus plant. The soil of citrus orchards has many communities of AMF instead of a single species. So, the root of the citrus plant may be colonized by more than one AMF at the same point, which indicates that the soil of citrus orchard is rich in diversity of AMF, and this will improve the nutrient's uptake and growth of a citrus plant. Citrus plants inoculated with mycorrhiza have more resistance against soil-borne pathogens and diseases than nonmycorrhizal plants. The mycorrhizal association releases the antibiotics in the soil which can control soil microorganisms such as nematodes and pathogenic fungi so the chances of infection reduced.

5.1 INTRODUCTION

To maintain the interaction between nutrients level and environmental factors tree plants have some adaptive and regulatory mechanisms which demand the whole plant nutritional requirement and regulation of above ground mechanisms like photosynthesis (Lukac et al., 2010). To improve the nutrient's uptake efficiency the tree plants have the ability to utilize the new resources just like the modification of the rhizosphere environment and the symbiotic association with mycorrhizal fungi. It is reported that the most important microorganisms that are symbiotically associated with the terrestrial plants are mycorrhizal fungi help in the uptake of important macro and micro nutrients, beneficial for normal plant growth and development. The AMF form arbuscules and vesicles on the penetration of the root cells of the plant without causing any harmful effect to them. Evolutionary studies reported that these associations have been found 460 million years ago (Redecker et al., 2000). The mycorrhiza is of many types especially ECM and AM. AMF is the major type of mycorrhizal fungi which form the symbiotic association with 90% of terrestrial plants which include citrus plants also (Gadkar et al., 2001). AMF have a broad variety of host species; according to an estimate about 150 AM species form association with 225,000 plant species (Wu et al., 2009). There are many other mycorrhizal species found in the rhizosphere of citrus plants, but the major AMF found is *Glomus* species. Through various species, this is proved that the predominant species in the rhizosphere of a citrus plant is *Glomus*. So this chapter provides the role of AMF when forming a symbiotic association with a citrus plant.

5.2 MYCORRHIZAL EFFECT ON THE GROWTH OF CITRUS

It is reported through various studies that mycorrhiza shows various responses on the growth of the citrus plant. The roots of mostly agronomic crops have association with the collection of many AMF that are differentiated in their functional benefits that include the transfer of nutrients, protection from pathogens, and water uptake (Verbruggen and Kiers, 2010). The most important benefit to the host plant physiology is the enhanced nutrients and water uptake. The main purpose of mycorrhizal association is when the Phosphorous is limited in the soil this association improved the uptake of P; hence, the growth of the plant is improved. When plants grow in nutrients depleted soil the mycorrhizal association helps in improving the nutrient's uptake efficiency of the plant from native sources (Marschner, 1994). The roles of AMF during this association are: (1) improving the nutrient's uptake, especially phosphorus, zinc, copper, ammonium, etc., (2) improving water uptake, (3) improving the growth of the plant, (4) improving fruit quality, (5) increasing the resistance against stresses, and (6) resistance against soil-borne diseases. Citrus plants are much selected to their environment and soil condition; they prefer to propagate in dry and light soil. When it grows under these circumstances plants acquire maximum benefit from the mycorrhizal association. The importance of mycorrhizal association in the citrus plant was first reported by Read and Fremont (1935). In less fertile and coarse textured soil condition mycorrhizal association is very beneficial to the plants. When methyl bromide is used in soil having nutrients deficiency, it eliminates the mycorrhizal fungi as it has the ability to eliminate soil-borne microorganisms which results in reduction of growth of the plants. When soils are fumigated for the cultivation of citrus, there is decrease in nutrients and stunned growth was reported. It is because of toxic effect of fumigation which eliminates the mycorrhizal fungi (Menge et al., 1978). As the soil-borne microorganisms are not good for citrus so fumigation and sterilization of soil are required, but it eliminates the mycorrhizal fungi as well and through various studies the positive role of mycorrhizae on plant growth has been reported.

5.3 THE ROLE OF INDIGENOUS MYCORRHIZA ON GROWTH OF CITRUS

The mycorrhizal association is also helpful in maintaining yield of the citrus plant. It is reported that the soil of citrus orchards has many communities of AMF instead of a single species. So, the root of the citrus plant may be colonized by more than one AMF at the same point, which indicates that the soil of citrus orchard is rich in diversity of AMF, and this will improve the nutrient's uptake and growth of a citrus plant. The same AMF but isolated with different geographical locations are varied in their ability to colonize the plant and improving growth rate (Graham et al., 1996). The quantification of spores of AMF in the soil of 12 orchards and 10 nurseries of citrus shows the great diversity of AMF species and the descending order of occurrence of these species was *Glomus macrocarpum* > *Scutellospora heterogama* > *Acaulospora scrobiculata* = *Acaulospora birreticulata* > *Glomus invermaium* = *Glomus occultum* = *Entrophospora colombiana* > *Glomus claroideum* = *Glomus constrictum* > *Scutellospora persica* (de Souza et al., 2002). In the same root area of a citrus plant there is a great diversity of AMF. It is reported that about 13 indigenous AMF have been isolated from the orchard of citrus rhizosphere in Thailand (Watanarojanaporn et al., 2011). Most of the studies based on the role of a single isolated AMF on the physiology and growth of the citrus plant. The inoculation of indigenous mycorrhizal fungi has the high response rate on citrus than the single AMF inoculation to the citrus plant. The significance of the species diversity to the field application and function of AMF is not very clear because of the lack of comparison data of various communities of AMF on the growth and physiology of the plant (Graham, 1986). It is reported that, when the citrus plant is colonized by the different numbers of AMF species then increase in height, diameter, and the length of root and leaf area has been noted (Sharma et al., 2009) (Table 5.1).

TABLE 5.1 Various Roles of AMF in Citrus Beyond Soil Fertility and Plant Nutrition.

AMF species	Response on plant	Reference
	Disease suppression	
Acaulospora tuberculata, Glomus etunicatum	Tiogun and Tangerine produced highest mycorrhizal efficiency index, P-uptake, *Phytophthora* suppression.	Watanarojanaporn et al. (2011)
Glomus fasciculatus	Suppression in *Phytophthora* infestation	Davis et al. (1980)
Glomus mosseae	Significant suppression of root rot of mandarin (*Citrus reticulata*) caused by *Fusarium oxysporum* when co-inoculated with *B. pumilus.*	Chakraborty et al. (2011)
	Root system architecture	
G. margarita, G. mosseae, G. versiform, G. caledonium	Formation of lateral roots of high order, colonization of this fungus induced more fine roots, and less coarse roots	Yao et al. (2009)
Glomus occultum	Suppression of root growth of volkamer lemon, less root dry weight, and total root length under continuously moist condition (–0.01 MPa) due to higher carbon costs to the plant	Fidelibus et al. (2000)
Glomus species	Stimulation in root growth of volkamer, lemon	Fidelius et al. (2000)
	Abiotic stress tolerance	
G. mosseae	Increased tolerance of citrus (*Citrus tangerina*) seedlings to soil water deficit through its higher root infection and antioxidant enzymatic activity in plants	Ni et al. (2013)
Glomus versiforme	Enhance growth and Mg distribution in Newhall seedlings under low Mg concentration of soil	Xiao et al. (2014)
Glomus intraradices, Glomus mosseae	Improved phosphorous nutrition of *Citrus volkameriana*	Haghighatnia et al. (2011)
	Flooding tolerance	
Glomus intraradices	Affecting root than hyphal nutrient uptake characterized by reduction in nitrogen and P content in Carrizo citrange and sour orange	Hartmond et al. (1987)

TABLE 5.1 *(Continued)*

AMF species	Response on plant	Reference
	Salinity tolerance	
Glomus intraradices, *G. versiforme*, *G. etunicatum*	EC of 162 ds/m with spore density of 100 spores/10 g soil due to stimulation of sporulation due to salt stress	Aliasgharzadeh et al. (2001)
Archaeospora, *Acaulospora*, *Glomus*	Increased sporulation at soil EC of 40.2 ds/m	Wang et al. (2004)
Rhizophagus irregularis, *Funneliformis massaceae*	Seedlings of cleopatra mandarin better response due to higher accumulation of P, K, Fe, and Cu in index values	Navarro et al. (2014)
Glomus intraradices	Improved acquisition of P, K, Mg, and Zn nutrients in leaves of sour orange and volkamer rootstocks	Khalil et al. (2011)
Glomus, *Gigasopora* species	Karna khatta, Troyer citrange performed better due to proline accumulation and higher Ca and Mg in leaves	Murkute et al. (2006)

Source: Adapted from Qiang-Sheng et al. (2017)

5.4 THE EFFECT OF MYCORRHIZAE ON GROWTH OF CITRUS SEEDLINGS

Mycorrhizal association has many beneficial impacts on the plants having horticultural importance. AMF improved the growth rate, quality of seedlings, the survival rate, height, and the diameter of stem, leaf numbers and area, length of shoot, dry mass, and production of the flower even on transferring to the field from the greenhouse (Wang and Xia, 2009). Through various studies it is reported that AMF colonization has the positive effect on the overall growth rate of the citrus plant under variable water content conditions. When in comparison with the nonmycorrhizal citrus plant the root and shoot growth, especially the fibrous root growth, increased in the mycorrhizal associated citrus plant (Tong et al., 2006). As previously discussed, the mycorrhizal association is beneficial when the soil is less fertile and has coarse texture, but when mycorrhizal fungi colonized citrus with high level of P use in the soil it decreased the development of plant (Jifon et al., 2002). The nutrient uptake efficiency of the mycorrhizal citrus plant is also more than the nonmycorrhizal one (Srivastava et al., 2002). The AMF also improve the mineralization thus enhancing the accessibility of iron content in the soil (Wang et al., 2009). It is reported that when AMF colonized the citrus plant, when growing at different pH levels, the efficiency of nutrient uptake and plant growth rate increased which also include plant height, stem diameter, the dry mass, and number of leaves; the accumulation of chlorophyll, iron uptake, and chelate reductase activity also improved.

5.5 NUTRIENTS UPTAKE AND ROLE OF MYCORRHIZAE

The response of citrus plant to AMF colonization is unpredictable specifically in the productive agriculture system. The only relationship studied is between the improved growth performance of orange and the density of inoculum of AMF. When inoculated with AMF the growth rate and nutrient's uptake of the plant increased. Before inoculation of AMF several species of mycorrhizae has been screened for the search for the best inoculum for the orange plant. It is reported through experiments that dry weight of shoot and root also improved in mycorrhizal citrus plants. All of this improved growth rate leads to the improvement in nutrition, increased plant height, root area, diameter of stem and biomass and also having more Phosphorous, Zinc, Cu, and Fe contents (Ortas et al., 2002a, 2002b).

The essential and important nutrients for the proper growth of the plant are C, H, O, N, P, K, Ca, Mg, S, Fe, Zn, Mn, B, Cu, Mo, Cl, and Ni. Other important mineral nutrients are obtained from water and soil. The nutrient deficiency in citrus is because of low nutrient's availability in the soil or the efficiency of the root to absorb these nutrients. The most suitable soil condition for the growth and development of citrus is the high pH value of soil, little nutrient's accessibility, and salinity, but the nutrition value of citrus is of great concern for the farmers when grown in such soil conditions. Because of different ecological situations it is difficult to analyze the nutritional level of a plant so the nutrient content of the plant is known before the management of nutritional level. The most common practice by the farmer is using high amounts of fertilizers which is unbalancing the nutrient content and also causes the reduction in the development of mycorrhizae (Ortas et al., 2002a). When colonized by mycorrhizae it is still important to know the optimum requirements of nutrients and the rate of uptake of nutrients. The colonization of AMF is predicted when the increased uptake of P is detected. Because of the high cost and low availability of P, it is suggested to use mycorrhizal association, crops with modified genotype, and highly resistant or renewable natural resources instead of fertilizers. The definition of uptake of P is the total rate of absorption of P in certain situations, but the demand of P is defined as the amount of P required for proper growth and fitness for the proper performance of a plant. The 1/3rd population around the world only received limited Fe and Zn by food chains and P, Zn, and Fe availability is also very limited in the soil resulting in the limited growth rate of the citrus, which is the reason for using chemical fertilizers to meet the nutrients demand of the plant. The nutrients-deficit soil will produce plants with low nutritional values, which affect human nutrition because plants are the main source through which nutrients enter into the food chain. As the citrus plant is consumed on the large scale in all over the world, to improve the mineral uptake of human for improving human health it is very important to improve the nutrient's level of the citrus plant. Mostly, citrus plants are in association with AMF which results in increase in the growth rate by increasing P content to the plant when the soil P content is limited. When colonized by AMF high P, Zn, Fe, N, K, and Cu content is found than in nonmycorrhizal citrus plants when grown under greenhouse environment. Through various studies it is reported that other important minerals also uptake efficiently than in nonmycorrhizal plants so to increase the nutritional level of citrus it is important to inoculate the citrus with AMF. The vegetative growth is also improved by AMF. AMF help in the increase in the Fe content in acidic soil, but not in the alkaline soil so AMF can also be used for bioremediation and to translocate the Fe to the host plant (Fig. 5.1).

5.6 ROOTSTOCKS AND MYCORRHIZAE

The difference in rootstock is based on their nutrients, requirements, and dependency on mycorrhizae. If the efficiency of nutrient's uptake of rootstock is not good and the mycorrhizal inoculation is not present in the soil, as fungicides and chemical fertilizers kill the AMF, the plant will not grow properly because of the deficiency of nutrients. The dependency of rootstock on mycorrhiza significantly varies and when inoculated by AMF increased root and shoot growth is reported (Dutra et al., 1996). The pattern of growth of root of rootstock affects the uptake of nutrients. As the root pattern of root stock is different from one another, the inoculation of mycorrhizae also varies in rootstocks. The nutrient's uptake is directly proportional to the surface area of root so it is suggested to choose the rootstock that can develop the better root system which also gives an idea that every rootstock of citrus plant should be tested individually. So rootstocks can differentiate to one another in nutrient's concentrations, ability of uptake of nutrients and

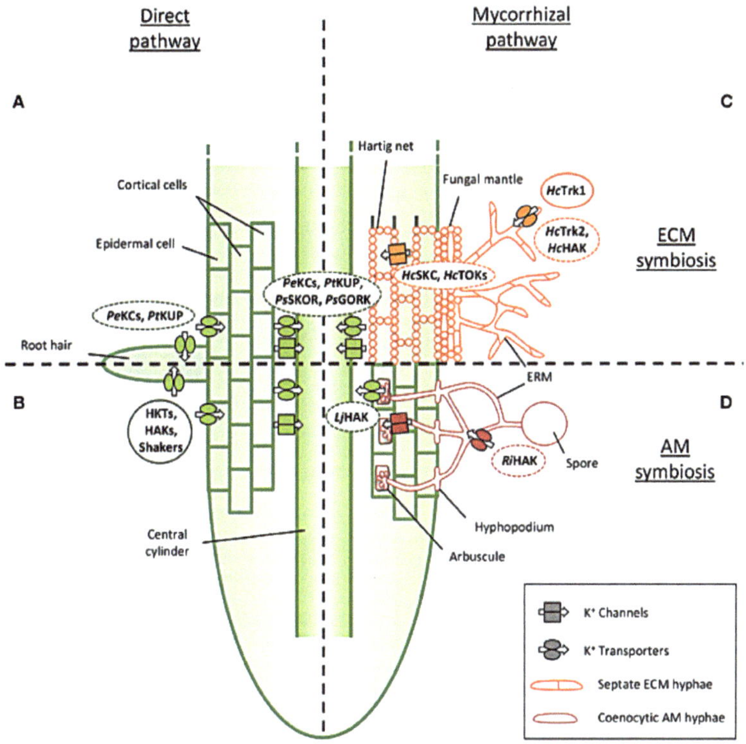

FIGURE 5.1 Current knowledge about potassium transport systems in mycorrhizal associations. (A, B) The transfer of potassium (K^+) from the soil to plant cells by the direct pathway needs plant transport systems in root uptake and release sites. (A) Shaker type channels from *Populus euphratica* (PeKCs) and a KUP transporter of *Populus trichocarpa* (PtKUP) could be involved in K^+ uptake in poplar trees (Zhang et al., 2010). (B) Members of HKT, HAK, and Shaker families were identified and characterized in various AM plants as transport systems involved in K nutrition. (C) To our knowledge, only K^+ transport systems of the fungus *Hebeloma cylindrosporum* are currently analyzed in ECM symbiosis, and just one, HcTrk1, is already characterized (Corratgé et al., 2007; Garcia et al., 2014). Few transcriptomic data suggest the role of some plant proteins allowing K^+ uptake from the apoplasm. (D) EST library of *Rhizophagus irregularis* allows the identification of several K^+ transport related proteins (Casieri et al., 2013), and transcriptomic analysis revealed an high overexpression of a K^+ transporter in inoculated *Lotus japonicus* (Guether et al., 2009). Full lines indicate transport systems whose capability in K^+ transport was verified. Dotted lines indicate transport systems whose involvement in K^+ transport during mycorrhizal symbioses is suggested. ECM fungal structures: Extraradical hyphae (ERM), fungal mantle, and Hartig net. AM fungal structures: ERM, spore, hyphopodium, and arbuscule. Plant root cells: roots hair, epidermal cells, and cortical cells, central cylinder. Hc, *Hebeloma cylindrosporum;* Lj, *Lotus japonicus;* Pe, *Populus euphratica;* Ps, *Pinus sylvestris;* Pt, *Populus tremula;* Ri, *Rhizophagus irregularis.*

Source: Reprinted from Garcia and Zimmerman (2014). https://creativecommons.org/licenses/by/4.0/

utilizing of these mineral nutrients and K/Mg ratio which is critical for nutritional balance. Another strategy to select citrus is the efficiency of high yield in the soil with nutrient's deficiency that is very prominent and common practice in the developing countries that spent high amounts on fertilizers. Another important selection criterion of citrus rootstock is their ability to adapt to soil conditions. It is reported through various studies that when citrus rootstocks inoculated with different important mycorrhizal species like *Glomus* and grown in unsterilized soil, improved nutrient's uptake and growth rate are reported which leads to the increased plant root, height and leaf area, diameter of stem and biomass, and also having more P, Zn, Cu, and Fe contents. The high amount of P content has been reported when the citrus plant treated with AMF and rock phosphate used as a P source (Srivastava et al., 2002). It is reported that the most effective AMF for improved citrus development which used as rootstocks in nurseries are *Glomus* species. These cultivars also have resistance to different stresses like drought, salinity, soil-borne diseases, and are highly adapted to soil and environmental conditions.

5.7 DEPENDENCE ON MYCORRHIZA

Plants are reliant on the mycorrhizal association for their improved growth and yield under natural habitat (Gerdemann, 1975), especially the tree plants are obligatory reliant on mycorrhizae (Ortas et al., 2002b). So mycorrhizae are required for the establishment of the plants that are depending on this association. The growth rate of the plant is based on soil fertility, P content availability, type of soil, and species of AMF (Graham et al., 1982). Rootstocks also vary in their dependency on the AMF; it is reported that rootstocks are less dependent when having long and more root hairs than with small and a small number of root hairs (Wu et al., 2009). It is assumed that the closely related genotypes the less dependent species has greater tendency to reduce the frequency but the degree of establishment is undisturbed even when high P content is found in the soil (Graham and Eissenstat, 1994). It is noted that the rootstocks having less reliance on mycorrhiza have increased hydraulic conductivity of root, improved transpiration, and CO_2 assimilation rates. To select AM-dependent rootstock, the use of water is also considered especially in the Middle East counties where the water is inadequate. The reduction in growth is caused by mycorrhizal fungi because of the high expenditure of carbon when growing in high P presence in the soil. Drought stress can affect the mycorrhizal-dependent citrus. The *Glomus* species considered more efficient AMF improved transpiration and stomatal conductance and water relation of citrus plant (Wu et al., 2007b). The most important reason for the dependency of citrus on AMF is increase in acquisition of phosphorous from the soil with limited P content. Several experiments have been designed to explain the dependency of citrus on mycorrhizal fungi (Ortas et al., 2002b). The dependency of citrus plant on mycorrhizal fungi also varies according to the type of rootstock as well (Ortas et al., 2002a). It is also reported that the mycorrhizae are less effected by the Zn nutrition than the P nutrition when citrus plant grows under different Zn and P supply (Ortas et al., 2002b). When the Zn supply is increased the dependency is slightly decreased, but if the supply of P is decreased it highly affects the dependency. As it is noted that there is no effect of Zn and the P supply on nonmycorrhizal plants, it is concluded that citrus plants are highly reliant on the mycorrhizae. The high Zn content is found in the mycorrhizal-associated plant, but no evidence has been found about whether mycorrhiza plays any role in increasing Zn content or it is just because of plant growth rate. As mycorrhizal fungi increases the Zn content it is astonished that mycorrhiza is depending on Zn content instead of P. Citrus plant when inoculated with AMF improved vegetative growth and is depending on mycorrhiza for nutrient supply. The genotype of citrus varies according to their dependency on mycorrhiza because of fertility of soil so as to select mycorrhizal-dependent rootstocks; we should consider P and Zn content. The dependency is an inherited character and for this the important parameters are nutrient's requirement and uptake efficiency more importantly P requirement. As the plant survival is also based on their dependency on mycorrhiza, it is significant to arrange plants according to their dependency (Wu et al., 2006a). The mycorrhizal association has the positive relation with the spore density, but the negative relation with the P content so it concludes that high spore density and low phosphorous content in the soil leads to the better mycorrhizal association development with citrus roots.

5.8 MYCORRHIZAE EFFECT ON ROOT GROWTH

The mycorrhiza has a significant impact on the morphology of the root. Mycorrhizal colonization can modify the branching pattern of the roots by producing plant growth hormones. The difference in acquisition of nutrients among plant species is based on growth and colonization of the roots by AMF (Marschner, 1994). The morphology of plant may be affected by the AMF colonization results in the better root length and root-to-shoot ratios.

It is difficult to identify the effects of potential AMF on plant function if there is no improvement in mineral nutrition. Several studies proved that AMF improved the plant root growth independent of P nutrition. The citrus plant when inoculated with AMF improved the growth parameters, height, diameter, and root length, number of lateral roots and fibrous roots, length of the seedlings, fresh weight of root and shoot, chlorophyll content, and leaf area than in nonmycorrhizal plants (Tong et al., 2006; Cho et al., 2009; Sharma et al., 2009). Citrus with small roots are susceptible to the withering, but when inoculated with mycorrhiza it improved the survival rate of the nursery seedlings when transplanted to the field. The dramatic difference in the growth rate of citrus has been reported when inoculated with different mycorrhizal fungi (Ortas et al., 2002a). As we know that the soil was partially sterilized before planting citrus seedlings so the difference in growth may be due to the limited availability of P and high C cost. It is reported through various studies that mycorrhizal infection improved the root length, lateral, and fibrous root numbers (Tong et al., 2006). The mycorrhizal root shows the lower value of SLR, less root/soil respiration, and also 10% less mortality of fine roots in citrus than nonmycorrhizal citrus. As shoot growth is less sensitive to the dry soil conditions than root growth, mostly root growth is limited. Thus, the root shoot ratio shows that when root production increased the dependency on the rootstock decreased (Nemec, 1979).

5.9 MYCORRHIZA AND NUTRITION OF CITRUS

The hyphae of AMF penetrate the intracellular spaces of the soil and break down to increase the fertility of the soil. The external hyphal network is used for the uptake and transport of minerals from the soil. Thus plant when associated with AMF formed large mycorhizosphere for the better uptake of mineral and nutrients from the soil and also increased their resistance against the various stresses. There are several useful functions of the external hyphal network of the AMF. Once inoculated the hyphae penetrates and spreads to all the available spaces in the roots of citrus plant. Because of the branching pattern of the hyphae the surface area has increased; thus, more nutrients and water can be absorbed. The hyphal networks are actually the link between the plant and soil, by absorbing important minerals like Phosphorus, Zinc, Copper, and Ammonium from the soil and transporting them to the plants. The uptake of nutrients by mycorrhiza involves several mechanisms. One of the important mechanisms is transporting nutrients from the soil solution across the hyphal membrane. The link between hyphae and roots is based on soil and environmental factors. The presence of Phosphorous content in the soil is inversely linked with AMF colonization. The P level is differently related to various plant communities. The population of AMF depends on analyzing the resting spores. It is significant to study the form of mycorrhizal association under various ecological situations linked with the adaptation mechanisms of the plants. The nutrient's uptake, especially P uptake by AMF, has been studied widely, but the dependency on the AMF differentiated according to the P level in the soil, plant species and AMF type studied less. The major role of the hyphae is to absorb the less mobile nutrients from the soil. The hyphal capacity of uptake of nutrients varies among fungal species, but the hyphal capacity is not studied under different Zn applications and farming strategies. Also, it is not explained yet why mycorrhizal fungi have the high nutrient's uptake rate under greenhouse situation than in the field. So this can address what is the impact of the soil biota on acquisition of nutrients under field conditions. The physical and biological quality of the soil is also affected by the mycorrhizal fungi. The hyphal network binds soil particles into the aggregates that are water stable (Tisdall and Oades, 1979) which varies among fungal species. The formation of these aggregates results in the production of immune-reactive glycoprotein named as glomalin.

5.10 PLANT PROTECTION AND ROLE OF MYCORRHIZA

Mycorrhiza not only improved the water and nutrient's uptake of the plant but also increased their resistance against diseases and pathogens. Citrus plants inoculated with mycorrhiza have more resistance against soil-borne pathogens and diseases than nonmycorrhizal plants. The mycorrhizal association releases the antibiotics in the soil which can control soil microorganisms such as nematodes and pathogenic fungi so the chances of infection are reduced. The mycorrhizal colonization reduced the damage caused by pests by improving plant growth. When the mycorrhizal association is well developed it helped the plant by developing resistance against nematodes and pathogens (Sikora, 1992). When nematodes are present in the soil then the plant has all kinds of improved growth parameters. The pathogenic damage reduced when the mycorrhizal development increased. The presence of nematodes reduced the plant yield up to 45%, but when the plant associated with AMF the yield is reduced up to 25% only. It is also reported that in mycorrhizal plants the phytopathological viruses can replicate in the better way. The health of the plant is directly related to the level of inoculation of AMF. So to prevent the plant from the soil pathogens the mycorrhizal association is very beneficial because it can increase the resistance against soil-borne pathogens.

5.11 MYCORRHIZA AND WATER RELATIONSHIP

The citrus plant growth is adversely effected by drought stress; it can inhibit the growth of the plant and damage the fruit quality (Levy et al., 1978). The mycorrhizal association alleviates the drought stress when it inoculated the citrus plant (Wu et al., 2009). Through several studies it was proved that drought stress reduced the colonization of the root. The mycorrhizal association not only improved the mineral uptake, but also has the positive impact on water uptake (Wu and Xia, 2004) and also helps in regulating the carbohydrate's metabolism and water relationship of citrus plant (Srivastava et al., 2002). While studying the relation between mycorrhiza and water uptake all the experiments proved that the mycorrhizal association improved the stomatal conductance, translation, and photosynthesis rate. The relation between water uptake and mycorrhiza and its impact on the growth of citrus with antioxidants was studied in detail and revealed that mycorrhiza improved shoot dry weight, diameter, and height of the citrus plant (Wu et al., 2006b) regardless of the level of water. This association also improved the photosynthesis, translation rate, root's hydraulic conductivity, transpiration, growth rate, resistance to soil conditions, and osmotic adjustment as well. Developing resistance against drought in citrus plants is independent of the nutrient's uptake, especially the P content. The mycorrhizal association helps in the soil moisture retention because of the formation of water stable aggregates by releasing glomalin that results in the enhanced growth of the citrus plant under drought stress. The plant's ability to resist the drought stress has been enhanced by AMF by improving mineral uptake, especially P, Zn, and Cu (Auge, 2001). The mycorrhizal association decreased the level of superoxide anion radical in leaf under drought stress. The establishment of the mycorrhizal association is high under normal water conditions (Wu et al., 2007b). Under stress conditions mycorrhiza influenced the secondary metabolites of the plant so it increased the solute accumulation to improve the osmotic adjustment of AM seedlings of citrus. The production of enzymatic and nonenzymatic antioxidants is giving protection against oxidative damage result in improved tolerance against drought (Wu et al., 2006a).

Mycorrhizal seedlings have more concentration of soluble sugars, starch, and nonstructural carbohydrates in roots and leaves than nonmycorrhizal seedlings. Many studies revealed that whenever mycorrhiza increased tolerance against drought is actually the drought avoidance (Auge, 2001).

5.12 MYCORRHIZA AND OTHER STRESS FACTORS

The major problems, which plants are facing all around the world especially in arid and semiarid regions, are drought, salinity, heavy metals, and heat stresses. AMF play an important role in alleviating the various stresses for example heavy metals, salts, drought, and temperature. At moderate temperature AMF will improve the photosynthesis, root morphology, uptake of nutrients, and growth, but the role of mycorrhiza failed at low temperature (Wu and Zou, 2010). Mycorrhizal seedlings improved diameter of stem, height of the plant, leaf area, root and total dry weights, higher photosynthetic and transpiration rate, and stomatal conductance, higher root volume, and increase in P, Ca, and Mg translocation than in nonmycorrhizal seedlings.

5.13 CONCLUSION

To improve the nutrient's uptake efficiency the tree plants have the ability to utilize the new resources just like the modification of the rhizosphere environment and the mycorrhizal association. The interaction between plant and fungi under certain circumstances is beneficial for both partners. Plant root and fungus combined, results in a single structure through which exchange of nutrients takes place. The most important mycorrhizal fungi are *Glomus* sp. which inoculated citrus plant very efficiently. Mycorrhizal association helps plants in uptake of mineral nutrients, promotes growth and photosynthates and nutrient's uptake. AMF has the ability in alleviating various stresses. The selection of rootstock, presence of fungal inoculum of AMF, and soil conditions for the cultivation of citrus plant are very important. It is also noticed that application of fungicides and herbicides to the soil can kill the AMF strains in the soil. The role of temperature for normal growth and improved yield is very important for both citrus plant and mycorrhizal fungi. At moderate temperature AMF play an important role in improving photosynthesis, root morphology, nutrient's uptake, and growth, but the role of mycorrhiza failed at low temperature. Citrus plants inoculated with mycorrhiza have more resistance against soil-borne pathogens and diseases than nonmycorrhizal plants. The mycorrhizal association releases the antibiotics in the soil which can control soil microorganisms such as nematodes and pathogenic fungi so the chances of infection are reduced.

KEYWORDS

- **AMF**
- **antibiotics**
- **citrus**
- **fungal plant interaction**
- ***Glomus* sp.**
- **nutrients uptake**
- **plant resistance**
- **rootstock**

REFERENCES

Aliasgharzadeh, N.; Rastin, S. N.; Towfighi, H.; Aizadeh, A. Occurence of of arbuscular mycorrhizal fungi in saline soils of the Tabriz plain of Iran on chemical properties of soil. *Mycorrhiza* **2001**, *11*, 119–122.

Auge, R. M. Water relations, drought and vesicular-arbuscular mycorrhizal symbiosis. *Mycorrhiza* **2001**, *11*(1), 3–42.

Casieri, L.; Lahmidi, N. A.; Doidy, J.; Veneault-Fourrey, C.; Migeon, A.; Bonneau, L. et al. Biotrophic transportome in mutualistic plant fungal interactions. *Mycorrhiza* **2013**, *23*, 597–625. doi: 10.1007/s00572-013-0496-9.

Chakraborty, U.; Chakraborty, B. N.; Allay, S.; De, U.; Chakraborty, A. P. Dual application of *Bacillus pumilus* and *Glomus mosseae* for improvement of health status of mandarin plants. *Acta Hort.* **2011**, *892*, 215–230.

Cho, E. J.; Lee, D. J.; Wee, C. D.; Kim, K. L.; Cheong, Y -H.; Cho, J -S.; Sohn, B. K. Effects of AMF inoculation on growth of *Panax ginseng* CA Meyer seedlings and on soil structures in mycorrhizosphere. *Sci. Hort.* **2009**, *122*(4), 633–637.

Corratgé, C.; Zimmermann, S.; Lambilliotte, R.; Plassard, C.; Marmeisse, R.; Thibaud, J. B. et al. Molecular and functional characterization of a Na^+-K^+ transporter from the Trk family in the ectomycorrhizal fungus *Hebeloma cylindrosporum*. *J. Biol. Chem.* **2007**, *282*, 26057–26066. doi: 10.1074/jbc.M611613200.

Davis, R. M.; Menge, J. A. Influence of *Glomus fasciculatus* and soil phosphorus on Phytophthora root rot of citrus. *Phytopathology* **1980**, *70,* 447–452.

de Souza, P. V. D.; Schmitz, J. A. K.; de Freitas, R. S.; Carniel, E. & Carrenho, R. Identification and quantification of native arbuscular mycorrhizae fungi of citrus in the State of Rio Grande do Sul, Brazil. *Pesquisa Agropecuaria Brasileira* **2002**, *37*(4), 553–558.

Dutra, P. V.; Abad, M.; Almela, V.; Agusti, M. Auxin interaction with the vesicular-arbuscular mycorrhizal fungus *Glomus intraradices* Schenck & Smith improves vegetative growth of two citrus rootstocks. *Sci. Hort.* **1996**, *66*(1–2), 77–83.

Fidelibus, M. W.; Martin, C. A.; Wright, G. C.; Stutz, J. C. Geographic isolates of *Glomus* increase root growth and whole plant transpiration of citrus seedlings grown with high phosphorous. *Mycorrhiza* **2000**, *10*: 231–236.

Gadkar, V.; David-Schwartz, R.; Kunik, T.; Kapulnik, Y. Arbuscular mycorrhizal fungal colonization factors involved in host recognition. *Plant Physiol.* **2001**, *127*(4), 1493–1499.

Garcia, K.; Delteil, A.; Conéjéro, G.; Becquer, A.; Plassard, C.; Sentenac, H. et al. Potassium nutrition of ectomycorrhizal *Pinus pinaster*: overexpression of the *Hebeloma cylindrosporum* *Hc*Trk1 transporter affects the translocation of both K^+ and P in the host plant. *New Phytol.* **2014**, *201*, 951–960. doi: 10.1111/nph.12603.

Gerdemann, J. W. VA mycorrhizae. In: *Development and function of roots*; Torrey, J. G., Clarkson, D. T., Eds.; Academic Press: London, 1975.

Graham, J. H.; Eissenstat, D. M. Host genotype and the formation and function of VA mycorrhizae. *Plant and Soil* **1994**, *159*(1), 179–185.

Graham, J. H. Citrus mycorrhizae - potential benefits and interactions with pathogens. *Hortscience* **1986**, *21*(6), 1302–1306.

Graham, J. H.; Drouillard, D. L.; Hodge, N. C. Carbon economy of sour orange in response to different *Glomus* spp. *Tree Physiology* **1996**, *16*(11–12), 1023–1029.

Graham, J. H.; Leonard, R. T.; Menge, J. A. Interaction of light-intensity and soil-temperature with phosphorus inhibition of vesicular arbuscular mycorrhiza formation. *New Phytol.* **1982**, *91*(4), 683–690.

Guether, M.; Balestrini, R.; Hannah, M.; He, J.; Udvardi, M. K.; Bonfante, P. Genome-wide reprogramming of regulatory networks, transport, cell wall and membrane biogenesis during arbuscular mycorrhizal symbiosis in *Lotus japonicus*. *New Phytol.* **2009**, *182*, 200–212. doi: 10.1111/j.1469-8137.2008.02725.x.

Haghighatnia, H.; Nadian, H. A.; Rejali, F. Effects of mycorrhizal colonization on growth, nutrients uptake and some other characteristics of *Citrus volkameriana* rootstock under drought stress. *World Appl. Sci. J.* **2011**, *13*(5), 1 077–84.

Hartmond, U.; Schaesberg, N. V.; Graham, J. H.; Syvetsen, J. P. Salinity and flooding stress effects on mycorrhizal and non mycorrhizal citrus rootstock seedlings. *Plant Soil* **1987**, 104, 37–43.

Jifon, J. L.; Graham, J. H.; Drouillard, D. L.; Syvertsen, J. P. Growth depression of mycorrhizal Citrus seedlings grown at high phosphorus supply is mitigated by elevated CO_2. *New Phytol* **2002**, *153*(1), 133–142.

Kevin, G.; Sabine, D. Z. The role of mycorrhizal associations in plant potassium nutrition. *Front. Plant Sci.* **2014**, *5*:337

Khalil, H. A.; Eirra, A. M.; El-Shazly, S. M.; Aboul Nasr, A. M. Improved growth of salinity stressed citrus after inoculation with mycorrhizal fungi. *Sci. Hort.* **2011**, *130*, 624–32.

Levy, Y.; Bielorai, H.; Shalhevet, J. Long-term effects of different irrigation regimes on grapefruit tree development and yield. *J. Am. Soc. Hort. Sci.* **1978**, *103*(5), 680–683.

Lukac, M.; Calfapietra, C.; Lagomarsino, A.; Loreto, F. Global climate change and tree nutrition: effects of elevated CO_2 and temperature. *Tree Physiol.* **2010**, *30*(9), 1209–1220.

Marschner, H.; Dell, B. Nutrient-uptake in mycorrhizal symbiosis. *Plant Soil* **1994**, *159*(1), 89–102.

Marschner, H. *Mineral nutrition of higher plants*. Academic Press: London, 1995.

Menge, J. A.; Johnson, E. L. V.; Platt, R. G. Mycorrhizal dependency of several citrus cultivars under three nutrient regimes. *New Phytol.* **1978**, *81*(3), 553–559.

Murkute, A. A.; Sharma, S.; Singh, S. K. Studies on salt stress tolerance of citrus rootstock genotypes with arbuscular mycorrhizal fungi. *Hort. Sci.* **2006**, *33*, 70–76.

Navarro, J. M.; Tornero, P.; Morte, A. Alleviations of salt stress in citrus seedlings inoculated with arbuscular mycorrhizal fungi depends on the rootstock salt tolerance. *J. Plant Physiol.* **2014**, *171*, 76–85.

Nemec, S. Response of six citrus rootstocks to three species of *Glomus*, a mycorrhizal fungus. *Citrus Ind.* **1979**, *60*(5), 5.

Ni, Q. D, Zou, Y. N.; Wu, Q. S.; Huang, Y. M. Increased tolerance of citrus (*Citrus tangerina*) seedlings to soil water deficit after mycorrhizal inoculation: Changes in antioxidant enzyme defense system. *Notulae Botanicae Horti Agrobotanici Cluj-Napoca* **2013**, *41*, 524–9.

Ortas, I.; Ortakci, D.; Kaya, Z. Various mycorrhizal fungi propagated on different hosts have different effect on citrus growth and nutrient uptake. *Commun. Soil Sci. Plant Analy.* **2002a**, *33*(1–2), 259–272.

Ortas, I.; Ortakci, D.; Kaya, Z.; Cinar, A.; Onelge, N. Mycorrhizal dependency of sour orange in relation to phosphorus and zinc nutrition. *J. Plant Nutr.* **2002b**, *25*(6), 1263–1279.

Qiang-Sheng, W.; Srivastava, A. K.; Ying-Ning, Z.; Malhotra, S. K. Mycorrhizas in citrus: Beyond soil fertility and plant nutrition. *Indian J. Agric. Sci.* **2017**, *87*(4), 427–443.

Read, H. S.; Fremont, T. Factors that influence the formation and development of mycorrhizal associations in citrus roots. *Phytophatology* **1935**, *25*, 645–647.

Redecker, D.; Kodner, R.; Graham, L. E. Glomalean fungi from the Ordovician. *Science* **2000**, *289*, 1920–1921.

Sharma, S. D.; Kumar, P.; Singh, S. K.; Patel, V. B. Indigenous AM fungi and *Azotobacter* isolates, and their screening from citrus seedlings at different levels of inorganic fertilizers application. *Indian J. Hort.* **2009**, *66*(2), 183–189.

Sikora, R. A. Management of the antagonistic potential in agricultural ecosystems for the biological-control of plant parasitic nematodes. *Ann. Rev. Phytopathol.* **1992**, *30*, 245–270.

Srivastava, A. K.; Singh, S.; Marathe, R. A. Organic citrus: Soil fertility and plant nutrition. *J. Sustain. Agric.* **2002**, *19*(3), 5–29.

Tisdall, J. M.; Oades, J. M. Stabilization of soil aggregates by the root systems of ryegrass. *Aust. J. Soil Res.* **1979**, *17*(3), 429–441.

Tong, R.; Yang, X.; Li, D. Effects of interspecies difference of arbuscular mycorrhizal fungi on *Citrus grandis* cv. Changshou Shatian You seedlings vegetative growth and mineral contents. *Yingyong Shengtai Xuebao* **2006**, *17*(7), 1229–1233.

Verbruggen, E.; Kiers, E. T. Evolutionary ecology of mycorrhizal functional diversity in agricultural systems. *Evol. App.* **2010**, *3*(5–6), 547–560.

Wang, F. Y, Liu, R. J.; Li, X. G.; Zhou, J. M. Abuscular mycorrhizal status of wild plant in saline alkaline soils of the Yellow River Delta. *Mycorrhiza* **2004**, *14*, 133–137.

Wang, M.; Xia, R. Effects of arbuscular mycorrhizal fungi on growth and iron uptake of *Poncirus trifoliata* under different pH. *Weishengwu Xuebao* **2009**, *49*(10), 1374–1379.

Wang, M.; Xia, R.; Wang, P. Effects of arbuscular mycorrhizal fungi on different iron species in *Poncirus trifoliata* rhizospheric soil. *Weishengwu Xuebao* **2009**, *49*(10), 1347–1352.

Watanarojanaporn, N.; Boonkerd, N.; Wongkaew, S.; Prommanop, P.; Teaumroong, N. Selection of arbuscular mycorrhizal fungi for citrus growth promotion and Phytophthora suppression. *Sci. Hort.* **2011**, *128*(4), 423–433.

Wu, Q. S.; Xia, R. X. Effects of arbuscular mycorrhizal fungi on plant growth and osmotic adjustment matter content of trifoliate orange seedling under water stress. *Zhiwu Shengli yu Fenzi Shengwuxue Xuebao* **2004**, *30*(5), 583–588.

Wu, Q. S.; Zou, Y. Arbuscular mycorrhizal symbiosis improves growth and root nutrient status of citrus subjected to salt stress. *Scienceasia* **2009**, *35*(4), 388–391.

Wu, Q. S.; Levy, Y.; Zou, Y. N. Arbuscular mycorrhizae and water relations in Citrus. In: *Tree and forestry science and biotechnology*. Global Science Books, London, 2009; pp 105–112.

Wu, Q.; Xia, R.; Zou, Y. Arbuscular mycorrhizal fungal growth on citrus roots and its correlations with soil available phosphorus content and phosphatase activity. *Yingyong Shengtai Xuebao*, **2006a**, *17*(4), 685–689.

Wu, Q. S.; Xia, R. X.; Zou, Y. N. Reactive oxygen metabolism in mycorrhizal and nonmycorrhizal citrus (*Poncirus trifoliata*) seedlings subjected to water stress. *J. Plant Physiol.* **2006b**, *163*(11), 1101–1110.

Wu, Q. S.; Zou, Y. N.; Xia, R. X.; Wang, M.Y. Five *Glomus* species affect water relations of Citrus tangerine during drought stress. *Botan. Stud.* **2007,** *48* (2), 147–154.

Wu, Q. S.; Zou, Y. N. Beneficial roles of arbuscular mycorrhizas in citrus seedlings at temperature stress. *Sci. Hort.* **2010**, *125*(3), 289–293.

Xiao, J. X.; Hu, C. Y.; Yang, B.; Hua, J. Effects of low magnesium and an arbuscular mycorrhizal fungus on the growth, magnesium distribution and photosynthesis of two citrus cultivars. *Sci. Hort.* **2014**, *177*, 14–20.

Yao, Q.; Wang, L. R.; Zhu, H. H.; Chen, J. Z. Effect of arbuscular mycorrhizal fungal inoculation on root system architecture of trifoliate orange (*Poncirus trifoliata* L. Raf.) seedlings. *Sci. Hort.* **2009**, *121*, 458–461.

Zhang, H.; Yin, W.; Xia, X. Shaker-like potassium channels in *Populus*, regulated by the CBL-CIPK signal transduction pathway, increase tolerance to low-K^+ stress. *Plant Cell Rep.* **2010**, *29*, 1007–1012. doi: 10.1007/s00299-010-0886-9.

CHAPTER 6

ARBUSCULAR MYCORRHIZAL BIOTECHNOLOGY AND ITS APPLICATIONS

SAMIA KHANUM[1], ABDEL RAHMAN MOHAMMAD AL-TAWAHA[2], YOUSEF ABU-ZAITOON[2], ABDEL RAZZAQ AL-TAWAHA[3], HIBA ALATRASH[4], ABDUR RAUF[5], ARUN KARNWAL[6], ABHIJIT DEY[7], MOHAMAD SHATNAWI[8], DEVARAJAN THANGADURAI[9], JEYABALAN SANGEETHA[10], SAHER ISLAM[11], IMRAN[12], AMANULLAH[12], SHAH KHALID[12], PALANI SARANRAJ[13], and ELIF GUNAL[14]

[1]*Department of Botany, University of the Punjab, Lahore, Pakistan*

[2]*Department of Biological Sciences, Al Hussein Bin Talal University, Maan, Jordan*

[3]*Department of Crop Science, Faculty of Agriculture, University Putra Malaysia, Serdang, Selangor, Malaysia*

[4]*General Commission for Scientific Agricultural Research, Syria*

[5]*Department of Chemistry, University of Swabi, Anbar, Khyber Pakhtunkhwa, Pakistan*

[6]*Department of Microbiology, School of Bioengineering and BioSciences, Lovely Professional University, Phagwara, Punjab, India*

[7]*Department of Life Sciences, Presidency University, Kolkata, India*

[8]*Department of Biotechnology, Faculty of Agricultural Technology, Al-Balqa Applied University, Al-Salt, Jordan*

Mycorrhizal Technology: Managing Plant Stress and Mitigating Climate Change using Mycorrhizae.
Jeyabalan Sangeetha, Abdel Rahman Mohammad Al-Tawaha, & Devarajan Thangadurai (Eds.)

[9]*Department of Botany, Karnatak University, Dharwad, Karnataka, India*

[10]*Department of Environmental Science, Central University of Kerala, Kasaragod, Kerala, India*

[11]*Institute of Biochemistry and Biotechnology, University of Veterinary and Animal Sciences, Lahore, Pakistan*

[12]*Department of Agronomy, The University of Agriculture, Peshawar, Pakistan*

[13]*Department of Microbiology, Sacred Heart College (Autonomous), Tirupattur, Tamil Nadu, India*

[14]*Department of Soil Science and Plant Nutrition, Faculty of Agriculture, Harran University, Sanliurfa, Turkey*

ABSTRACT

Roots of higher plants interact with the hyphae of some fungi and form a mutualistic association called mycorrhizal association. Mycorrhizal association helps in improving growth and development of plants by various ways such as improved nutrients and water uptake, improved defense mechanisms, and developed resistance against various biotic and abiotic stresses. Mycorrhizal biotechnology is very important for the better crop, forest, and the nursery establishment. The use of mycorrhizal inoculum for improved crop performance is very common and the focus of recent studies. The selection of inoculum in this regard is very critical. Inoculum must be ecologically and physiologically suitable for the planting site. Several commercial inoculation methods and techniques have been designed for inoculation of plant seedlings on a large scale. Recent studies focus on the role of mycorrhizal association in improvement and better growth of nursery seedlings and tree production. It also emphasizes on the establishment of new forests and nurseries.

6.1 INTRODUCTION

In the ectomycorrhizae system, the fungus forms a thick mantle around short lateral roots and grows over their surface (Haselwandter and Bowen, 1996). The ECM is usually white or brightly colored and therefore can be seen easily with a hand lens. In the same way, if ectomycorrhizae are excessive in the soil they give the appearance of dense mold like the fungus when seedlings are lifted (Molina and Trappe, 1984).

Basidiomycetes or ascomycetes form a very large variety of ECM systems (*Cenococcum geophilum* member of ascomycetes is very common); Phycomycetes (Endogone species) also form ectomycorrhizae (Haselwandter and Bowen, 1996).

Ectomycorrhizae are formed by all members of Family Pinaceae, including pine (*Pinus*), Douglas-fir (*Pseudotsuga*), larch (*Larix*), true fir (*Abies*), hemlock (*Tsuga*), and spruce (*Picea*) species. Members of the family Fagaceae like oak (*Quercus*) and beech (*Fagus*) species and from family Betulaceae, alder (*Alnus*) and birch (*Betula*) species also form ectomycorrhizae (Molina and Trappe, 1984).

6.2 THE BENEFITS OF MYCORRHIZA

1. Mycorrhiza significantly improved uptake of nutrients, predominantly (P) phosphorus.
2. They can take up water (Duddridge et al., 1980).
3. They increase drought tolerance of young saplings (Parke et al., 1983).
4. Mycorrhizal fungi detoxification of soil pollutants (Zak, 1971).
5. They facilitate plantlets to survive extreme acidity (Marx, 1976) or elevated soil temperatures (Marx and Bryan, 1971).
6. Mycorrhiza is also practically important for nursery management because these fungi could defend roots against particular pathogens (Marx, 1973). For example, the mycorrhizal fungi *Laccaria laccata* help in protecting the feeder roots from the damage caused by *Fusarium* (Sinclair et al., 1982) (Fig. 6.1).

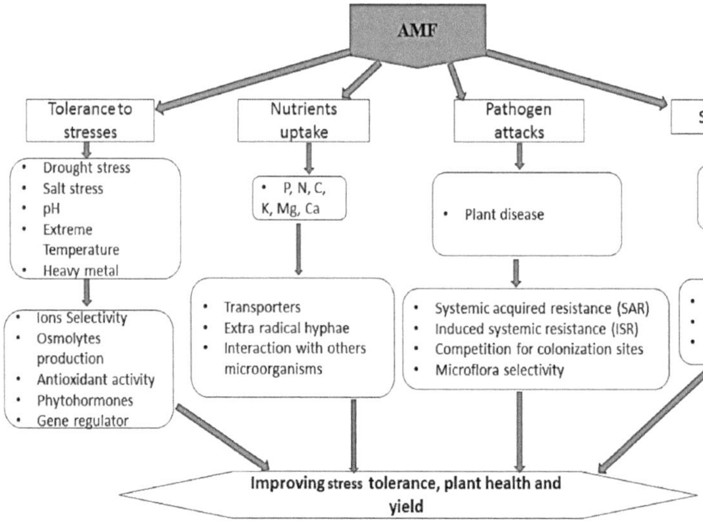

FIGURE 6.1 Role of AMF in plant growth.

Source: Adapted from Diagne et al. (2020), https://creativecommons.org/licenses/by/4.0/

6.3 MYCORRHIZAL INOCULATION

To improve crop performance the use of mycorrhizal inoculum which is ecologically and physiologically suitable for the planting site was first carried out by Moser (1958) in Austria, Takacs (1967) in Argentina, and Theodorou and Bowen (1973) in Australia.

6.3.1 SELECTION CRITERIA FOR INOCULUM

The fungal inoculum must stand against all kinds of biological, environmental, chemical, and physical stresses; then, it will be compatible with the crop and improved the crop. The selected fungal inoculum also has the ability to survive during its production and then storage conditions. It also has the ability to survive under all kinds of soil and site characteristics first in the forest and then to the site where the seedlings are planted (Kropp and Langlois, 1990; Grove and Le Tacon, 1993). For this, the criteria for selecting the fungal inoculum have been designed for testing it at the small or large scale to meet the nursery goals. The key criteria of selection have been summarized by Molina:

1. **Improved growth in culture:** The growth of ectomycorrhizal fungi is usually slow so the fungal inoculum has the ability to grow faster as preferred.
2. **Effective in the formation of mycorrhizae:** For the production of inoculum mostly fungi can grow easily but only few are useful as vegetative inoculum.
3. **Adapted to ecological situations:** The fungi that have greater ability to stand against the ecological stresses like drought are preferred.
4. **Competitive ability:** The fungal inoculum has more ability to compete already present mycorrhizal fungi in the soil and dominated in nature to colonize the roots of the plant seedlings. It also has the ability to protect the roots from the soil-borne pathogens as well.
5. **Range of host:** Most of the mycorrhizal fungi form association with multiple hosts and are not host specific, but some fungi are host specific so while selecting the inoculum it must be kept in mind that mycorrhizal fungi has compatibility with the host plant.
6. **Improving performance of seedlings in plantations:** This is the most important criterion to be kept in mind while selecting the inoculum for the productions of nursery inoculum on the wide range (Trappe, 1977; Molina, 1982).

6.3.2 *PISOLITHUS TINCTORIUS*

The most important ECM inoculum used successfully and on the large scale is *Pisolithus tinctorius* because of its geographically widespread distribution and varieties of plant host range (Marx, 1977). It also has the ability to survive the various adverse soil conditions like drought, high temperature, and contaminated with heavy metals and toxic compounds (Schramm, 1966).

In the USA and other regions of the world the inoculum of Basidiospores of *P. tinctorius* has been used in both conditions like sprayed on soil in the form of suspension or encapsulation of the seed with clay. The result of this experiment shows improved growth of the seedlings, but the production of mycorrhiza is lower than the production of mycorrhiza by vegetative inoculum as super strain so this method is considered not much effective (Marx et al., 1984b, 1991).

6.4 REINVASION AND FUMIGATION OF *THELEPHORA TERRESTRIS*

The most common practice to control the soil-borne microorganisms, weeds, and pathogenic fungi is to fumigate the soil, seed bed, and mixture of pots before inoculation. But still after fumigation, the soil is colonized by the spores of ECM fungi within few days just like *Thelephora terrestris* which is one of the dominant ECM fungi present in the soil of the forest nursery all around the world (Mikola, 1970; Ivory, 1980; Marx et al., 1984a). The experiments are designed to understand the influence of inoculated mycorrhizal fungi which is very critical as most of the plant seedlings and control plants also are colonized by the already present mycorrhizal fungi.

6.4.1 POOR AND GOOD INOCULUM

It is reported that *Rhizopogon* and *Suillus* species are not considered good inoculum because they have low ability to colonize the plants (Perry et al., 1987). Many fungi, for example, different strains of *Laccaria* species have greater ability to produce mycorrhiza on the large scale in plants, grown in containers (Molina, 1982). The most common example of *Laccaria* species for most beneficial inoculum is *L. bicolor* S238. Many strains of *Laccaria* and *Hebeloma* are produced on the large scale to use them as an inoculum on the commercial level on *Pseudostuga menziesii* under nursery conditions in containers (Hung and Molina, 1986).

6.4.2 INOCULATION TECHNIQUES

Several commercial inoculation methods and techniques have been designed for inoculation of plant seedlings on large scale (Marx and Bryan, 1975; Marx, 1980, 1991; Marx and Kenney, 1982; Sieverding, 1990; Bethlenfalvay and Linderman, 1992; Brundrett et al., 1996).

6.4.2.1 SOIL INOCULUM

The most suitable and preferred inoculum that is commonly used is the soil of forest taken from the host of ECM. The soil inoculum is about 10% by volume which is assimilated about 10 cm of the nursery bed before transplanting the seedlings (Mikola, 1973). One side effect of such an application of inoculum is the seeds of weed, rhizomes, and other pathogens also introduced with beneficial fungi (Molina and Trappe, 1984).

6.4.2.2 NURSE SEEDLINGS

The seedlings that are colonized by mycorrhizal fungi and used for the colonization of new seedlings are known as nurse seedlings. Their plantation or incorporation into the roots of ECM plant hosts in the nursery bed can produce the source of ECM fungal inoculum (Molina and Trappe, 1984).

6.4.2.3 SPORE-BASED INOCULUM (PELLETS AND SEED COATINGS)

The spores of ECM are widely used in the nurseries of forest (Castellano, 1994). The sporocarps are produced on large numbers and easily collected. The most commonly and widely used fungi are puffballs *Pisolithus* and *Scleroderma* because of their fruiting bodies having more spores and being abundantly present in the forest. Truffle fungi also produced the large number of spore inoculum, but they produce spore-bearing tissues under the soil so it will be challenging to dig out the enough inoculum of these fungal inoculum.

6.4.2.4 MYCELIAL INOCULUM CONSISTS OF SLURRIES AND ALGINATE BEAD FORMULATIONS

The most popular inoculum technique is based on the vegetative mycelium growth in the mixture of vermiculite and peat and further moistens with liquid nutrient medium (Marx and Kenney, 1982). Vermiculite offers a coated substrate that is well aerated which helps in the protection of the mycelium and the pH range also maintained 4.8 to 5.5 with the application of peat. The nutrient solutions usually consist of C to N ratio and range from 50 to 60. This solution is applied in the quantity so that mycorrhiza can consume all available C, which is required for its proper growth and development. Mycorrhizal fungi can be replaced by the other saprotrophs if C is available in the medium at the time of inoculation (Brundrett et al., 1996).

6.5 FIELD APPLICATION OF ECTOMYCORRHIZAE

6.5.1 FOREST PRODUCTION

The importance of ectomycorrhizae as the essential part of the forest and for the better growth and nutrition of trees has been reported since the first experiments are carried out while managing the forest for the better production of timber (Frank, 1885). Later on the role of mycorrhiza in better development of plants and the dependency of these forests on ECM have been reported repeatedly, especially when ECM Pinaceae has been planted in the Southern Hemisphere. These plants only grow and survive better when associated with their compatible mycorrhizal fungi (Mikola, 1970, 1973). The first attempt to establish nurseries of pine seedling in Australia and New Zealand was unsuccessful. However, the pine seedlings colonized by some fungal species result in increased survival rate and growth were observed. Now the most well developed pine trees are found in New Zealand and Australia among the other important world's forests. The similar result has been reported during the afforestation of treeless grasslands of the US (McComb, 1938) and Russia showing the dependency on mycorrhiza. The productivity of forest has been susceptible by global change, for example, nitrogen or sulfur direct application to the soil or the imbalanced pH range which adversely affects the mycorrhizal association and on both partners. These effects are believed to be contributing factors in the forest decline syndrome particularly observed in Europe and north-eastern USA.

6.5.1.1 PLANTATION OF FORESTRY PROGRAMS

The increase in plantation forestry, with the increase in world demand of timber and pulp, has been reported in several regions of the world in reaction to increased requisitioning of CO_2 by increasing forest cover. In the result special strategies have been designed based on colonization of plants before out-planting of these seedlings. Using such techniques increased the crop performance in different regions of the world (Marx et al., 2002). It is reported that about 1.5 billion *Pinus* seedlings have been produced in nurseries in southern USA annually (Marx, 1985). After that, plantation forestry and re-vegetation programs have been increased worldwide which results in the increase in demand for stocks of various conifers and broadleaf species. The Forestry Department of New South Wales of Australia developed 7.5 million seedlings and 2 million cuttings in 2006.

6.5.1.2 AFFORESTATION PROGRAMS IN SOUTH EAST ASIA

When eucalyptus has been planted in the sites having the low level of P and absence of proper ECM fungi in China, the whole experiment of establishing the early plantation was failed. Thus for the better growth and development of the nursery plants the presence of the adequate amount of P and ECM fungi is necessary (Xu et al., 2001). When superphosphate was applied to the seedlings it increases the survival and volume of the stand within 3 years. When the nursery plants were colonized by ECM fungi, no differences have been reported in the development of inoculated and noninoculated seedlings at the time of out planting. So it shows that inoculation with ECM has the positive impact on the seedling growth and it totally depends on the presence of the adequate amount of P and ECM fungal strains. The *Eucalyptus europhylla*, when colonized by *Hebeloma westraliense*, results in decrease in the stand volume of the seedlings after 3 years' growth in the field, but when colonized by *Laccaria lateritia* or *Pisolithus albus* it results in the increase in the stand volume when the same level of P has been applied.

6.5.1.3 AGROFORESTRY AND LAND REHABILITATION

The performance of mycorrhiza can be affected by factors such as soil, plant, and type of fungi. The maximum response from low rooting plant trees and less fertile soil is obtained by regular inoculation with mycorrhizal fungi which are selected for producing large populations of indigenous fungi because of their high effectiveness but present in less in numbers. Mycorrhizal fungi may improve harmful soil conditions other than nutrient deficiency and is also helpful to improve the symbiotic nitrogen fixation.

6.5.1.4 TIMBER EXTRACTION, CLEAR-CUT LOGGING, AND FOREST FIRE IMPACT ON FOREST ECOSYSTEMS

The major effects of forests causing extraction of timber and clear-cut logging will affect the forest ecosystems, plants that can regenerate and ECM fungi that can colonize the plant species which can regenerate or can be transplanted (Jones et al., 2003). The conifer forests of western America are adapted to the fire. The pine trees depend on the spores of ECM fungi because they can survive from the fire and help the plants to regenerate and establish the forest ecosystem again after severe wildfires. Researches used these natural forest fires to study the role of ECM in building new forest ecosystems. It is reported that the spores of ECM fungi are fire resistant and can survive the wildfire, but there are also some fungi that are now adapted to fire *Rhizopogon olivaceotinctus* that increased in abundance after the fire. The experiments show that the ECM spores still survive the high intensity fire, but the reduction in ECM spores has been observed as well (Glassman et al., 2015).

6.5.1.5 COMMERCIAL FOREST MANAGEMENT

It is critical to understand how long mycorrhizal fungi remain in the roots as existing fungi are replaced by the already present fungi in the soil. The two species *Hebeloma* and *Laccaria* when introduced in arid regions of western US were replaced by unknown native fungi just after 6 months of planting out (Bledsoe et al., 1989). Another example revealed that the newly grown roots are colonized by native fungi while introduced fungi still remain in the older roots (Ruehle, 1983).

6.6 ECTOMYCORRHIZAE AND NURSERY PRODUCTION

It has been recognized that the well-developed and the long root system is very important for the production of resilient planting stock. Recent studies revealed the role of mycorrhiza for the better growth and development of the nursery seedling and also this association is necessary for the better performance of the seedlings after planting out. The main objection of inoculation of nursery seedlings is for the production of healthy seedlings which can survive when planting in the field while colonized by selected fungal strains. In the US and Canada when the seed beds and containers of nursery seedlings of conifers and hardwood plants are artificially mixed with the inoculum of selected fungus like *Pisolithus tinctorius* (Pt), it improved the growth and quality of the seedlings. In correlated field out plantings that have the maximum number of ECM fungi improved growth and the survival rate of trees has been reported in a wide range of conifers and on various planting sites (Marx and Cordell, 1984). The species belonging to the genus *Thelephora* is the most commonly found ECM fungi in the nurseries having bare roots. The ECM of *Thelephora terrestris* is the very smooth and pale cream brown in color and found worldwide and produces the fruit at the base of the seedlings. Other ECM fungi are also common like *Laccaria laccata* and *Inocybe lacera*, but truffle fungi of genus *Rhizopogon*, *Boletes*, and *Suillus* are only reported in few.

6.6.1 NURSERIES DEVELOPED IN AND AWAY FROM FOREST ZONES

The nurseries having a large number of ECM hosts and established in forest zones are able to produce seedlings colonized with plentiful and different ectomycorrhizae. It is reported that the seedling produced from soil which is not fumigated can form the mycorrhizal association in their early stages of development. On the other hand, it is also reported that when soil is completely fumigated the spores from the fruiting bodies of neighboring forests are enough for the establishment of ECM associations. So when nurseries established in the region with no mycorrhizal species and even with no neighbor source for ECM have no mycorrhizal association, it leads to financial and reforestation setbacks. Such nurseries produced seedlings having stunned growth, nutrients deficiencies, and chlorotic. The symptoms of mycorrhizal deficiency may continue to the next year and even when colonized or recovered the size of seedlings is still different in a same seedbed.

6.7 ECTOMYCORRHIZAE AND GLASSHOUSE EXPERIMENTATION

The screening trials of glasshouse help in establishing the mycorrhizal association with the host plant.

1. In experimentation of the glasshouse the performance of fungal species can be analyzed by evaluating survival and the growth rate of the plant species associated with ECM by comparing them with control plants.
2. The measurement of nutrient content of plants can be used for comparison of fungal uptake capacity and also for measuring the capacity of fungi to exclude toxic elements from the soil.
3. The study of anatomy of the hartig net of ECM roots helps in confirmation of the establishment of mycorrhizal association and also explained the compatibility of host and fungal species. It also helps in measuring the potential of nutrient exchange between two partners.
4. The mycorrhizal fungi ability to compete soil-borne microorganisms can be determined by analyzing the ECM fungi ability to establish and enhance plant growth in pasteurized and unsterile soil (Aggangan et al., 1995). But, field trials can give the chance to estimate the persistence and performance of fungi in the field.

6.7.1 OBJECTIVES OF GLASSHOUSE EXPERIMENTS

1. *Mycorrhizal synthesis*: Compatibility of the plant and mycorrhizal fungi and also based on the mycorrhizal status.

2. *Mycorrhizal dependency*: The difference in the growth rate of the mycorrhizal and nonmycorrhizal plants when inoculated at various fertility levels of the soil.
3. *Bioassays*: Assessment of the inoculum mycorrhizal fungi which is compatible in the certain soil conditions.
4. *Isolate screening*: The inoculum of mycorrhizal fungi selected on the basis of their growth rate in certain soil situations.
5. *Micronutrient research*: Calculate the specific amount of trace elements in mycorrhizal and nonmycorrhizal plant seedlings.
6. *Histology*: Study the roots of mycorrhizal plant seedlings under the microscope.
7. *Physiology*: To consent testing the response of the host plant or fungal partner under different soil conditions.

6.8 EDIBLE ECTOMYCORRHIZAL FUNGI

The most of the important concern of application of ECM is for improved trees production, but recently researchers also emphasize on using the fruiting bodies of ECM as medicine and food. In most countries of the world, the fruiting bodies of ECM fungi are collected and used as a major food source so also play an important role in increasing income. In Mexico, it is estimated that tons of fresh mushrooms are retailed per annum in markets and even in rural areas major portion of income of a family is based on sales of mushroom in the season. The possibilities of evaluation of commercial market value for the fruiting bodies of ECM are being explored in many countries. Currently, a small number of mycorrhizal species are particularly prized for their gastronomic quality and are hence of very high value. Black truffles (*Tuber melanosporum*) price was increased to £1100 per kg in 2005 (Table 6.1).

TABLE 6.1 Price List of the Edible Fungi Selected for the Market.

Common name	Latin name	Selling form	Price/kg (£)	USD/kg
Black truffle	*Tuber melanosporum*	Fresh	425	900
Branchetto truffle	*Tuber albidum*	Fresh	700	–
Summer truffle	*T. aestivum*	Fresh	1700	–
Porcini	*Boletus edulis*	Fresh	105	–
Morel	*Morcella elata*	Sliced and dried	95	–
Chanterelle	*Cantherellus cibarius*	Sliced and shredded	255	–
Matsutake	*Tricholoma matsutaki*	Dried	15	76

6.9 CONCLUSION

Mycorrhizal association helps plants in uptake of mineral nutrients, promotes growth, inhibits soil-borne pathogen, competes with pathogen for space, photosynthates and nutrients, and increases resistance of plant against such pathogens. Mycorrhizal biotechnology is very important for the better crop, forest, and the nursery establishment. The use of mycorrhizal inoculum for improved crop performance is very common and the focus of recent studies. The selection of inoculum in this regard is very critical. Inoculum must be ecologically and physiologically suitable for the planting site. Several commercial inoculation methods and techniques have been designed for inoculation of plant seedlings on a large scale. The importance of ectomycorrhizae as the essential part of forest and for the better growth and nutrition of trees has been reported since the first experiments carried out while managing the forest for the better production of timber. Recent studies revealed the role of mycorrhiza for the better growth and development of the nursery seedling and also this association is necessary for the better performance of the seedlings after out planting. The most of the important concern of application of ECM is for improved trees production, but recently researchers also emphasize on using the fruiting bodies of ECM as medicine and food.

KEYWORDS

- **afforestation**
- **bioassay**
- **biotechnology**
- **ECM**
- **fungal inoculum**
- **nursery production**
- **symbiotic association**

REFERENCES

Bledsoe, C.; Brown, D.; Coleman, M.; Littke, W.; Rygiewicz, P.; Sangwanit, U.; Rogers, S.; Ammirati, J. Physiology and metabolism of ectomycorrhizae. *Annales des Sciences Forestières, EDP Sci.* **1989**, *46,* 697S–705S.

Diagne, N.; Ngom, M.; Djighaly, P. I.; Fall, D.; Hocher, V.; Svistoonoff, S. Roles of arbuscular mycorrhizal fungi on plant growth and performance: Importance in biotic and abiotic stressed regulation. *Diversity* **2020**, *12*(10), 370. https://doi.org/10.3390/d12100370.

Duddridge, J. A.; Malibari, A.; Read, D. J. Structure and function of mycorrhizal rhizomorphs with special reference to their role in water transport. *Nature* **1980**, *287*, 834–836.

Marx, D. H.; Bryan, W. C. Influence of ectomycorrhizae on survival and growth of aseptic seedlings of loblolly pine at high temperatures. *Forest Sci.* **1971**, *17*, 37–41.

Marx, D. H. Mycorrhizae and feeder root diseases. In: *Ectomycorrhizae—Their ecology and physiology*; Marks, G. C., Kozlowski, T. T., Eds.; Academic Press: New York, 1973; pp 351–382.

Marx, D. H. Use of specific mycorrhizal fungi on tree roots for forestation on disturbed surface areas. In: *Conference on forestation of disturbed areas*; Utz, K. A., Ed.; Birmingham, Alabama, USDA Forest Service: Atlanta, Georgia, 1976; pp 47–65.

McComb, A. L. The relation between mycorrhizae and the development and nutrient absorption of pine seedlings in a prairie nursery. *J. Forestry* **1938**, *36*, 1148–1153.

Mikola, P. Mycorrhizal inoculation in afforestation. *Int. Rev. Fores. Res.* **1970**, *3*, 123–196.

Mikola, P. Application of mycorrhizal symbiosis in forestry practices. In: *Ectomycorrhizae—Their ecology and physiology*; Marks, G. C., Kozlowski, T. T., Eds.; Academic Press: New York, 1973; pp 383–411.

Molina, R.; Trappe, J. Mycorrhiza management in bareroot nurseries. *Forestry Nursery Manual: Prod. Bareroot Seedlings*, *11*, 211–223.

Molina, R. Mycorrhizal inoculation and its potential impact on seedling survival and growth in southwest Oregon. In: *Pacific Northwest forest and range experiment 1S station*; Hobbs, S. D., Helgerson, O. T., Eds.; Forest Research Laboratory, Oregon State University, Corvallis: Oregon, 1982, pp 86–91.

Parke, J. L.; Linderman, R. G.; Black, C. H. The role of ectomycorrhizae in drought tolerance of Douglas-fir seedlings. *New Phytol.* **1983**, *95*, 83–95.

Ruehle, J. L. The relationship between lateral-root development and spread of *Pisolithus tinctorius* ectomycorrhizae after planting of container-grown loblolly pine seedlings. *Forest Sci.* **1983**, *29*(3), 519–526.

Sinclair, W. A.; Sylvia, D. M.; Larsen, A. O. Disease suppression and growth promotion in Douglas-fir seedlings by the ectomycorrhizal fungus *Laccaria laccata*. *Forest Sci.* **1982**, *28*, 191–201.

Trappe, J. M. Selection of fungi for ectomycorrhizal inoculation in nurseries. *Annu. Rev. Phytopathol.* **1977**, *15*, 203–222.

Zak, B. *Detoxification of autoclaved soil by a mycorrhizal fungus*; USDA Forest Service, Pacific Northwest Forest and Range Experimental Station: Portland, 1971.

PART III

MYCORRHIZAL FUNGI FOR MANAGING BIOTIC AND ABIOTIC STRESSES IN CROP PLANTS

CHAPTER 7

ARBUSCULAR MYCORRHIZA UNDER BIOTIC AND ABIOTIC STRESSES

SAMIA KHANUM[1], ABDEL RAHMAN MOHAMMAD AL-TAWAHA[2], ABDEL RAZZAQ AL-TAWAHA[3], HIBA ALATRASH[4], ABDUR RAUF[5], ARUN KARNWAL[6], ABHIJIT DEY[7], NUJOUD ALIMAD[8], DEVARAJAN THANGADURAI[9], JEYABALAN SANGEETHA[10], SAHER ISLAM[11], IMRAN[12], AMANULLAH[12], SHAH KHALID[12], PALANI SARANRAJ[13], and ELIF GUNAL[14]

[1]*Department of Botany, University of the Punjab, Lahore, Pakistan*

[2]*Department of Biological Sciences, Al Hussein Bin Talal University, Maan, Jordan*

[3]*Department of Crop Science, Faculty of Agriculture, University Putra Malaysia, Serdang, Selangor, Malaysia*

[4]*General Commission for Scientific Agricultural Research, Syria*

[5]*Department of Chemistry, University of Swabi, Anbar, Khyber Pakhtunkhwa, Pakistan*

[6]*Department of Microbiology, School of Bioengineering and BioSciences, Lovely Professional University, Phagwara, Punjab, India*

[7]*Department of Life Sciences, Presidency University, Kolkata, India*

[8]*Damascus University, Faculty of Agriculture, Damascus, Syria*

[9]*Department of Botany, Karnatak University, Dharwad, Karnataka, India*

Mycorrhizal Technology: Managing Plant Stress and Mitigating Climate Change using Mycorrhizae.
Jeyabalan Sangeetha, Abdel Rahman Mohammad Al-Tawaha, & Devarajan Thangadurai (Eds.)

[10]*Department of Environmental Science, Central University of Kerala, Kasaragod, Kerala, India*

[11]*Institute of Biochemistry and Biotechnology, University of Veterinary and Animal Sciences, Lahore, Pakistan*

[12]*Department of Agronomy, The University of Agriculture, Peshawar, Pakistan*

[13]*Department of Microbiology, Sacred Heart College (Autonomous), Tirupattur, Tamil Nadu, India*

[14]*Department of Soil Science and Plant Nutrition, Faculty of Agriculture, Harran University, Sanliurfa, Turkey*

ABSTRACT

Plants have the ability to interact with their environment. Different biotic and abiotic stresses, for example drought, salinity, heat, cold, metals and nutrient imbalance, affect the development and growth of plants. Plants modify themselves morphologically, physiologically, and modulating their biochemistry to suppress impact of various stresses. Plant roots develop association with soil microbes for better growth in every situation. Roots of higher plants interact with hyphae of some fungi and form a mutualistic association known as mycorrhizal association. There are many types of this mycorrhizal association, but the most important ones are ectomycorrhiza and arbuscular mycorrhiza. Mycorrhizal association helps in improving plant growth and yield by increasing nutrients uptake and water absorption from fine particles of soil. The role of mycorrhizal association is not only beneficial for enhancing plant tolerance to various stresses but this interaction also responds to many other stresses of their host plant.

7.1 INTRODUCTION

Plants have the ability to interact with their environment. Different stresses, for example drought, salinity, heat, cold, metals and nutrient imbalance, affect the development and growth of plants and the yield of their growth reduced up to 70% (Turk et al., 2003a, 2003b, 2003c; Nikus et al., 2004a, 2004b; Tawaha and Turk, 2004; Turk and Tawaha, 2004; Saxena et al., 2013; Hayyawi et al., 2018; Aludatt et al., 2019; Amanullah et al., 2019; Hani et al., 2019). Plants modify themselves morphologically, physiologically, and modulating their biochemistry to reduce the impacts of stresses. Plants also adopt several stress tolerance strategies to assist the mending of disturbed systems (Al-Tawaha et al., 2005; Lee et al., 2005; Sulpanjan et al., 2005; Tawaha et al., 2005; Assaf et al., 2006; Turk et al., 2006; Sharifi et al., 2007; Al-Kiyyam et al., 2008; Al-Ajlouni et al., 2009; Al-Tawaha et al., 2010; Souza et al., 2012; Al-Tawaha et al., 2018a, 2018b). Plant roots develop association with soil microbes and recover growth and development of plants in every situation (Talavera et al., 2001; Sailo and Bagyaraj, 2005). Arbuscular mycorrhiza is filamentous-shaped microscopic fungi that make association with 90% of all land plants; these filaments colonized the roots and rhizosphere. These branched filamentous microscopic fungi then widely spread out over several centimeters. Mycorrhizal association significantly contributes several useful benefits to its host plants.

To survive it is necessary to adapt to the changes occurring in the relative environment and the most important and abrupt adaptation is relocation, but plants being sessile are unable to relocate themselves; that is why plants adopt different strategies to avoid biotic and abiotic stresses. Salinity is the most common abiotic stress; others are high temperature, low precipitation, because it leads to increasing the quantity of salts in the soil and affecting plant growth, establishment, development, delayed root growth, and cell toxicity which leads to root necrosis (Beltrano et al., 2013).

To minimize the detrimental effect of abiotic stresses many strategies have been proposed like new salt-tolerant crops, genetically modified plants, desalinization, and removal of excessive salts from the soil. These strategies are very useful but costly so these are not commonly practiced in developing countries.

In such condition, the mutualistic association between soil microbes and roots of plants is considered very important in agricultural research and development of these associations promotes growth rate in plants and offers an alternative biological option. The omnipresence of arbuscular mycorrhizal fungi at the boundary among the roots of plant and soil marks them an important microorganism of soil biota because of their nutritive and non-nutritive actions that strongly affect the ecosystem phenomenon. Because of these characteristics they are also helpful for ecosystem services in agro-ecology. We discuss here how this symbiotic association is helpful and benefits to improve productivity of plants under various stresses in an ecosystem. The overall impact of various stresses can be controlled by mycorrhizal association by proper administration. It will also help in conservation of natural resources and also their consumption with an outward clear incentive on human population.

7.2 ABIOTIC STRESSES AND MYCORRHIZA

The agricultural production rate has been reduced because of drought and salinity and several other abiotic stresses (Hameed et al., 2008; Al-Tawaha et al., 2010; Al-Tawaha et al., 2017; Ananthi et al., 2017; Abu Obaid et al., 2018; Amanullah et al., 2019a, 2019b; Hani et al., 2019) (Tables 7.1 and 7.2). In contrast, the reduction of mineral, soil salinity, water stress, alkalinity, temperature fluctuation, especially high temperature and existence of heavy metals are also affecting the agricultural crops. All of these above-mentioned abiotic stresses are severe issues in different areas in all over the world, predominantly in semiarid and arid areas. It is also expected that per resident agricultural land area will reduce up to 0.15 ha in the year 2050. Another alarming situation is observed that about one-third of the country land area of US and Spain are under degradation.

It is already reported since long that mycorrhizal fungus has the ability to increase tolerance against abiotic stresses in plants. The management and applying agriculture system will be implausible and remarkably important for improving the quality of soil and production of crops under edapho-climatic circumstances (Lal, 2009). The useful results of co-inoculation of bacteria with mycorrhizal fungi collected from arid environments on plants which are growing under drought stress studied and these studies emphasized the increased interest of influencing mycorrhiza which are isolated from the dry regions for replanting and rebuilding of the disturbed land areas to improve the quality of soil, productivity, and also it will help to cope up the degradation of Mediterranean ecosystems. This will explain with the example when *Glomus intraradices* is in association with soil-borne bacteria; it will reduce the requirement of water for the normal growth of *Retamas phaerocarpa* up to 42% (Marulanda et al., 2006). In another example reported from arid regions of North Africa and Spain fungi with mycorrhizal association improve salinity stress in olive trees and where palm grove yields are considerably affected by conditions like salinity and drought (Table 7.2).

TABLE 7.1 Impact of AM Fungi on Management of Thermal Stress in Vegetables.

Stress	Family of host plant	Host plant	Fungus species	Observed response	Reference
Cold	Solanaceae	*Solanum lycopersicum*	*Funneliformis mosseae*	Higher fresh and dry weight reduced level of MDA, H_2O_2, and O_2- and increased Ca precipitates in apoplast, enhanced enzyme activity, and reduced redox state in root cells	Liu et al. (2016)
Cold	Solanaceae	*Solanum lycopersicum*	*Funneliformis mosseae*	Reduced cell membrane injuries, more efficient photosystem II, improved growth of seedlings	Caradonia et al. (2019)
Cold	Solanaceae	*Solanum melongena*	*Funneliformis mosseae*, *Claroideoglomus etunicatum*, *Rhizophagus irregularis*, and *Diversispora versiformis*	Improved photochemical reactions and activation of antioxidant defense mechanisms, accumulation of protecting molecules, and reduction of membrane damages	Pasbani et al. (2020)
Cold	Cucurbitaceae	*Cucumis sativus*	*Funneliformis mosseae*	Improved dry weight and fresh weight of seedlings, enhanced secondary metabolites content and enzymatic activity, reduced hydrogen peroxide content	Chen et al. (2013)
Cold	Cucurbitaceae	*Cucumis sativus*	*Funneliformis mosseae*	Increased total fresh and dry weight, reduced H_2O_2 accumulation in cell walls, and increased ATPase concentration and its activity, enhanced protein content in plasma membrane	Liu et al. (2014)
Cold	Cucurbitaceae	*Cucumis sativus*	*Glomus mosseae*	Increased root and shoot fresh weight, enhanced antioxidant activity and phenol content	Haghighi et al. (2015)
Cold	Cucurbitaceae	*Cucumis sativus*	*Rhizophagus irregularis*	Improved efficiency of photosynthesis and enhanced carbon sink strength, sugar content of leaves, and nonphotochemical quenching	Ma et al. (2019)
Heat	Solanaceae	*Cucumis sativus*	*Penicillium resedanum* LK6	Increased number of leaves and biomass per plant, higher nutrient content in plants, significant proline accumulation, higher amount of flavonoids synthesis	Khan et al. (2013)

TABLE 7.1 *(Continued)*

Stress	Family of host plant	Host plant	Fungus species	Observed response	Reference
Heat	Solanaceae	*Solanum lycopersicum*	*Septoglomus deserticola* and *Septoglomus constrictum*	Increased leaf water potential, stomatal conductance, and enhanced biomass production	Duc et al. (2013)
Heat	Cucurbitaceae	*Cucumis sativus*	*Glomus mosseae*	Increased root and shoot fresh weight, enhanced antioxidant activity and phenol content	Haghighi et al. (2015)

Source: Reproduced with permission from Malhi et al. (2021), Copyright © Elsevier B.V. (2021).

TABLE 7.2 Impact of AM Fungi on Management of Drought Stress and Salinity Stress in Vegetables.

Family of host plant	Host plant	Fungus species	Observed response			
			Drought stress	References	Salinity stress	References
Solanaceae	*Solanum lycopersicum*	*G. mosseae*, *G. etunicatum*, *G. claroideum*, *G. microaggregatum*, *G. geosporum*, and *R. irregularis*	Enhanced fruit yield, water use efficiency, biomass, and leaf water potential	Bakr et al. (2016)	–	–
Solanaceae	*Lycopersicon esculentum* (L.)	*Glomus mosseae*	–	–	Increased dry matter and leaf area, enhance enzyme activity	Huang et al. (2010)
Solanaceae	*Lycopersicon esculentum* (L.)	*Glomus fasciculatum*	–	–	Improved P, phosphatase enzyme activity, and reduced pH	Shivakumar et al. (2019)

TABLE 7.2 *(Continued)*

Family of host plant	Host plant	Fungus species	Observed response			
			Drought stress	**References**	**Salinity stress**	**References**
Solanaceae	*Capsicum annuum* (L.)	*Glomus intraradices*	–	–	Improved root and shoot biomass, leaf area, and cell membrane integrity	Beltrano et al. (2013)
Solanaceae	*Lycopersicon esculentum* (L.)	*Glomus intraradices*, *Glomus mosseae*, *Glomus aggregatum*, *Glomus clarum*, *Glomus monosporus*, *Glomus deserticola*, *Glomus brasilianum*, *Glomus etunicatum*, *Gigaspora margarita*	–	–	Increased plant growth, fresh stem weight, enhanced chlorophyll a and b content	Demir et al. (2011)
Solanaceae	*Solanum lycopersicum*	*Funneliformis mosseae*, *Rhizophagus intraradices*	Improved water use efficiency and improved growth	Chitarra et al. (2016)	–	–
Solanaceae	*Solanum lycopersicum*	*R. irregularis* strain EEZ 58	Improved rate of growth and more efficient photosystem II, increased production of strigolactone for symbiosis promotion	Ruiz-Lozano et al. (2016)	–	–
Solanaceae	*Solanum lycopersicum*	*Funneliformis mosseae*, *Funneliformis geosporum*, *Claroideoglomus etunicatum*, *Claroideoglomus claroideum*, *Glomus microaggregatum*, and *Rhizophagus irregularis*	Improved water use efficiency, stomatal conductance, more efficient photosystem II, and higher water potential of leaves	Bakr et al. (2018)	–	–

TABLE 7.2 *(Continued)*

Family of host plant	Host plant	Fungus species	Observed response			
			Drought stress	**References**	**Salinity stress**	**References**
Solanaceae	*Solanum lycopersicum*	*Septoglomus deserticola*, *Septoglomus constrictum*	Improved stomatal conductance, relative water content, and leaf water potential	Duc et al. (2018)	–	–
Solanaceae	*Solanum melongena*	*Glomus fasciculatum*	Enhanced crop growth, dry biomass, high water use efficiency, and increased number of fruits	Badr et al. (2020)	–	–
Solanaceae	*Capsicum annuum*, *Solanum lycopersicum*	*Rhizophagus intraradices*, *Rhizophagus fasciculatum*	Increased root and shoot length, higher biomass, and chlorophyll content	Tallapragada et al. (2016)	–	–
Solanaceae	*Capsicum annuum* (L.)	*Glomus clarum*	–	–	Regulation of ions (K, Mg, Ca, Zn, Na) uptake, increased nutrient uptake and growth	Altuntas et al. (2016)
Solanaceae	*Solanum melogena*	*Glomus deserticola*	–	–	Enhanced growth and aggressive colonization in roots	Mohammad and Mittra (2013)
Amaryllidaceae	*Allium sativum* (L.)	*Glomus fasciculatum*	Improved shoot length and plant growth, increased biomass, increased water use efficiency	Borde et al. (2012)	–	–
Amaryllidaceae	*Allium cepa* (L.)	*Glomus fasciculatum*	–	–	Increased plant height, fresh weight and dry weight, and reduced electrical conductivity	Suhail and Mahdi (2013)

TABLE 7.2 *(Continued)*

Family of host plant	Host plant	Fungus species	Observed response			
			Drought stress	**References**	**Salinity stress**	**References**
Cucurbitaceae	*Cucumis melo* (L.)	*Glomus* spp.	Improved plant height, root length, enzyme activity, biomass production and soluble sugar content, higher net photosynthetic rate, and water use efficiency	Huang et al. (2010)	–	–
Cucurbitaceae	*Citrullus lanatus* (L.)	*Funneliformis mosseae*	–	–	Improved photosynthetic activity, elemental, and water uptake	Ye et al. (2019)
Cucurbitaceae	*Cucumis sativus* (L.)	*Glomus versiforme*	–	–	Enhanced growth, improved photosynthesis and growth, and reduced Na toxicity	Ahmad et al. (2018)
Cucurbitaceae	*Cucumis sativus* (L.)	*Glomus etunicatum*, *Glomus intraradices*, *Glomus mosseae*	–	–	Enhancement of pigment synthesis, antioxidant enzyme activity, lower lipid peroxidation and electrolyte leakage, accumulation of K, Ca, Zn, Fe, Mg, Mn, Cu, and reduction of Na^+ ions	Hashem et al. (2018)
Cucurbitaceae	*Cucumis melo* (L.)	*Glomus intraradices* and *Glomus* spp.	Improved physiological parameters and photosynthesis, increased chlorophyll content	Cakmakci et al. (2017)	–	–

TABLE 7.2 *(Continued)*

Family of host plant	Host plant	Fungus species	Observed response			
			Drought stress	**References**	**Salinity stress**	**References**
Asteraceae	*Lactuca sativa*	*Glomus intraradices*, *Glomus mosseae*	Improved antioxidant compounds accumulation and chlorophylls and phenolics content in leaves, improved quality	Baslam and Goicoechea (2012)	–	–
Asteraceae	*Lactuca sativa*	*Rhizophagus intraradices*	Higher photosynthetic pigments, stomatal conductance, and improved antioxidant enzyme level (catalase, ascorbate peroxidase, and glutathione reductase)	Durán et al. (2016)	–	–
Asteraceae	*Lactuca sativa* (L.)	*Claroideoglomus claroideum*	–	–	Reduction of oxidative damage and enhanced growth, development, and yield	Santander et al. (2019)
Asteraceae	*Lactuca sativa*	*R. irregularis* strain EEZ 58	Improved rate of growth and more efficient photosystem II, increased production of strigolactone for symbiosis promotion	Ruiz-Lozano et al. (2016)	–	–
Asteraceae	*Lactuca sativa* (L.)	*Glomus iranicum*	–	–	Enhanced net photosynthesis, plant biomass, root mycorrhizal colonization, leaf ion content, plant biomass, and stomatal conductance	Vicente-Sánchez et al. (2013)

TABLE 7.2 *(Continued)*

Family of host plant	Host plant	Fungus species	Observed response			
			Drought stress	**References**	**Salinity stress**	**References**
Fabaceae	*Phaseolus vulgaris* (L.)	*Glomus etunicatum*, *G. intraradices* and *G. monosporum*	Improved vegetative growth, proteins, and sugar concentration in leaves, reduced proline content and yield	Salim and Abou El-yazied (2015)	–	–
Fabaceae	*Pisum sativum* (L.)	*Rhizoglomus intraradices*, *Funneliformis mosseae*, *Gigaspora* sp.	–	–	Enhanced nutrient uptake, increased biomass, synthesis of chlorophyll, compatible osmolytes accumulation higher growth attributes, and yield	Parihar et al. (2020)

Source: Reproduced and modified with permission from Malhi et al. (2021), Copyright © Elsevier B.V. (2021).

Inoculation of mycorrhizal fungi has also developed as an alternative source for increasing tolerance in plants exposed to different environmental stresses, especially in the rebuilding of soils that are contaminated with metals due to natural or industrial activities. It is also proposed through various studies that mycorrhizal association may help in phytoremediation, because high accumulation of Ni has been found in plants growing in soil enriched with metals (Gamalero et al., 2009). Also, phosphate fertilizers are main cause of soil pollution by Cd in cultivated land areas that decrease crop dependence on phosphate fertilizers. It is studied that the mycelium increased the uptake of inorganic phosphate but also acted as a shield to Cd uptake, that is how this association reduces the toxic effect of Cd on production and growth of plant.

7.3 ROLE OF MYCORRHIZA IN DROUGHT

Drought is defined as the insufficient amount of underground water compulsory for normal growth and development of plants; drought is also defined as insufficiency of water or water stress (Turk and Tawaha, 2001; Tawaha and Turk, 2002a, 2002b; Turk and Tawaha, 2002a, 2002b; Tawaha et al., 2003; Abebe et al., 2005; Abera et al., 2005; Al-Tawaha et al., 2005; Al-Kiyyam et al., 2008; Hameed et al., 2008; Marulanda and Barea, 2009; Al-Tawaha et al., 2010; Al-Tawaha et al., 2017; Ananthi et al., 2017; Abu Obaid et al., 2018). The actual reasons for the effect of drought in plants can be the unreachability of water to the plant roots system, stem transpiration rate, and subsequent instigation of oxidative stress in plants (Barzana et al., 2012). The symbiotic association between plant and mycorrhizal fungi increases root length and its thickness, the area of leaf, plant biomass, uptake of nutrients, and their transport in different drought conditions and that is how this association can increase the plant growth (Davies et al., 2002). Inoculation of mycorrhizal fungi increases plant growth by forming the extensive net of mycorrhizal hyphae and excreting glomalin, this in return improves the uptake of essential nutrients, including micro and macronutrients and water by the plants (Gholamhoseini et al., 2013). This symbiotic association participates in several biochemical and physiological processes including

- Mycorrhizal fungi directly uptake and transport the water and nutrients to plants
- Increased osmotic regulation
- Improving exchange of gases and efficient use of water

- In the literature, it is also reported that the defensive mechanisms against oxidative damage are also improved (Ruiz-Sanchez et al., 2010).

The mycorrhizal association led to improving water potential in leaf, exchange of gasses; stomatal conductance, photosynthesis, and transpiration also improved in plants which are inoculated with mycorrhiza growing in drought stress (Mena-Violante et al., 2006). Mycorrhizal fungi also adjust regulation of water in the plants that are in symbiotic association by adjusting their signaling balance of hormones or by initiating the activity of osmolytes in comparison with plants having no mycorrhizal association. The stomatal conductance and physiological phenomenon can also be stimulated by abscisic acid in mycorrhizal plants growing under drought stress (Porcel et al., 2006; Ruiz-Lozano et al., 2006).

7.4 ROLE OF MYCORRHIZA UNDER NUTRIENTS STRESS

The insufficient amount of micro- and macronutrients affects plant growth by changing the pattern of growth, chemical structure, and defense capacity of antioxidants in plants which leads to reduced plant liability to various stresses (Hajiboland, 2012). The macro and micronutrients, including iron, copper, and zinc having low mobility rate and their uptake, can also improved by mycorrhizal association (Baslam et al., 2013). AMF is important in increasing the rate of mineral uptake for the proper growth and development of plants. Due to increased rate of uptake of mineral, plants have the ability to stand against the pathogen and reduce uptake of compounds which are toxic in nature to plant growth. AMF association increased the surface area of roots which cover the better volume of soil and these roots penetrate in the depletion zone for better water and nutrients uptake (Smith and Read, 2008).

The lettuce plant that inoculated with mycorrhiza growing in soil with high concentration of nitrogen and phosphorous results in reduced macro and micronutrients in plant tissues (Azcon et al., 2003). Mycorrhiza also improved the uptake of nutrients from the soil when there are limited fertilizers of nitrogen and phosphorous (Xie et al., 2014). In another research, chickpea plant in association with mycorrhiza shows maximum uptake of micro and macronutrients (Calcium, Potassium, Magnesium, Phosphorous, Iron, and Silicon) (Yaseen et al., 2012). Due to the establishment of the broad hyphal system in the soil the phosphorus uptake improved; that is how it will help to improve the effect of very low value of pH (Muthukumar et al., 2014).

It is reported that about 80% of total amount of phosphorous required by plant for normal growth is carried by the mycelium network which radically grow outward from the plants. The diameter of hyphae is less than the root hair; that is why these hyphae even penetrate the pores of soil where root hairs are unable to reach. AMF increased the nutrients uptake, most importantly phosphorous uptake (Jakobsen, 1995). Absorption of phosphorous from soil is faster in mycorrhizal plants than in the non-mycorrhizal one because the diffusion distance of the HPO_4^{-2} and H_2PO_4 ions is shorter for fungal hyphae than the roots of plant. The increased rate of mineral uptake results in many physiological changes in the roots (Akhtar and Siddiqui, 2008). Also, it is studied that, when maize plant is inhabited by *Gigaspora margarita* (mycorrhizal fungi), it increases the uptake and transport of phosphate under acidic conditions.

7.5 MYCORRHIZA UNDER BIOTIC STRESSES

One of the most common and widely used agricultural practices to control the harmful effect of bacteria, nematodes, viruses, and fungi is using different types of pesticides. These pests cause the great damage to common cultivated crops. The genetically modified plants are also having the ability to show resistance against different pathogens. Many scientists are working on the plant-breeding system to develop such cultivars. However, the effect of pesticides against potential soil-borne diseases is only partly or moderately effective. Moreover, pesticides are detrimental and unfavorable to the environment and human health and consequently an accumulative number of pesticides are being taken off the market. Furthermore, cultivars with modified genome developed from plant-breeding process can easily deteriorate by changing in the biodiversity of the pathogenic organisms or agents because only a single-plant gene is used to get these cultivars with disease resistance. We can develop different stable strategies that can assure effective resistance in plants to the potential pathogens.

Through various studies the importance and benefits of mycorrhizal fungi have been approved. Soil-borne potential pathogens that are related to many plant species causing biotic stress and mycorrhizal fungi induce tolerance against this biotic stress in plants. There are different kinds of fungi of group Oomycetes which are pathogenic in nature including *Fusarium*, *Rhizoctonia*, *Verticillium*, *Thievalopsis*, *Aphanomyces*, *Phytophthora*, *Pythium*, and nematodes from another group named Heterodera has been constantly demonstrated (Hao et al., 2009). Many researches have been performed under very accurate and strategic circumstances. These researches performed at the initial growth stages of plant both in the agriculture field and also under controlled environmental conditions in greenhouses which are favorable for the good production rate approve these results (Utkhede, 2006).

It is unfeasible to mention all the research recognized related to this topic. This can be explained by the output achieved from the extensively cultivated and one of the expensive vegetable tomato crops in all over the world. However, tomato is susceptible to many pathogens affecting about 34% decline in production of fruit, if cultivated and grown in recent cultivation strategies. The tomato fruit is not very suitable for mycorrhizal association if we consider in relation to the crop production (Smith et al., 2009). It is also reported that mycorrhizal association combined with pathogens that affect the roots of plants can recompense these losses. It is also reported that when plant roots are in association with mycorrhiza, infections of roots and the intensity of the diseases caused by pathogens can primarily decrease. It also increased fresh plant weight up to 198% and fruit yield by 14.3% compared to the plants that are not inoculated by mycorrhiza and infected with pathogens.

7.6 MYCORRHIZOSPHERE: A BIOCONTROL ZONE

Mycorrhizosphere is defined as the zone of soil associated or in interaction with mycorrhizal fungi (Oswald and Ferchau, 1968). It is observed that the rhizosphere soil extract of mycorrhizal plants reduced the formation of sporangia of pathogenic fungus than the rhizosphere soil extract of nonmycorrhizal. The reason behind this is the absence of sporulation-inducing microorganisms or the number of sporulation microorganisms increased. Microbial communities around the roots are affected by changes in exudates of roots, the reason for the establishment of mycorrhizosphere (Varma et al., 1999). The mycorrhiza influences the microbes of rhizosphere. The microflora in mycorrhizosphere differs quantitatively as well as qualitatively as compared with nonmycorrhizal. Mycorrhizosphere has two components: a soil layer surrounding the roots of mycorrhiza and a soil layer surrounding the arbuscular mycorrhizal hyphae in the soil known as hyphosphere. There are two possible interactions between soil-borne pathogens and AMF; these interactions may be positive (synergistic) or negative (antagonism) (Mukerji et al., 2002). In positive interaction, AMF interacts with plant growth promoting bacteria, nitrogen-fixing bacteria, phosphorous-solubilizing bacteria that enhance the germination of AMF spores (Mayo et al., 1986). AMF when negatively interacting can reduce the soil pathogens (Dehne, 1982).

The mycorrhizosphere is characterized as the environment that is favorable to microbes and plants, but is unfriendly to the potential pathogens present in soil, their existence, and their propagation. Surely, to reduce the effect of the disease caused by *Fusarium oxysporum dianthi* in the *D. caryophyllus* co-inoculation of the nonmycorrhizal species *Dianthus caryophyllus* with the mycorrhizal species *Tagetes patula* and *G. intraradices* is very useful; this inoculation patently decreased the disease by reducing the growth of pathogens and its growth in mycorrhizosphere (St-Arnaud et al., 1997). Additionally, it is also noticed that the tomato roots which are previously colonized with *G. mosseae* when inoculated with *P. nicotianae* zoospores cause the decrease in the number of infection loci which also leads to disturbed infiltration and root penetration of pathogen in mycorrhizosphere (Vigo et al., 2000).

Three types of interactions have been reported which influenced the mycorhizosphere; these interactions included rhizobacteria, mycorrhiza, and plant roots. This three-party association represents precise characteristics and every character influencing one another growth, development, and health. They are producing glycoproteins especially glomalin; also mycorrhizal fungi play an important role in providing stable microsites that are useful for the establishment of microbes and plant roots by helping in the formation of aggregates (Rillig and Mummey, 2006).

On the other hand, the extra radical network of mycorrhizal mycelium also provides microsites for the establishment of different bacteria. The nitrogen-fixing and phosphate-solubilizing bacteria are categorized as the plant growth promoting rhizobacteria because these bacteria can interconnect and collaborate with the fungi results in increase in the accessibility of nitrogen and phosphorous to plants. That is why these bacteria are very helpful in supporting the growth of plants and also improve the ability of plants to reduce the harmful effects of pathogens on growth, development, and production of plants (Lioussanne et al., 2009).

7.7 MYCORRHIZA AND PHYTOPHAGOUS INSECTS

Plants with mycorrhizal association also affect the performance of insects and herbivores, but the extent and trend of effect depends directly on their mode of feeding and regime of pathogenic insects (Koricheva et al., 2009). Many different researches on the three-party associations (mycorrhiza-plant-insects) under controlled, natural, or field conditions have been conducted and upon the wide range of review data it concluded that usually mycorrhizal association has the negative effect on rhizophagous insects. It is also noted that the effect on insects that are getting their food from shoot of the plants is fragile and very irregular. Irrespective of this irregularity, some common ways are studied: one category that is named generalist insects are mostly badly affected by mycorrhizal fungi, but on the other hand, the specialist insects mostly take advantage of it. Furthermore, the mycorrhizal association negatively affects the insects that are used to chew the leaves of plants but aphids generally perform better on mycorrhizal plants.

The above-mentioned pattern may be observed due to the difference in nourishing and defense aspects of pathogenic insects. Though insects show sensitivity to the defense mechanism, but herbivores are able to avoid the plant defense mechanism and remain unnoticed. Consequently, when plants improve their dense mechanism by developing the mycorrhizal association the common and general insects effected by this improved defense capacity but the specialist insects are unnoticed by improved defense mechanism and enjoy the improved nutritional status of the host plant. Plants that are inoculated with mycorrhizal fungi produce jasmonate or jasmonic acid; these chemicals are used as defense mechanism against leaf chewer insects because these chemicals are susceptible to these insects and having negative effects on them (Pena-Cortes et al., 2004). Also, to control the negative effect of herbivores on plant, mycorrhizal fungi can manipulate the performance and function of their predators and parasitoids. This phenomenon is explained by an example of tomato plant when in mycorrhizal interaction producing blend of volatile chemicals which attracts the aphid parasitoids and this process is not observed in the nonmycorrhizal tomato plant (Guerrieri et al., 2004).

7.8 CONCLUSION

From all the studies, we concluded that mycorrhizal association is not only beneficial for enhancing plant tolerance to biotic and abiotic stresses, but this interaction also responds to many other stresses of their host plant. All above-mentioned stresses have some influence on richness and mycorrhizal community structure. The change of mycorrhizal fungal diversity will directly relate to change in the variety of leading plant species of that community because changes in the mycorrhizal community will feedback into the host plant of that community. Other factors that speed up such kind of responses are changes in climate, agricultural practices, and other plants that invaded the same community. Mycorrhizal fungi are efficient in adapting to any changes in their biotic and abiotic environment. But this may or may not improve their mutualistic association.

The effect of evolutionary and ecological responses to biotic and abiotic stresses of mycorrhizal fungi is now becoming more commanding for agricultural system such as the invasion of new species that are not native to the community considered under the aspect of change in climate and biotic stresses. To understand the mycorrhizal role in conferring drought stress to plants considerable progress has been made, but different characteristics still need attention to identify different metabolites and also unknown metabolic pathways. The various omic techniques, their procedure, and combinations will benefit physiological, biochemical, and molecular data techniques established on standard resources and methods. For detailed analysis, we can use different progressive techniques to simplify and explain the mechanisms of tolerance and avoidance of drought induced by mycorrhizal association. It also helps in differentiation of the biotic and abiotic stresses induced by the processes of the defensive mechanisms controlled by mycorrhizal association. To increase our info about how plants, modify their physiology when in association with mycorrhizal fungi along with information and biology of the potential insects attacked the plants are very important to describe markers to stimulate resistance. It will also help for the generation of analytical models of the outcome of specific interactions between mycorrhiza and insects. Another task is to identify the connections how plants responded to abiotic and biotic stresses. It is significant to consider mycorrhizal fungi as a part of multi-trophic environment because other organisms present in relevant ecosystem may affect the interactions of plants colonized by mycorrhizal fungi.

KEYWORDS

- **abiotic stresses**
- **arbuscular mycorrhiza**
- **biotic stresses**
- **tolerance**
- **host plant**

REFERENCES

Abebe, G.; Hattar, B.; Al-Tawah, A. M. Nutrient availability as affected by manure application to cowpea (*Vigna unguiculata* L. Walp.) on calacarious soils. *J. Agric. Soc. Sci.* **2005**, *1*(1), 1–6.

Abera, T.; Feyisa, D.; Yusuf, H.; Nikus, O.; Al-Tawaha, A. R. Grain yield of maize as affected by biogass slurry and NP fertilizer rate at Bako, Western Oromiya, Ethopia. *Biosci. Res.* **2005**, *2*, 31–38.

Ahmad, H.; Hayat, S.; Ali, M.; Liu, T.; Cheng, Z. The combination of arbuscular mycorrhizal fungi inoculation (*Glomus versiforme*) and 28-homobrassinolide spraying intervals improves growth by enhancing photosynthesis, nutrient absorption, and antioxidant system in cucumber (*Cucumis sativus* L.) under salinity. *Ecol. Evol.* **2018**, *8*, 5724–5740. https://doi.org/10.1002/ ec3.4112.

Al Zoubi, O. M.; Al-Tawaha, A. R. Allelopathic effect of beetroot (*Beta vulgaris* L.) on germination and growth of *Zea mays* and *Vigna umbellata*. *Int. J. Bot. Stud.* **2019**, *4*(4), 47–51.

Al-Ajlouni, M. M.; Al-Ghzawi, A. L. A.; Al-Tawaha, A. R. Crop rotation and fertilization effect on barley yield grown in arid conditions. *J. Food Agric. Environ.* **2010**, *88*(3), 869–872.

Al-Juthery, H. W.; Habeeb, K. H.; Jawad, F.; Altaee, K.; AL-Taey, D. K. A.; Al-Tawaha, A. R. M. Effect of foliar application of different sources of nano-fertilizers on growth and yield of wheat. *Biosci. Res.* **2018**, *15*(4), 3988–3997.

Al-Kiyyam, M. A.; Turk, M.; Al-Mahmoud, M. and Al-Tawaha, A. R. Effect of plant density and nitrogen rate on herbage yields of marjoram under Mediterranean conditions. *J. Am.-Eurasian J. Agric. Environ. Sci.* **2008**, *3*, 153–158.

Al-Tawaha, A. M.; Turk, M.; Hameed, K.; Assaf, T.; Tahhan, R.; Al-Jamali, A. F. Effects of application of olive mill by-products on lentil yield and their symbiosis with mycorrhizal fungi under arid conditions. *Crop Res.* **2010**, *40*(3), 66–75.

AL-Tawaha, A. R. M.; Nidal, O. Use of sorghum and maize allelopathic properties to inhibit germination and growth of wild barley (*Hordeum spontaneum*). *Notulae Botanicae Horti Agrobotanici Cluj-Napoca* **2010**, *38*(3), 124–127.

Al-Tawaha, A. R.; Al-Karaki, G.; Al-Tawaha, A. R.; Sirajuddin, S. N.; Makhadmeh, I.; Wahab, P. E. M.; Youssef, R. A.; Al Sultan, W.; Massadeh, A. Effect of water flow rate on quantity and quality of lettuce (*Lactuca sativa* L.) in nutrient film technique (NFT) under hydroponics conditions. *Bulgarian J. Agric. Sci.* **2018a**, *245*, 791–798.

Al-Tawaha, A. R.; Al-Tawaha, A. R.; Alu'datt, M.; Al-Ghzawi, A. L.; Wedyan, M.; Al-Obaidy, S. D. A.; Al-Ramamneh, E. A. D. Effects of soil type and rainwater harvesting treatments in the growth, productivity and morphological trains of barley plants cultivated in semi-arid environment. *Aust. J. Crop Sci.* **2018b**, *12*(6), 975–979.

Al-Tawaha, A. R.; McNeil, D. L.; Yadav, S. S.; Turk, M.; Ajlouni, M.; Abu-Darwish, M. S.; Al-Ghzawi, A. L. A.; Al-udatt, M.; Aladaileh, S. Integrated legume crops production and management technology. In *Climate change and management of cool season grain legume crops;* Yadav, S. S., Redden, R., Eds., Springer: Dordrecht, 2010; pp 325–349.

Al-Tawaha, A. R.; Turk, M. A.; Abu-Zaitoon, Y. M.; Aladaileh, S. H.; Al-Rawashdeh, I. M.; Alnaimat, S.; Al-Tawaha, A. R. M.; Aludatt, M. H.; Wedyan, M. Plants adaptation to drought environment. *Bulgarian J. Agric. Sci.* **2017**, *23*(3), 381–388.

Altuntas, O.; Dasgan, H. Y.; Akhondnejad, Y. Effect of mycorrhiza on alleviating salt stress of *Capsicum annum* L. by ion regulation. Proceedings of XVIth EUCARPIA *Capsicum* and eggplant working Group meeting in memoriam Dr. Alain Pallox, 2016.

Aludatt, M. H.; Rababah, T.; Alhamad, M. N.; Al-Tawaha, A.; Al-Tawaha, A. R.; Gammoh, S.; Ereifej, K. I.; Al-Karaki, G.; Hamasha, H. R.; Tranchant, C. C.; Kubow, S. Herbal yield, nutritive composition, phenolic contents and antioxidant activity of purslane (*Portulaca oleracea* L.) grown in different soilless media in a closed system. *Industrial Crops and Products* **2019**, *141*, 111746.

Ananthi, T.; Amanullah, M. M.; Al-Tawaha, A. R. M. S. A review on maize-legume intercropping for enhancing the productivity and soil fertility for sustainable agriculture in India. *Advances in Environmental Biology* **2017**, *11*(5), 49–63.

Azcon, R.; Ambrosano, E.; Charest, C. Nutrient acquisition in mycorrhizal lettuce plants under different phosphorus and nitrogen concentration. *Plant Sci.* **2003**, *165*, 1137–1145.

Badr, M. A.; EI-Tohamy, W. A.; Abou-Hussein, S. D.; Gruda, N. S. Deflict irrigation and arbuscular mycorrhiza as a water-saving stratergy for eggplant production. *Horticulturae* **2020**, *6*, 45. http://doi.org/10.3390/horticulturae6030045.

Bakr, J.; Daood, H.; Pék, Z.; Helyes, L.; Posta, K. Yield and quality of mycorrhized processing tomato under water scarcity. *Appl. Ecol. Environ. Res.* **2016**, *15*, 401–413.

Bakr, J.; Pek, Z.; Helyes, L.; Posta, K. Mycorrhizal inoculation alleviastes water deflict impact on field-grown processing tomato. *Polish J. Environ. Stud.* **2018**, *27*, 1949–1958. http://doi. Org/10.15244/pjoes/78624.

Barzana, G.; Aroca, R.; Paz, J. A. Arbuscular mycorrhizal symbiosis increases relative apoplastic water flow in roots of the host plant under both well-watered and drought stress conditions. *Ann Bot.* **2012**, *109*, 1009–1017.

Baslam, M.; Goicoechea, N. Water deflict improved the capacity of arbuscular mycorrhizal fungi (AMF) for including the accumulation of antioxidant compounds in lettuce leaves. *Mycorrhiza* **2012**, *22*, 347–359.

Baslam, M.; Garmendia, I.; Goicoechea, N. Enhanced accumulation of vitamins, nutraceuticals and minerals in lettuces associated with arbuscular mycorrhizal fungi (AMF): A question of interest for both vegetables and humans. *Agriculture* **2013**, *3*,188–209.

Beltrano, J.; Ruscitti, M.; Arango, M. C.; Ronco, M. Effects of arbuscolar mycorrhiza inoculation on plant growth, biological and physiological parameters and mineral nutrition in pepper grown under different salinity and P levels. *J. Soil Sci. Plant Nutr.* **2013**, *13*, 123–141.

Beltrano, J.; Ruscitti, M.; Arango, M. C.; Ronco, M. Effects of arbuscular mycorrhiza inoculation on plant growth, biological and physiological parameters and mineral nutrition in pepper grown under different salinity and P levels. *J Soil Sci Plant Nutr*. *13*, 123–141.

Borde, M.; Dudhane, M.; Jite, P. Growth water use efficiency and antioxidant defense responses of mycorrhizal and non mycorrizal *Allium sativum* L. under drought stress condition. *Ann. Plant Sci.* **2012**, *1*, 6–11.

Cakmakci, O.; Cakmakci, T.; Durak, E. D.; Demir, S.; Sensoy, S. Effects of arbuscular mycorrhizal fungi in melon (*Cucumis melo* L.) seedling under deflict irrigation. *FEB-Fresenius Environ. Bull.* **2017**, 7513.

Caradonia, F.; Francia, E.; Morcia, C.; Ghizzoni, R.; Moulin, L.; Terzi, V.; Ronga, D. Arbuscular mycorrhizal fungi and plant growth promoting rhizobacteria avoid processing tomato leaf damage during chilling stress. *Agronomy* **2019**, *9*, 1–20.

Chen, G.; Yan, W.; Yang, L.; Gai, J.; Zhu, Y. Overexpression of StNHX1, a novel vacuolar Na^+/H^+ antiporter gene from *Solanum torvum*, enhances salt tolerance in transgenic vegetables soyabean. *Hort. Environ. Biotechnol.* **2014**, *55*, 213–221.

Chitarra, W.; Pagliarani, C.; Maserti, B.; Lumini, E.; Siciliano, I.; Cascone, P.; Schubert, A.; Gambino, G.; Balestrini, R.; Guerrieri, E. Insights on the impact of arbuscular mycorrhizal symbiosis on tomato tolerance to water stress. *Plant Physiol.* **2016**, *171*, 1009–1023.

Davies, F. T. Jr.; Olalde-Portugal, V.; Aguilera-Gomez, L. Alleviation of drought stress of Chile ancho pepper (*Capsicum annuum* L. cv. San Luis) with arbuscular mycorrhiza indigenous to Mexico. *SciHortic (Amst)* **2002**, *92*, 347–359.

Dehne, H. W. Interaction between vesicular-arbuscular mycorrhizal fungi and plant pathogens. *Phytopathology* **1982**, *72*, 1115–1119.

Demir, K.; Başak, H.; Okay, F. Y.; Kasim, R. The effect of endo-mycorrhiza (VAM) treatment on growth of tomato seedling grown under saline conditions. *Afr. J. Agric. Res.* **2011**, *6*, 3326–3332.

Duc, N. H.; Csintalan, Z.; Posta, K. Arbuscular mycorrhizal fungi mitigate negative effects of combined drought and heat stress on tomato plants. *Plant Physiol. Biochem.* **2018**, *132*, 297–307.

Duran, P.; Acuna, J. J.; Armada, E.; Lopez-Caornejo, P.; Mora, M. L.; Azcon, R. Inoculation with selenobacteria and arbuscular mycorrhizal fungi to enhance selenium content in lettuce plants and improve tolerance against drought stress. *J. Soil Sci. Plant Nutr.* **2016**, *16*, 347–359.

Gamalero, E.; Lingua, G.; Berta, G.; Glick, B. R. Beneficial role of plant growth promoting bacteria and arbuscular mycorrhizal fungi on plant responses to heavy metal stress. *Can J Microbiol.* **2009**, *55*, 501–514.

Gholamhoseini, M.; Ghalavand, A.; Dolatabadian, A. Effects of arbuscular mycorrhizal inoculation on growth, yield, nutrient uptake and irrigation water productivity of sunflowers grown under drought stress. *Agric Water Manag.* **2013**, *117*, 106–114.

Guerrieri, E.; Lingua, G.; Digilio, M. C.; Massa, N.; Berta, G. Do interactions between plant roots and the rhizosphere affect parasitoid behaviour? *Ecol. Entomol.* **2004**, *29*, 753–756.

Gurdeep, S. M.; Manpreet, K.; Prashant, K.; Mohammed, N. A.; Abdulaziz, A. A.; Parvaiz, A. Arbuscular mycorrhiza in combating abiotic stresses in vegetables: An eco-friendly approach. *Saudi J. Biol. Sci.* **2021**, *28*(2), 1465–1476.

Haghighi, M.; Mozafariyan, M.; Abdolahipour, B. Effect of cucumber mycorrhiza inoculation under low and high root temperature grown on hydroponic conditions. *J. Crop Sci. Biotechnol.* **2015**, *18*, 89–96. https://doi.org/ 10.1007/s12892–014–0083–4.

Hajiboland, R. Effect of micronutrient deficiencies on plants stress responses. In: *Abiotic stress responses in plants: Metabolism, productivity and sustainability*; Ahmad, P., Prasad, M. N. V., Eds.; Springer Science + Business Media, LLC: New York, 2012; pp 283–329.

Hani, N. B.; Al-Ramamneh, E. A. D.; Haddad, M.; Al-Tawaha, A. R.; Al-Satari, Y. The impact of cattle manure on the content of major minerals and nitrogen uptake from ^{15}N isotope-labeled ammonium sulphate fertilizer in maize (*Zea mays* L.) plants. *Pak. J. Bot.* **2019**, *51*(1), 185–189.

Hao, Z.; Fayolle, L.; van Tuinen, D.; Gianinazzi-Pearson, V.; Gianinazzi, S. Mycorrhiza reduce development of nematode vector of Grapevine fanleaf virus in soils and root systems. In: *Extended Abstract 16th meeting of ICVG*; Boudon-Padfieu, E., Ed.; Dijon: France, 2009; pp 100–101.

Hashem, A.; Alqarawi, A. A.; Radhakrishnan, R.; Al-Arjani, A. F.; Aldehaish, H. A.; Eganberdieva, D.; Abd-allah, E. F. Arbuscular mycorrhizal fungi regulate the oxidative system, hormones and ionic equilibrium to trigger salt stress tolerance in *Cucumis sativus* L. *Saudi J. Biol. Sci.* **2018**, *25*, 1102–1114.

Huang, Z.; Zou, Z.; He, C.; He, Z.; Zhang, Z.; Li, J. Physiological and photosynthetic responses of melon (*Cucumis melo* L.) seedlings to three *Glomus* species under water deficit. *Plant Soil* **2010**, *339*, 391–399. https://doi.org/ 10.1007/s11104–010–0591-z.

Khan, A. L.; Kang, S. -M.; Dhakal, K. H.; Hussain, J.; Adnan, M.; Kim, J. -G.; Lee, I. J. Flavonoids and amino acid regulation in *Capsicum annuum* L. by endophytic fungi under different heat stress regimes. *Sci. Hortic.* **2013**, *155*, 1–7. https://doi.org/ 10.1016/j.scienta.2013.02.028.

Koricheva, J.; Gange, A. C.; Jones, T. Effects of mycorrhizal fungi on insect herbivores: A meta-analysis. *Ecology* **2009**, *90*, 2088–2097.

Lal, R. Soil degradation as a reason for inadequate human nutrition. *Food Secur.* **2009**, *1*, 45–57.

Li, X.; Ahammed, G. J.; Zhang, G. Q.; Sun, Z. H.; Zhou, J.; Zhou, Y. H.; Xia, X. J.; Yu, J. Q.; Shi, K. Carbon dioxide enrichment alleviates heat stress by improving cellular redox homeostasis through an ABA-independent process in tomato plants. *Plant Biol.* **2014**, *17*. https://doi.org/10.1111/plb.12211.

Lioussanne, L.; Beauregard, M. S.; Hamel, C.; Jolicoeur, M.; St-Arnaud, M. Interactions between arbuscular mycorrhiza and soil microorganisms. In: *Advances in Mycorhizal biotechnology: A Canadian perspective*; Khasa, D., Piche, Y., Coughlan, A., Eds.; NRC Press: Ottawa, 2009.

Liu, A.; Chen, S.; Wang, M.; Liu, D.; Chang, R.; Wang, Z.; Lin, X.; Bai, B.; Ahammed, G. J. Arbuscular mycorrhizal fungus alleviates chilling stress by boosting redox poise and antioxidant potential of tomato seedlings. *J. Plant Growth Regul.* **2016**, *35*, 109–120. https://doi.org/10.1007/s00344–015–9511-z.

Ma, J.; Janouskova, M.; Ye, L.; Bai, L. Q.; Dong, R. R.; Yan, Y.; Yu, X. C.; Zou, Z. R.; Li, Y. S.; He, C. X. Role of arbuscular mycorrhiza in alleviating the effect of cold on the photosynthesis of cucumber seedlings. *Photosynthetica* **2019**, *57*, 86–95. https:// doi.org/10.32615/ps.2019.001.

Marulanda, A.; Barea, J. M. Stimulation of plant growth and drought tolerance by native microorganisms (AM fungi and bacteria) from dry environments: mechanisms related to bacterial effectiveness. *J. Plant Growth Regul.* *28*, 115–124.

Marulanda, A.; Barea, J. M.; Azcon, R. An indigenous drought tolerant strain of *Glomus intraradices* associated with a native bacterium improves water transport and root development in *Retama sphaerocarpa*. *Microb. Ecol.* **2006**, *52*, 670–678.

Mayo, K.; Davis, R. E.; Motta, J. Stimulation of germination of spores of *Glomus versiforme* by spore-associated bacteria. *Mycologia* **1986**, *78*, 426–431.

Mena-Violante, H. G.; Ocampo-Jimenez, O.; Dendooven, L. Arbuscular mycorrhizal fungi enhance fruit growth and quality of chileancho (*Capsicum annuum* L. cv San Luis) plants exposed to drought. *Mycorrhiza* **2006**, *16*, 261–267.

Mohammad, A.; Mittra, B. Effect of inoculation with stress-adapted arbascular mycorrhizal fungus *Glomus deserticola* on growth of *Solanum melongena* L. and *Sorghum sudanses* Staph. seedlings under salinity and heavy metal stress conditions. *Arch. Agron. Soil Sci.* **2013**, *59*, 173–183. http://doi.org/10.1080/03650340.2011.610029.

Mukerji, K. G.; Manoharachary, C.; Chamola, B. P. *Techniques in mycorrhizal studies*; Kluwer Academic Publishers, Netherlands, 2002; pp 285–296.

Muthukumar, T.; Priyadharsini, P.; Uma, E.; Jaison, S.; Pandey, R. R. Role of arbuscular mycorrhizal fungi in alleviation of acidity stress on plant growth. In: *Use of microbes for the alleviation of soil stresses*; Miransari, M., Ed.; Springer Science + Business Media: New York, 2014; pp 43–71.

Nikus, O.; Al-Tawaha, A. M.; Turk, M. A. Effect of manure supplemented with phosphate fertilizer on the fodder yield and quality of two sorghum cultivars (*Sorghum bicolor* L.). *Biosci. Res.* **2004a**, *1*, 1–7.

Nikus, O.; Al-Tawaha, A. M.; Turk, M. A. Yield response of Sorghum (*Sorghum bicolor* L.) to manufacture supplemented with phosphate fertilizer under semi-arid Mediterranean conditions. *Int. J. Agric. Biol.* **2004b**, *6*, 889–893.

Oswald, E. T.; Ferchau, H. A. Bacterial associations of coniferous mycorrhizae. *Plant Soil* **1968**, *28*, 187–192.

Parihar, M.; Rakshit, A.; Rana, K.; Tiwari, G.; Jatav, S. S. Colonization with arbuscular mycorrhizal fungi inoculation in mitigating salt stress of Pea (*Pisum sativum* L.). *Commun. Soil Sci. Plant Anal.* **2020**, *51*, 1545–1559.

Pasbani, B.; Salimi, A.; Aliasgharzad, N.; Hajiboland, R. Colonization with arbuscular mycorrhizal fungi mitigates cold stress through improvement of antioxidant defense and accumulation of protecting molecules in eggplants. *Sci. Hortic.* **2020**, *272*. https://doi.org/10.1016/j.scienta.2020.109575

Pena-Cortes, H.; Barrios, P.; Dorta, F.; Polanco, V.; Sanchez, C.; Sanchez, E.; Ramirez, I. Involvement of jasmonic acid and derivatives in plant response to pathogen and insects and in fruit ripening. *J. Plant Growth Regul.* **2004**, *23*, 246–260.

Porcel, R.; Aroca, R.; Cano, C.; Ruiz-Lozano, J. M. Identification of a gene from the arbuscular mycorrhizal fungus *Glomus intraradices* encoding for a 14-3-3 protein that is up-regulated by drought stress during the AM symbiosis. *Microb. Ecol.* **2006**, *52*(3), 575–582.

Rillig, M. C.; Mummey, D. L. Mycorrhizas and soil structure. *New Phytol.* **2006**, *171*, 41–53.

Ruiz- Lozano, J. M.; Aroca, R.; Zamarreho, Á. M.; Molina, S.; Andero-Jiménez, B.; Porcel, R.; García–Mina, J. M.; Ruyter-Sprira, C.; López-Ráez, J. A. Arbuscular mycorrhizal symbiosis induces strig-D-lactone biosynthesis under drought and improves drought tolerance in lettuce and tomato. *Plant. Cell Environ.* **2006**, *39*, 441–452. http://doi.org/10.1111/pce. 12631.

Ruiz-Lozano, J. M.; Porcel, R.; Aroca, R. Does the enhanced tolerance of arbuscular mycorrhizal plants to water deficit involve modulation of drought-induced plant genes? *New Phytol.* **2006**, *171*, 693–698.

Ruiz-Sanchez, M.; Aroca, R.; Munoz, Y.; Armada, E.; Polon, R.; Ruiz-Lozano, J. M. The arbuscular mycorrhizal symbiosis enhances the photosynthetic efficiency and the antioxidative response of rice plants subjected to drought stress. *J. Plant Physiol.* **2010**, *167*, 862–869.

Sailo, G. L.; Bagyaraj, D. J. Influence of different AM fungi on the growth, nutrition and forskolin content of *Coleus forskohlii*. *Mycol. Res.* **2005**, *109*, 795–798.

Salim, B.; Abou El-Yazied, A. Effect of mycorriza on growth, biochemical constituents and yield of snap bean plants under water deflict conditions. *J. Hortic. Sci. Ornam Plants.* **2015**, *7*, 131–140.

Santander, C.; Ruiz, A.; Garcia, S.; Aroca, R.; Cumming. J.; Cornejo, P. Efficiency of two arbuscular mycorrhizal fungi inocula to improve saline stress tolerance in lettuce plants by changes of antioxidant defense mechanism. *J. Sci. Food Agric.* **2019**, *100*, 1577–1587.

Saxena, S.; Kaur, H.; Verma, P.; Petla, B. P.; Andugula, V. R.; Majee, M. Osmoprotectants: potential for crop improvement under adverse conditions. In: *Plant acclimation to environmental stress*; Tuteja, N., Gill, S. S., Eds.; Springer Science + Business Media: New York, 2013; pp 197–232.

Sharifi, M.; Ghorbanli, M.; Ebrahimzadeh, H. Improved growth of salinity stressed soybean after inoculation with pre-treated mycorrhizal fungi. *J. Plant Physiol.* **2007**, *164*, 1144–1151.

Shivakumar, K.; Kumaresan, G.; Sugapriya, N. Arbuscular mycorrhizal fungi (AM fungi) and phosphate solubilizing bacteria (PSB) on tolerance of tomato under salt stress. *J. Pharmacochem. Phytochem.* **2019**, *8*, 4717–4721.

Smith, F. A.; Grace, E. J.; Smith, S. E. More than a carbon economy: nutrient trade and ecological sustainability in facultative arbuscular mycorrhizal symbioses. *New Phytol.* **2009**, *182*, 347–358.

Smith, S. E.; Read, D. J. *Mycorrhizal symbiosis*; Academic Press: London, 2008.

Souza, L. A.; Andrade, S. A. L.; Souza, S. C. R.; Schiavinato, M. A. Evaluation of mycorrhizal influence on the development and phytoremediation potential of *Canavalia gladiata* in Pb contaminated soils. *Int. J. Phytoremed.* **2012**, *15*, 465–476.

St-Arnaud, M.; Hamel, C.; Vimard, B.; Caron, M.; Fortin, J. A. Inhibition of *Fusarium oxysporum* f. sp. *dianthi* in the non-VAM species *Dianthus caryophyllus* by co-culture with *Tagetes patula* companion plants colonized by *Glomus intraradices*. *Can. J. Bot.* **1997**, *75*, 998–1005.

Suhail, F. M.; Mahdi, I. A. Test the efficiency of mycorrhizal fungi (*Glomus fasciculatum*) and magnetic water to reduce the effect of salinity on plant onion (*Allium cepa* L.). *Bull. Univ. Agric. Sci. Veterin. Med. Cluj-Napoca Agric.* **2013**, *70*, 325–333.

Talavera, M.; Itou, K.; Mizukubo, T. Reduction of nematode damage by root colonization with arbuscular mycorrhiza (*Glomus* spp.) in tomato *Meloidogyne incognita* (Tylenchida: Meloidogynidae) and carrot-*Pratylenchus penetrans* (Tylenchida: Pratylenchidae) pathosystems. *Appl Entomol Zool.* **2001**, *36*, 387–392.

Tallapragada, P.; Dikshit, R.; Seshagiri, S. Influence of *Rhizophagus* spp. and *Burkholderia seminalis* on the growth of tomato (*Lycopersicon esculentum*) and bell pepper (*Capsicum annuum*) under drought stress. *Common. Soil Sci. Plant Anal.* **2016**, *47*, 1975–1984.

Tawaha, A. M.; Turk, M. A. Lentil (*Lens culinaris* Medic.) productivity as influenced by rate and method of phosphate placement in a Mediterranean environment. *Acta Agronomica Hungarica* **2002**, *50*(2), 197–201.

Tawaha, A. M.; Turk, M. A. Field pea seeding management for semi-arid Mediterranean conditions. *J. Agron. Crop Sci.* **2004**, *190*, 86–92.

Tawaha, A. M.; Singh, V. P.; Turk, M. A. A review on growth, yield components and yield of barley as influenced by genotypes, herbicides and fertilizer application. *Res. Crop* **2003**, *4*(1), 1–9.

Tawaha, A. R. M.; Turk, M. A.; Lee, K. D.; Zheng, W. J.; Ababneh, M.; Abebe, G.; Musallam, I. W. Impact of fertilizer and herbicide application on performance of ten barley genotypes grown in northeastern part of Jordan. *Int. J. Agric. Biol.* **2005**, *7*(2), 162–166.

Turk, M. A.; Tawaha, A. M. Common vetch (*Vicia sativa* L.) productivity as influenced by rate and method of phosphate fertilization in a Mediterranean environment. *Agricultura Mediterranea* **2001,** *131*,108–111.

Turk, M. A.; Tawaha, A. M. Onion (*Allium cepa* L.) as influenced by rate and method of phosphorus placement. *Crop Res.* **2002a**, *23*(1), 105–107.

Turk, M. A.; Tawaha, A. M. Impact of seeding rate, seeding date and method of phosphorous application in faba bean (*Vicia faba* L. Minor) in the absence of moisture stress. *Biotechnol. Agron. Soc. Environ.* **2002b**, *6*(3), 171–178.

Turk, M. A.; Tawaha, A. M. Allelopathic effect of black mustard (*Brassica nigra* L.) on germination and growth of wild oat (*Avena fatua* L.). *Crop Protect.* **2003a**, *22*(4), 673.

Turk, M. A.; Tawaha, A. M. (2003b). Allelopathic effect of black mustard (*Brassica nigra* L.) on germination and growth of wild barley (*Hordeum spontaneum*). *J. Agron. Crop Sci.* **2003b**, *189*(5), 298–303.

Turk, M. A.; Tawaha, A. M. Effect of variable sowing ratios and sowing rates of bitter vetch on the herbage yield of oat-bitter vetch mixed cropping. In *Proceedings 7th International Oat Conference*; Peltonen-Sainio, P., Topi-Hulmi, M., Eds.; MTT, 2004.

Turk, M. A.; Lee, K. D.; Tawaha, A. M. Inhibitory effects of aqueous extracts of black mustard on germination and growth of radish. *Res. J. Agric. Biol. Sci.* **2005**, *1*(3), 227–231.

Turk, M. A.; Tawaha, A. M.; Shatnawi, M. Lentil (*Lens culinaris* Medik) response to plant density, sowing date, phosporous fertilization and Ethephon application in the absence of moisture stress. *J. Agron. Crop Sci.* **2003a**, *189*(1), 1–6.

Turk, M. A.; Tawaha, A. M.; Samara, N. Effects of seedling rate and date, and phosphorous application on growth and yield of carbon vetch (*Vicia narbonensis*). *Agronomie* **2003b**, *23*, 1–4.

Turk, M. A.; Hmaeed, K. M.; Aqeel, A. M.; Tawaha, A. M. Nutritional status of durum wheat grown in soil supplemented with olive mill by-products. *Agrochimica* **2003c**, 47(5–6), 209–219.

Utkhede, R. Increased growth and yield of hydroponically grown greenhouse tomato plants inoculated with arbuscular mycorrhizal fungi and *Fusarium oxysporum* f. sp. *radicislycopersici*. *Biocontrol* **2006**, *51*, 393–400.

Varma, A. Function and applications of arbuscular mycorrhizal fungi in arid and semi-arid soils. In: *Mycorrhiza: Structure, function, molecular biology and biotechnology*; Varma, A., Hock, B., Eds.; Springer-Verlag: Germany, 1999; pp 521–556.

Vicente- Sánchez, J.; Nicolás, E.; Pedrero, F.; Alarcón, J. J.; Maestre-Valero, J. F.; Fernández, F. Arbuscular mycorrizal symbiosis alleviates detrimental effects of saline reclaimed water in lettuce plants. *Mycorrhiza* **2013**, *24*, 339–348.

Vigo, C.; Norman, J. R.; Hooker, J. E. Biocontrol of the pathogen *Phytophthora parasitica* by arbuscular mycorrhizal fungi is a consequence of effects on infection loci. *Plant Pathol.* **2000**, *49*, 509–514.

Xie, X.; Weng, B.; Cai, B.; Dong, Y.; Yan, C. Effects of arbuscular mycorrhizal inoculation and phosphorus supply on the growth and nutrient uptake of *Kandelia obovata* (Sheue, Liu & Yong) seedlings in autoclaved soil. *Appl. Soil. Ecol.* **2014**, *75*, 162–171.

Yaseen, T.; Burni, T.; Hussain, F. Effect of arbuscular mycorrhizal inoculation on nutrient uptake, growth and productivity of chickpea (*Cicer arietinum* L.) varieties. *Int. J. Agron. Plant Prod.* **2012**, *3*, 334–345.

Ye, L.; Zhao, X.; Bao, E.; Cao, K.; Zou, Z. Effects of arbuscular mycorrhizal fungi on watermelon growth, elemental uptake, antioxidant and photosynthesis II activities and stress-response gene expression under salinity-alkalinity stresses. *Front. Plant Sci.* **2019**, *10*, 1–12.

CHAPTER 8

ROLE OF MYCORRHIZAL FUNGI IN PLANT GROWTH: IMPLICATIONS IN ABIOTIC STRESS TOLERANCE

MINA BAYANATI[1], ABDEL RAHMAN MOHAMMAD AL-TAWAHA[2], JEYABALAN SANGEETHA[3], DEVARAJAN THANGADURAI[4], and POOJASHREE NAGAPPA KUMMUR[4]

[1]*Department of Horticultural Science, Ferdowsi University of Mashhad, Mashhad, Iran*

[2]*Department of Biological Sciences, Al Hussein Bin Talal University, Maan, Jordan*

[3]*Department of Environmental Science, Central University of Kerala, Kasaragod, Kerala, India*

[4]*Department of Botany, Karnatak University, Dharwad, Karnataka, India*

ABSTRACT

Sustainable agricultural systems emphasize on reducing the consumption of agricultural inputs, renewable resources, observe plant rotation, and use nitrogen-fixing plants to maintain soil fertility, and use biological fertilizers to improve soil structure. Microorganisms taking advantage of the interactions between factors affecting production basically reduced production costs and maintained sustainable cropping systems, one of the main microorganisms in the environment. Mycorrhiza fungus provide many benefits to their host by coexisting with crops, orchards, and ornamental plants. Mycorrhiza fungus

Mycorrhizal Technology: Managing Plant Stress and Mitigating Climate Change using Mycorrhizae.
Jeyabalan Sangeetha, Abdel Rahman Mohammad Al-Tawaha, & Devarajan Thangadurai (Eds.)

increase tolerance to lack of water in the host plants through increasing water and nutrient uptake. Increased resistance to pathogens and salinity, production of plant hormones, and improvement of soil structure by facilitating the formation of aggregates are some of the benefits that the host plant enjoys in this coexistence. Mycorrhiza fungus have frequented and long-standing symbiotic relationships with the plant lineage. Selective mechanisms of this fungus to absorb ions improve photosynthesis by improving nutrient uptake and improving water uptake. The plant response to mycorrhiza is equivalent to the efficiency of a fungus in improving the growth status and absorption of plant elements. Therefore, the main way to measure the efficiency of the fungus is to determine the host growth response, which is called the mycorrhizal growth response (MGR). Mycorrhiza via long hyphae, which also have high growth power, can enter places where the roots are not able to penetrate and use limited water and nutrients. Fungal fibers move and grow to soil phosphates and are able to absorb it. Mycorrhizal arbuscular fungi are the largest group of mycorrhizal fungi that penetrate root skin cells and form structures called vesicles and arbuscules, which increase the level of metabolic content.

8.1 INTRODUCTION

Although the use of biological fertilizers in agriculture has a long history, but the scientific exploitation of such resources has little history. Today, due to the problems caused by the excessive use of chemical fertilizers in farms and soils, the use of biological fertilizer resources in agriculture is more important than ever (Ehteshami et al., 2012). Fertilizer consumption management is also an influential factor in the success of plant cultivation and in the meantime, identification and use of biological fertilizers compatible with nature and suitable for plant growth and development can have favorable effects on quantitative and qualitative indicators of the crop (Astaraei and Kouchaki, 1996). Biological fertilizers do not refer exclusively to organic materials obtained from animal manure, cement plant, etc., but products derived from the activity of microorganisms associated with nitrogen fixation or the availability of phosphorus and other nutrients in the soil (Saleh Rastin, 2001). The increasing food needs of the people due to the rapid growth of the population require an increase in agricultural production. Organic farming has emerged as an important global priority in terms of increasing demand for healthy food and long-term sustainability of food supply and concerns about environmental pollution associated with the indiscriminate use of chemicals in agriculture. Although the use of chemical inputs in agriculture is inevitable to increase yield crops, the production of organic products can be an incentive to export crops. Biofertilizers are an essential component of organic agriculture, which contains organisms or efficient species of nitrogen-fixing and phosphate-solubilizes for seeds, soil, or compost-containing sites with the aim of increasing the number of these microorganisms and accelerating microbial processes that provide access to nutrients for proper assimilation in plants (Azarnia et al., 2015). Mycorrhiza is made up of two words, Myco (meaning fungus) and Rhiza (meaning root) (Peterson et al., 2004). It was in the early 21st Century that this coexistence was recognized and now it is well understood that this relationship is one of the most common coexistence systems on earth. This type of coexistence is one of the oldest coexistence in plants and is the most ecologically important coexistence between microorganisms and plants (Siddiqui and Pichtel, 2008).

Mycorrhizal fungi show very high efficiency and ability in terms of supplying the plant with the required phosphorus, which plays a role in coexistence with different types of plants, so that other important benefits of mycorrhizal coexistence, such as greater absorption of some high-consumption elements and low consumption (e.g., increased uptake of nitrogen, potash, calcium, magnesium, manganese, iron), better water uptake, producing of growth-promoting hormones, and finally the potential to fight root pathogens, are overshadowed by the issue of phosphorus uptake by them. These fungi penetrate into the cells of the cortex and then multiply rapidly using the root photosynthetic material and increase the absorption of nutrients, especially phosphorus, by spreading their roots into the soil.

Achieving sustainable agriculture along with increasing the yield of agricultural products and ensuring community health is one of the goals of researchers in agriculture. The overuse of chemical inputs in agricultural lands has caused many environmental problems, including of water, air, and soil resources pollution, fertility loss, and balance of chemical elements in the soil. A new method to prevent these problems is the use of biological and chemical modifiers in the soil. Researches show that the application of mycorrhiza alone and in combination with macro and micro elements improves plant growth and soil fertility. This chapter explains the concept of mycorrhizal fungi development that is involved in improving soil, growth, quality, crop yield, and protection of them.

8.2 BASED ON ANATOMICAL AND MORPHOLOGICAL FEATURES

Currently, mycorrhiza is classified based on the relationship between fungus and plant and how it relates fungal mycelium to root cells.

8.2.1 ENDOMYCORRHIZAE

This fungus belongs to the endogamous family and the order Glomus and belongs to the group of phycomycetes. Host plants are mostly pteridophytes, angiosperms, and all gymnosperms. However, Pinaceae, Betulaceae, and Orchidaceae families do not have a symbiotic relationship with this fungus. These fungi grow around the roots and penetrate into the root tissue, forming mainly branching organs, which are called arbuscules. Few genera of the endogamous family create vesicular-arbuscular mycorrhiza (VAM) such as *Glomus*, *Gigaspora*, *Acaulospora*, and *Scetelospora*.

8.2.2 ECTOMYCORRHIZAE

This fungus has a sheath, which is connected to the deadly fibers. The mycelium penetrates the intercellular space and forms a mycelial network called the harting. In this case, intracellular penetration is very little or not seen at all. In many fungal species, they form ectomycorrhizae. These fungi contain basidiomycota and ascomycota. Some ectomycetes are specific and coexist with a specific host and some have a wide host.

8.2.3 ECTENDOMYCORRHIZAE

The Harting network is usually expanded but mycelium penetrates into the host cell. In these fungi, VAM is more important in agriculture due to its symbiotic relationship with crops. Studies show that fungi can coexist with a large number of plants, that the amount and intensity is different from each other (Peterson et al., 2004; Smith and Read, 2010).

8.3 MYCORRHIZA AS A BIO-FERTILIZER IN SUSTAINABLE AGRICULTURE

Agriculture, as the most important way of providing human food, has increased the area under cultivation and changed land use. Human exploitation of natural resources is always unilateral and uncontrolled and exploitation of these resources is done without observing its protective aspects and only on the basis of providing short-term benefits. Rising population growth and food crisis have led to over-exploitation of natural resources, resulting in biological imbalances. Improper use of chemicals in agriculture has threatened other organisms and ecosystems. Sustainable agriculture refers to the proper management of agricultural resources that are used to meet the changing needs of human beings, while maintaining and even improving natural resources and the quality of the environment. Sustainable agriculture relies on the conservation of natural resources. This agricultural method is the most economical and the most beneficial way to use energy and convert it into agricultural products, without destroying soil fertility and environmental quality. The benefits of sustainable agriculture can be summarized as follows: increasing soil organic matter and reducing the use of chemical fertilizers, erosion reduction and saving in mechanical operations of soil protection, reducing production costs and increasing economic efficiency, increasing production efficiency, improving quantity and quality of water and preventing its loss, creating a healthy environment for raising livestock, poultry, aquatic animals, and maintaining environmental order (Bhardwai et al., 2014; Berruti et al., 2016; Igiehon and Babalola, 2017).

The aim of sustainable agriculture is the use of biofertilizers to eliminate or significantly reduce the use of chemical fertilizers that are a good solution to overcome these problems. Arbuscular mycorrhizal fungi (AMF) creates a key link with plants, and interest in using them as natural fertilizers is growing. These fungi, phosphate solubilizing bacteria, growth-promoting bacteria and other biological fertilizers play an important role in improving plant nutrition and growth. They have salty conditions, as some call them bioremediations, which are called saline soils (Singh et al., 1997). This fungus surrounds the root layer and carbohydrates and mineral nutrients are exchanged between them and the plant, such as arbuscules. In addition to improving nutrient supply, they increase salinity, drought, and disease resistance and reduce the toxicity of heavy metals (Ruiz-Lozano et al., 2012; Calvo-Polanco et al., 2014; Saia et al., 2014; Sánchez-Romera et al., 2015; Berruti et al., 2016). However, the mechanism of tolerance to abiotic stress is not fully understood. Several reports have suggested that mycorrhizal fungi can directly and indirectly affect ecosystems. The effects of these fungi on greenhouse gases have been investigated. They can help to reduce NO_2 by increasing N uptake in plants and reducing consumption in soil solution so they can reduce climate change (Bender and van der Heijden, 2015). The effects of mycorrhiza fungi have been investigated in the development of sustainable agriculture as given below (Table 8.1; Fig. 8.1).

TABLE 8.1 Contributions of Biofertilizers to Plant Productivity.

Mycorrhizal fungi	Contributions toward plant productivity	References
Glomus etunicatum	Enhanced vegetative growth, chlorophyll, and nutrient level in maize more than nonmycorrhizal maize plant	Sadhana (2014)
Alien arbuscular mycorrhizal fungi	Enhanced root and shoot biomass of *Juniperus oxycedrus* than nonmycorrhizal fungi even in the presence of environmental stress	Alguacil et al. (2006)
Vigna radiata	The fungus had positive impact on the nitrogen, potassium, phosphorus, and protein content of the green grain	Bhat et al. (2010)
Necrotrophic mycorrhizal species	Increase the nutrient status and secure plant against herbivore (such as aphids) attack	Gilbert and Johnson (2015)
AM fungi	Reduce the concentration of cadmium in plant grown under cadmium-contaminated site, thereby enhancing its productivity	Janoušková et al. (2006)
Glomus intraradices	Mycoremediate zinc from *Medicago truncatula* environment with the resultant effect of improved growth and development	Hildebrandt et al. (2007)
Glomus mosseae	Improved wheat yield even under more efficient in enhancing crop yield under stress	Daei et al. (2009)
Glomus spp. inoculated with *Mortierella* sp.	*G. aggregatum* or *G. mosseae* when inoculated with *Mortierella* sp. had significant impacts on the dry weights of shoot and root of *Kosteletzkya virginica* in the presence of salt stress	Zhang et al. (2011)
Acaulospora lacunosa and *Glomus constrictum*	Improved foliar nutrient status of onions (*Allium porrum*)	Hart and Forsythe (2012)
AMF	Enhanced nutrient uptake capability of soybean (*Glycine max*) and consequently increased yield	Liu et al. (2012); Tian et al. (2013)

Source: Reproduced with permission from Igiehon and Babalola (2017) with slight modification, Copyright © Springer-Verlag (2017).

8.4 THE EFFECT OF MYCORRHIZAL FUNGI ON PLANT WATER RELATIONS

Mycorrhiza fungi affect plant moisture relationships, so that these fungi affect the amount of water and leaf gas exchanges increase the amount of water in plant tissues and also reduce evapotranspiration. Mycorrhizal fungi are still able to absorb water from soil particles at very low soil moisture levels where the plant is unable to absorb water, causing drought resistance. Askari et al. (2019) expressed that mycorrhizal plants consume less water

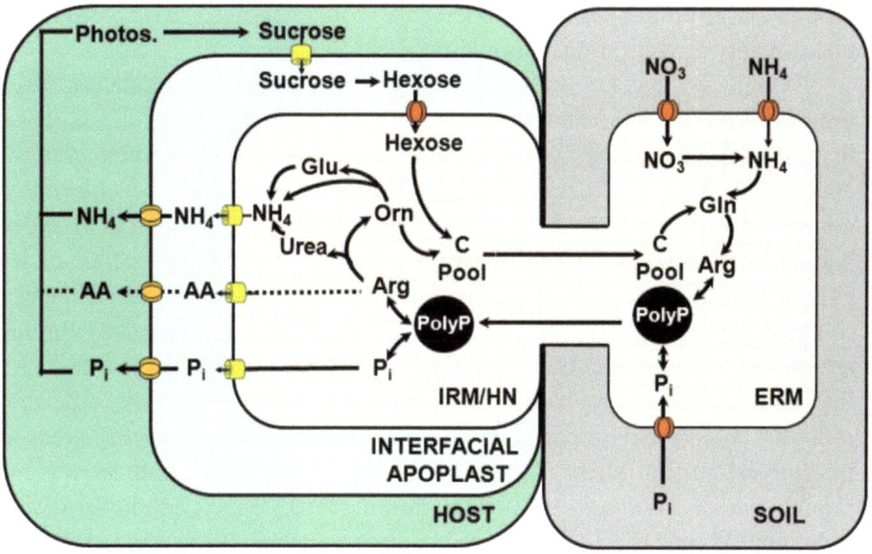

FIGURE 8.1 The process of uptake and transfer of nutrients between the fungal ERM and host plants by transporters, the anabolic arm of the urea cycle.

Source: Reprinted from Heike et al. (2011), https://creativecommons.org/licenses/by/3.0/

per unit of dry matter production. Abbaspour et al. (2012) studied the effect of mycorrhizal fungus on drought stress in pistachio seedlings. The results showed that mycorrhizal seedlings grew more than control and the amount of copper, zinc, potassium, and phosphorus in mycorrhizal shoots is higher than nonmycorrhizal in both optimal irrigation and water stress conditions also, nitrogen and calcium levels were also higher, while coexistence had no effect on magnesium levels. Mycorrhizal plants had higher levels of soluble sugars, proteins, flavonoids and proline, and inoculation with mycorrhiza also increased peroxidase activity. Therefore, mycorrhizae increased tolerance to drought in pistachio plant by raising the accumulation of osmotic and nutrient regulating compounds and antioxidant enzymatic activity.

8.4.1 MYCORRHIZAL FUNGI AND DROUGHT STRESS

Water stress is the lack of sufficient water and, also the presence of excess water around the plant. In the first case, which occurs as a result of drought or lack of water, it is known as water deficit stress, which is synonymous with drought stress. Rainfall that causes damage to the plant depends on the type of plant, water, soil, and atmospheric storage capacity, which affects the rate of evapotranspiration (Sarmadnia, 1993).

Drought stress is known as an environmental factor restricting photosynthesis (Kramer, 1986; Santaniello et al., 2017). Many researchers believe that moisture stress conditions are necessary for photosynthesis to continue and this trait is considered as an indicator of resistance of chlorophyll content of leaves of one of the plants to drought stress (Rangan et al., 2016; Niether et al., 2020). Durability of photosynthesis and maintenance of leaf chlorophyll under stress conditions are among the physiological indicators of stress resistance. Chloroplasts and their pigments are affected by drought stress. Drought stress causes hydrolysis of thylakoid proteins and decreases the amount of chlorophyll inhibiting plant photosynthesis, changes the chlorophyll content, and damages photosynthetic structures. One of the reasons that environmental stresses such as drought reduce the plant growth and photosynthetic ability by the accumulation of active oxygen molecules and induction of oxidative stress, and damage to species production (Hasanuzzaman et al., 2020). Mycorrhizal plants transfer about 20% of their stabilized pure carbon to their coexisting fungi, and the carbon stabilized by mycorrhizal coexistence is much more than the fungus needs. Mycorrhiza fungi increases the rate of photosynthesis in their symbiotic plant.

AMF has an effect on the activity of antioxidant enzymes, intrinsic water use efficiency (iWUE), and MDA content. Li et al. (2019) suggested that AMF has an effective role in determining the plant community, so that in his experiments, AMF increased the relative abundance of C3 (*Leymus chinensis*) to C4 (*Hemarthria altissima*) species in drought stress. C3 and C4 plants are different in two functional groups. C3 plants are suitable for growing in temperate conditions and C4 plants grow better in dry conditions than C3 plants. C4 plants are more resistant to drought esterification compared to C3 plants (Pearcy and Ehleringer, 1984; Nelson et al., 2004). Recent research shows that mycorrhizal colonization in C3 plants is higher and mycorrhizal dependence is stronger in C3 species (Lanfranco et al., 2018; Li et al., 2019). One of the functions of AMF against drought stress is to regulate biochemical mechanisms and reduce oxidative damage. In the first mechanism, water absorbs by hyphae and transfers to the host plant, and increased water in plant inhibits ROS, such as hydroxyl radicals (OH), hydrogen peroxide (H_2O_2), and superoxide anion radical (O). In the second mechanism, an increase in enzymatic and nonenzymatic production occurs. Enzymatic antioxidants include ascorbate peroxidases (APX), superoxide dismutase (SOD), catalase (CAT), glutathione reductase (GR), guaiacol peroxidase (G-POD), and glutathione peroxidase (GPX). The nonenzymatic antioxidants include glutathione (GSH), ascorbate (ASC), carotenoid, flavonoids, and tocopherol (Bahadur et al., 2019).

Osmotic regulation is the overall response of plants to moisture stresses, during which the solute content of cells increases. When plants are under stress, they keep their water potential at a lower level than the soil potential, usually by increasing the concentration of osmotic substances in the cell or by absorbing solutes from the soil solution or by increasing compatible solute synthesis (Ahmad and Sharma, 2008). The two most important osmolytes in plant are sucrose and proline in drought stress. AMF can regulate the osmotic activity of plants by regulating the activity of sucrose and proline metabolizing enzymes. The two most important plant osmolytes are sugar and proline in drought stress. Sugar can regulate the osmotic activity of plants by regulating the activity of sucrose and proline metabolizing enzymes. Sucrose converts to glucose and fructose by invertase enzymes, then glucose can be directly absorbed by mycorrhizae. AMF enhances invertase content in plant (Ravnskov et al., 2003; Wu et al., 2017). Proline is classified as an amino acid that has various roles in protecting plants against stress. This amino acid is vigorously associated with osmotic regulation, inhibition of free radicals and as a protective molecule to stabilize the structure of cellular proteins, thus protecting cells from the damaging effects of various environmental stresses (Ahmad and Sharma, 2008; De-Carvalho et al., 2013). Induction of increased proline was reported in *Medicago sativa* (Slama et al., 2011), *Triticum aestivum* (Jatav et al., 2012), *Oryza sativa* (Pandey et al., 2004), and *Arabidopsis thaliana* (Ju et al., 2013).

In general, mycorrhizal fungi can rise drought tolerance by improving the physical properties of the soil around the roots, such as aggregation and improving soil structure, increasing the level of root uptake, and thus increasing the water uptake efficiency, increased absorption of phosphorus and other nutrients, activate the host plant defense system, biochemical protection of plants from oxidative hazards produced by drought stress, stimulation of growth-related genes in the host plant (Song, 2005). However, the plant's ability to withstand stress in the presence of mycorrhiza fungi increases and the reaction of plants to fungi should be further investigated by RNA-seq (Fig. 8.2).

8.5 THE EFFECT OF MYCORRHIZA ON MORPHOLOGICAL AND PHYSIOLOGICAL CHARACTERISTICS OF PLANTS

Mycorrhizal fungi are able to coexist with many plants and in reduction absorption of heavy metals, the increase of plant resistance to root pathogens

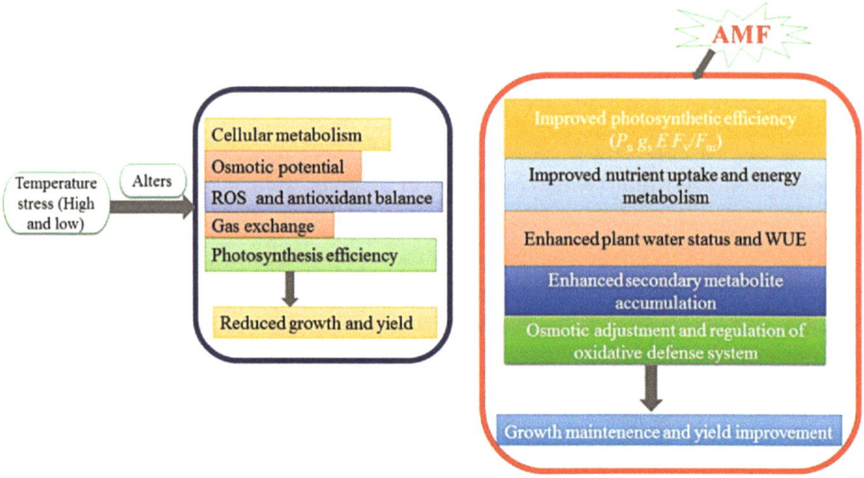

FIGURE 8.2 AMF reduces the oxidative effects of heat stress.

Source: Reprinted from Diagne et al. (2019), https://creativecommons.org/licenses/by/4.0/

and the increase of root growth capillaries play an effective role and increase the growth and development of the host plant by intensifying photosynthesis. Numerous studies have shown that mycorrhizal fungi grow on vegetative growth. Many plants have a symbiotic effect and increase growth. The results of a research show mycorrhiza has all the characteristics of vegetative growth of pistachio seedlings (stem height, stem diameter, the surface leaves increase the dry weight of leaves, stems, and roots compared to the control (without mycorrhiza) (James et al., 2008; Wu et al., 2017). One of the most important mechanisms related to the effect of mycorrhiza on the vegetative growth of plants is the effect of mycorrhiza on the uptake of nutrients such as macro elements in the soil. The presence of mycorrhizal biofertilizer improves soil properties such as organic matter content and increases access to nitrogen, potassium phosphorus, and micronutrients. Water and nutrients affect plant height by increasing the number of nodes and the length of internodes, normally (Eydizadeh et al., 2010). Mycorrhizal fungi increase absorption and transport of substances through the expansion of extra-root hyphae networks are added nutrients to the roots, thus improving plant growth and yield. Mycorrhizae increase height by the increasing leaf area, increasing photosynthesis, and the fabrication and transport of materials stated that mycorrhizal fungi produced. Mycorrhizal fungi with the production of growth hormones, especially gibberellin, caused a significant increase in leaf numbers, diameter, and the dry weight of organs (Khan et al., 2019).

Mycorrhizal fungi can increase the quality of products, so that in strawberries, by increasing secondary metabolites, it has improved its antioxidant properties (Castellanos-Morales et al., 2010). In addition, mycorrhizal fungi are shown beneficial effects on the quality of citrus fruit and tomato by increasing vitamin C, carotenoids, and flavonoids (Baslam et al., 2011; Zeng et al., 2014). Mycorrhizal fungi can also accelerate fruit ripening. The acceleration of the maturation process in fungus-treated plants attributes to increased ABA and ethylene biosynthesis (Palumbo et al., 2020) (Fig. 8.3).

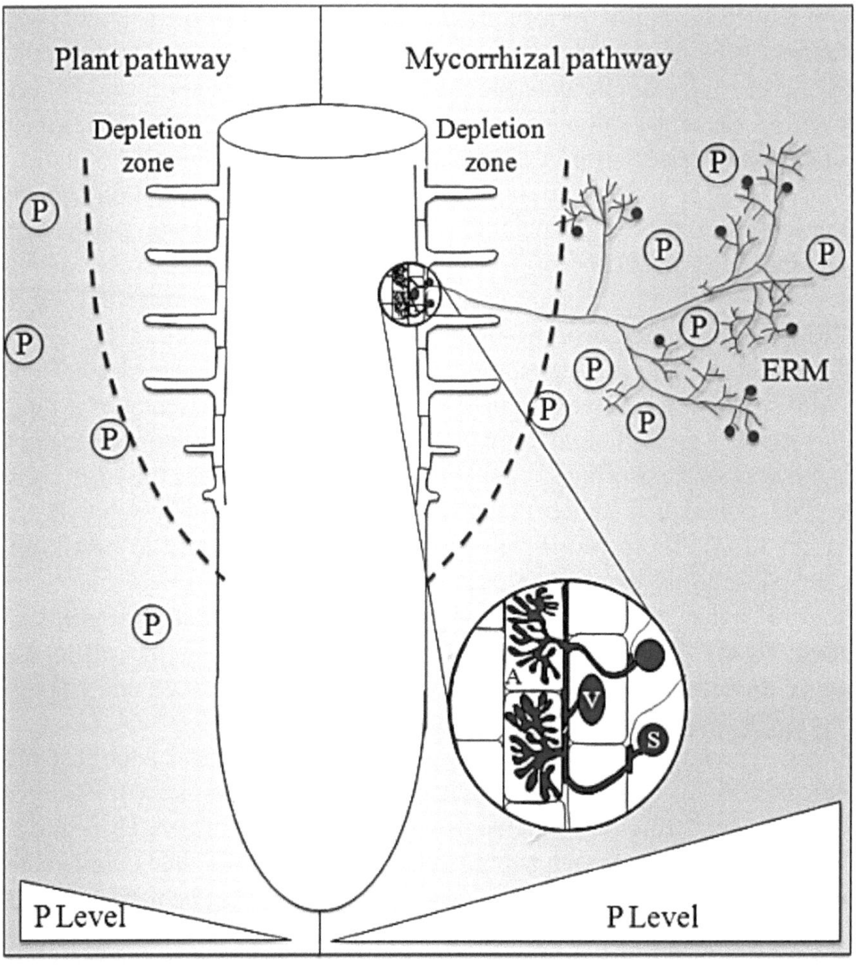

FIGURE 8.3 The process of a macro element such as phosphorus coexists through fungi. Abbreviations: Extraradical mycelium of the fungus (ERM), vesicles (V) and spores (S) of the arbuscular mycorrhizal fungus.

Source: Reprinted from Heike et al. (2011), https://creativecommons.org/licenses/by/3.0

8.6 MYCORRHIZAL FUNGI AND NUTRIENT UPTAKE

Food absorption can reduce input costs, nutrient loss rates, and increase crop yields. Genetics, morphology, physiology, and their interaction with external factors such as soil moisture, temperature and light, biological management methods of soil, fertilizers, and nutrients evaluate and improve plant nutrient efficiency. The production potential of many soils in the world depends little on the supply of nutrients due to the physical and chemical limitations of the soil. Changes in root morphology due to unfavorable physical properties of soil reduce the ability of root growth and movement in the soil and reduce nutrient penetration and water availability and lower yields. Organic matter is a large source of nutrients and intensifies biological activity and helps the cycle of elements, soil adsorption capacity, and acidity. These mechanisms of absorption and transport of nutrients in plants are all function of moisture in the roots and soil. In general, it is responsible for the absorption of mobile nutrients such as nitrate, calcium and magnesium, evaporation, transpiration, and mass movement.

The effect of different species of mycorrhiza is different on the growth, yield, and absorption of elements. The efficiency of Mycorrhiza fungus also changes, depending on the species fungi (Janos, 2007). The plant's response to mycorrhiza is equivalent to the efficiency of a fungus in recovery growth status and absorption of plant elements. Hence, the main way of measuring fungal efficiency is determining the host growth response, which is called mycorrhizal growth response (MGR) (Hetrick et al., 1993). The mycorrhizal response is directly related to host growth rate phosphorus demand (Kodie et al., 1996). Mycorrhizal plants help to ionic balance of the cytoplasm with the accumulation of potassium in their organs and excretion of sodium to the outside of the cell, or they place it inside the vacuole and the above retention of potassium to sodium during salinity (Giri et al., 2007), thus salinity tolerance increases in the plant. The importance of these physiological processes is such that some researchers have identified mechanisms that improve growth of mycorrhizae plants which may be sensitive to stress that affects absorption of elements such as nitrogen and phosphorus (Tian et al., 2004). Soil salinity reduces the uptake of minerals, especially phosphorus; because phosphate ions react rapidly with calcium ions and magnesium and zinc in saline soils it becomes immobile and out of reach for the plant (Munns, 1993). AMF protects plants by the immobilization of trace elements in its mycelium against the phytotoxicity (Christie et al., 2004; Whitfield et al., 2004).

AMF has biological effects in the soil by preventing reduced absorption of sodium and chlorine or their transfer to aerial parts. Trace elements such as Fe, Zn, Co, Cu, Mo, Mn, and Ni act more as an enzymatic cofactor in maintaining functional metabolism, and many studies have shown that mycorrhizal fungi can increase the absorption of these elements in the roots (Cabral et al., 2015).

Different genera of mycorrhiza fungi have increased the adsorption capacity in different wheat cultivars for N, Ca, Mn, Fe in saline water conditions. In another study, the three mentioned species of mycorrhiza increase the uptake capacity of wheat. It is also said that the fungus was necessary for growing wheat in saline conditions and increasing the yield of the ecosystem. Researchers stated that different wheat cultivars can coexist with the fungus, and lead to better root yield, high nutrient uptake, and increased yield. Mechanisms affecting the yield of different wheat cultivars to nutrient uptake are increasing of root nutrients and the rate of transmission by the wall of the uptake-resistant cell and the higher concentration of ions. Mardukhi et al. (2011) showed that *Glomus mosseae* and *G. intraradices* are the most efficient and *Glomus mosseae* absorbs high amounts of K, P, Zn and has a high root growth rate and yield. They stated that *G. mosseae* had a greater effect than *G. intraradices* in terms of increasing the intake of water and nutrients required by the plant. Also, in *Triticum* L., the effects of inoculation increased the root colony, grain yield, and water use efficiency. Al-Karaki (2006) reported that inoculation of tomato seeds with *Glomus mosseae* increased dry matter of plant as well as P, K, Fe, Zn, and Cu concentrations. Phosphorus uptake, node number, dry matter, and nitrogenase activity increased with chickpea inoculation. The root colonized with mycorrhizae is responsible for plant growth, nutrient uptake, and node support during the plant aging under stress. Mycorrhiza improves soil structure by coating a viscous glycoprotein called glomalin. The glomalin plays a key role in the formation of aggregates and the prevention of major hazards for better hyphae growth; this will improve the penetration of the climate and prevent soil erosion.

Hamzei and Sadeghi (2013) showed that phosphatase activity in wheat rhizosphere with higher *Glomus mosseae* and *Glomus geosporum* is higher than that of controlled plants. Studies showed that different strains had different efficiencies in colonization of alfalfa roots; so that the *musa* strain has the highest rate of colonization index, the highest root dry weight, the ability to absorb and transfer phosphorus, zinc, and potassium more than other strains. They also found that the use of inoculum separately and in combination with *Glomus mosseae* and *G. intraradices* increased potassium in leaves. Researches stated that only ectomycorrhizas (ectomycorrhizae) are able to increase nitrate uptake in their symbiotic plant. But, ammonium uptake is increased by ectomycorrhizae, ericoid mycorrhizae, and arbuscular vesicles. Transpiration and mass movement play an important role, and the larger this ratio, the more these elements are harvested, but in the uptake of immobile nutrients in soil such as phosphorus, zinc, and copper, plant root characteristics such as longitudinal velocity, root uptake rate, and total length will be effective. Researchers have found that mycorrhiza can improve host plant nutrition by slowly transporting it to the roots and by releasing nutrient ions (zinc and copper) to the roots, which account for 9% of Zn in shoots. AM can increase the concentration and delivery of zinc by 9–14% in roots, shoots, and flowers in low zinc soil compared to without mycorrhiza (Raju et al., 1990; Sabannavar and Lakshman, 2008; Watts-Williams et al., 2015). Mycorrhiza increases N, P, K, Ca, Mg, chlorophyll, carotene, and protein by improving stem nutrient uptake compared to without mycorrhiza. In mycorrhizal plants, the concentration of root metals is usually higher than in the stem of plants without mycorrhiza, and AMF increases chlorophyll formation because chlorophyll deficiency causes a lack of nutrients. AMF is beneficial for the development of sustainable agriculture in tropical soils and is an unknown source for sustainable soil management and conservation. The fungus helps to preserve zinc and cadmium in alfalfa and stabilizes the soil. Previously, studies have shown that the type of host plant, its genetic and morphological characteristics cause the uptake of any trace element in it to be different from that of other plants (Muleta et al., 2007; Chintala et al., 2015). The important mechanisms suggested by mycorrhizae include: (1) Mobilization and acquisition by the fungal mycelia, and (2) Crossing of absorbed nutrients from the fungal–root interface (Teotia et al., 2017). However, the effect of mycorrhiza on the nutrition of various elements in the plant is different and depends on the type and distribution of nutrients between the aerial and underground organs of the plant (Fig. 8.4).

8.7 MYCORRHIZAL FUNGI AND HEAVY METALS

The presence of various pollutants, including petroleum hydrocarbons and heavy metal in soil and water, is dangerous to life on Earth. Due to the

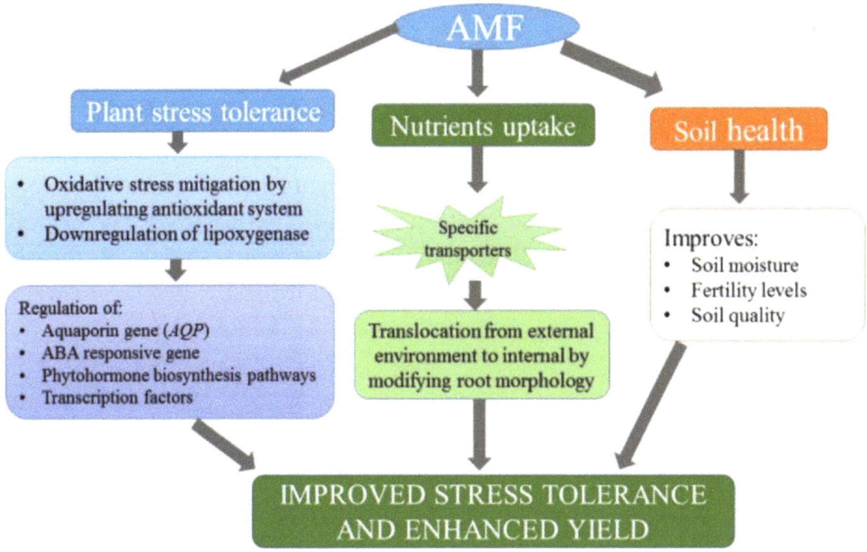

FIGURE 8.4 Schematic view of the effects of mycorrhiza on the induction of resistance to environmental stresses in plants.

Source: Reprinted from Diagne et al. (2019), https://creativecommons.org/licenses/by/4.0/

pollution of the environment by organic and inorganic pollutants during the last 85 years, many physical and chemical methods for ecosystem cleanup have been used, which are often costly. Hence today from the biological resources of the environment cleaning contaminated areas helps with a variety of contaminants. Recent research indicates the role of mycorrhiza fungi in regenerating contaminated areas by improving and accelerating plant growth in contaminated soils and increasing decomposition relative to contaminants. Inoculation of the plant with arbuscular mycorrhizal fungi is a suitable method for the regeneration and purification of contaminated soils. AMF establish a direct relationship between soil and roots; thus, the plant efficiency increases by affecting the availability of heavy metals and plant tolerance. Soil is a limited resource and its health is important for sustainable development; too much human activity causes soil loss, and in the case of agricultural soil it reduces exponential fertility (Alarcón et al., 2008). Soil quality depends not only on its physical and chemical properties, but also on the diversity and activity of living organisms or their biology. Beneficial interactions between plants and soil microorganisms in the rhizosphere play a decisive role in plant growth and health, and fertility (Jeffries et al., 2003). Mycorrhizal fungi are the most important microbial symbionts for most plants in all soils. Even algae create a symbiotic relationship (Bainard et al., 2011). This symbiosis occurs only in a limited number of plant families including Amaranthaceae, Pinaceae, Brassicaceae, Chenopodiaceae, Cyperaceae, Juncaceae, Proteaceae, and Plygonaceae (Smith and Read, 2010; Watts-Williams et al., 2012). AMF, as a microbial group in the soil, are effective in absorbing heavy metals by forming complexations of metals by glomalin, immobilizing heavy metals in roots and AMF structures, and promoting the adaptation of their host plants in stressful environments (Qiao et al., 2015; Zhang et al., 2019). Cadmium is an unnecessary element and has a high bioavailability and causes chlorosis, reduces root and stem growth. Garg and Bhandari (2014) and Zhang et al. (2019) reported that by binding cadmium metal (Cd) to fungal hyphae, its toxicity is reduced. Xianfeng et al. (2009) expressed those mycorrhizal fungi, increasing the rate of arsenic absorption in tobacco (*Nicotiana tabacum*) organs with increased phosphorus uptake. Among the three types of fungi, only *Glomus versiforme* was able to increase growth. Whitfield et al. (2004) observed that mycorrhiza fungus increases the absorption of zinc metal and improves plant growth (Thyme, *Thymus polytrichus*) under metal stress (Whitfield et al., 2004). In another study on *Cannabis sativa* it was stated that mycorrhiza in contaminated soil increases chromium (Cr), Nickel, and Cadmium in plant shoots, but mycorrhiza growth slightly decreased (Citterio et al., 2005).

However, the accumulation of mycorrhiza in contaminated soils pertains on the composition of heavy metal-fungi and plant and is subject to soil conditions. Many studies on mycorrhizal symbiosis with different plants have been performed to increase the absorption of nutrients as well as heavy metals. Each of these elements had different physicochemical properties and therefore their behavior was different in the soil environment (Table 8.2).

8.8 CONCLUSION

The importance of mycorrhiza fungi in sustainable agriculture shows the role of them as rhizosphere organisms between soil and plant. Using proper management and observing the principles of sustainable agriculture in the

TABLE 8.2 Synergistic Effect of AMF With Various Amendments on the Bioavailability of Trace Elements Under Different Heavy Metal Stresses.

Mycorrhizal species	Amendment	Host plant	Heavy metals	Mechanisms of HM alleviation	References
Funneliformis mosseae, *Rhizophagus aggregatus*, *Claroideoglomus etunicatum*, *Rhizophagus intraradices*	Lignin-derived biochar	*Hordeum vulgare* L.	Pb-acid battery effluents	AMF + lignin biochar (LBC) reduced bioavailable Pb, reduction of Pb concentration in barley grains	Khan et al. (2020)
Glomus aggregatum, *Glomus intraradices*, *Glomus elunicatum*, *Glomus versiforme*	Rice straw Biochar	*Medicago sativa* L.	Cd-contaminated soils	Increasing uptake of N and P in cadmium-contaminated soil plant, Cd content decreased	Zhang et al. (2019a)
Funneliformis mosseae (Fm), *Glomus versiforme* (Gv), *Rhizophagus intraradices* (Ri)	Steel slag	*Zea mays* L.	Cd, Pb	Single (SS) and AMF + SS reduced Cd and Pb contents, increased glomalin production	Hu et al. (2019)
Glomus intraradices	Wheat-straw biochar	*Zea mays* L.	Cd-contaminated soil	Synergistic effect of AM and biochar application in reducing cadmium and increasing corn growth	Liu et al. (2018c)
Rhizophagus clarus	Feedstock of common reed biochar	*Zea mays* L.	Cd-spiked soil	Biochar + MF1 treatment enhanced plant biomass, increased Cd uptake	Rafique et al. (2019)
Funneliformis mosseae (Fm), *Glomus versiforme* (Gv), *Rhizophagus intraradices* (Ri)	Tobacco stems biochar	*Zea mays* L.	Cd/Pb accumulation	Synergistic effect of AMF (Gv, Fm, Ri) and BC on growth and decreasing Cd/Pb accumulation	Zhuo et al. (2020)
Glomus intraradices, *Glomus etunicatum*	Rice straw biochar	*Trifolium repens*	Cd-contaminated soil	AMF, biochar, and P input increased plant growth	Xiao et al. (2020a)
Glomus aggregatum, *Glomus intraradices*, *Glomus elunicatum*, *Glomus versiforme*	Rice straw biochar	*Lolium multiflorum*	Acidic Cd contaminated soils	Effect of biochar on phosphorus and potassium uptake	Liu et al. (2019b)

TABLE 8.2 *(Continued)*

Mycorrhizal species	Amendment	Host plant	Heavy metals	Mechanisms of HM alleviation	References
Glomus aggregatum, *Glomus etunicatum*, *Glomus intraradices*, *Glomus tortuosum*, *Glomus versiforme*	Organic manure	*Trifolium repens*	Cd-contaminated sterilized soil	Synergistic effect AMF and organic manure on *Trifolium*	Xiao et al. (2020b)
Arbuscular mycorrhizal fungi	Cow dung, sawdust, and rice husk biochar (BC) and Selenium (Se)	*Vigna radiata* L.	As stress	Reducing oxidative stress, induce antioxidant resistance	Alam et al. (2019b)
Arbuscular mycorrhizal fungi	Rice husk and saw dust biochar	*Pisum sativum* L.	As stress	Reducing as content	Alam et al. (2019c)
Arbuscular mycorrhizae	Corn stalks biochar	*Zea mays* L.	Cd, Cr, Ni, Cu, Pb, Zn aged soil	Reducing heavy metals (Cd, Cr, Ni, Cu, Zn)	Qiao et al. (2015)

Source: Reproduced with permission from Muhammad et al. (2021), Copyright © Elsevier (2021).

use of mycorrhizal fungi leads to reducing the use of pesticides and chemical fertilizers and ultimately reducing production costs and, most importantly, preserving the environment. These fungi are one of the important biological factors in agricultural soils, which beneficial feature of coexistence with plants has increased the interest of researchers in the field of science and commercial use of these fertilizers. However, the type of plants, fungi, and soil are important in the impact of these biofertilizers.

KEYWORDS

- **glomalin**
- **heavy metals**
- **hyphae**
- **mineral uptake**
- **mycorrhizal fungi**
- **nitrogenase activity**
- **organic materials**
- **plant growth**
- **salt stress**

REFERENCES

Abbaspour, H.; Saeidi-Sar, S.; Afsharia, H.; Abdel-Wahhab, M. A. Tolerance of Mycorrhiza infected Pistachio (*Pistacia vera* L.) seedling to drought stress under glasshouse conditions. *J. Plant Physiol.* **2019**, *169*, 704–709.

Abdelhameed, R. E.; Rabab, A. M. Alleviation of cadmium stress by arbuscularmycorrhizal symbiosis. *Int. J. Phytoremed.* **2019**, *21*(7), 663–671.

Ahmad, P.; Sharma, S. Salt stress and phyto-biochemical responses of plants. *Plant Soil Environ.* **2008**, *54*(3), 89–99.

Alam, M. Z.; Hoque, M. A.; Ahammed, G. J.; Boggs, L. C. Arbuscular mycorrhizal fungi, selenium, sulfur, silica-gel and biochar reduce arsenic uptake in plant biomass and improve nutritional quality in *Pisum sativum*. *BioRxiv*. **2019**, *20*, 663120.

Alam, M. Z.; McGee, R.; Hoque, M. A.; Ahammed, G. J.; Carpenter-Boggs, L. Effect of arbuscular mycorrhizal fungi, Selenium and biochar on photosynthetic pigments and antioxidant enzyme activity under arsenic stress in Mung Bean (*Vigna radiata*). *Front. Physiol.* **2019**, *10*, 193.

Alarcón, A.; Davies Jr, F. T.; Autenrieth, R. L.; Zuberer, D. A. Arbuscular mycorrhiza and petroleum-degrading microorganisms enhance phytoremediation of petroleum-contaminated soil. *Int. J. Phytoremed.* **2008**, *10*, 251–263.

Alguacil, M.; Caravaca, F.; Díaz-Vivancos, P.; Hernández, J. A.; Roldán, A. Effect of arbuscular mycorrhizae and induced drought stress on antioxidant enzyme and nitrate reductase activities in *Juniperus oxycedrus* L. grown in a composted sewage sludge-amended semiarid soil. *Plant Soil* **2006**, *279*, 218–209.

Al-Karaki, G. N. Nursery inoculation of tomato with arbuscular mycorrhizal fungi and subsequent performance under irrigation with saline water. *Sci. Hortic.* **2006**, *109*(1), 1–7.

Amiri, R.; Nikbakht, A.; Etemadi, N. Alleviation of drought stress on rose geranium *Pelargonium graveolen* L. Herit. in terms of antioxidant activity and secondary metabolites by mycorrhizal inoculation. *Sci. Hort.* **2015**, *197*, 373–380.

Askari, A.; Ardakani, M.; Paknejad, F.; Hosseini, Y. Effects of mycorrhizal symbiosis and seed priming on yield and water use efficiency of sesame under drought stress condition. *Sci. Hort.* **2019**, *257*, 1–8.

Asrar, A. A.; Abdel-Fattah, G. M.; Elhindi, K. M. Improving growth, flower yield, and water relations of snapdragon *Antirhinum majus* L. plants grown under well-watered and water-stress conditions using arbuscular mycorrhizal fungi. *Photosynthetica* **2012**, *50*, 305–316.

Astaraei, A.; Koochaki, A. *Application of biologic fertilizers in sustainable agriculture*; Jahad Daneshgahi Publication, Mashhad (In Persian), 1996; pp 89–101.

Azarnia, M.; Safikhani, S.; Biabani, A. The effect of bio-fertilizer on crops yield, sustainable agriculture and organic farming. *J. Biosaf.* **2015**, *8*(2), 85–97.

Bahadur, A.; Batool, A.; Nasir, F.; Jiang, S.; Mingsen, Q.; Zhang, Q.; Pan, J.; Liu, Y.; Feng, H. Mechanistic insights into arbuscular mycorrhizal fungi-mediated drought stress tolerance in plants. *Int. J. Mol. Sci.* **2019**, *20* (17), 1–18. https://doi.org/10.3390/ijms20174199.

Bainard, L. D.; Klironomos, J. N.; Gordon, A. M. Arbuscular mycorrhizal fungi in tree-based intercropping systems: A review of their abundance and diversity. *Pedobiologia* **2011**, *54*(2), 57–61.

Baslam, M.; Garmendia, I.; Goicoechea, N. Arbuscular mycorrhizal fungi (AMF) improved growth and nutritional quality of greenhouse grown lettuce. *J. Agric. Food Chem.* **2011**, *59*, 5504–5515. https://doi.org/10.1021/jf200501c.

Bayani, R.; Saateyi, A.; Faghani, E. Influence of arbuscular mycorrhiza in phosphorus acquisition efficiency and drought-tolerance mechanisms in barley *Hordeum vulgare* L. *Int. J. Biosci.* **2015**, *7*, 86–94.

Bender, S. F.; van der Heijden, G. A. Soil biota enhance agricultural sustainability by improving crop yield, nutrient uptake and reducing nitrogen leaching losses. *J. Appl. Ecol.* **2015**, *52*, 228–239. https://doi.org/10.1111/1365–2664.12351.

Berruti, A.; Lumini, E.; Balestrini, R.; Bianciotto, V. Arbuscular mycorrhizal fungi as natural biofertilizers: let's benefit from past successes. *Front. Microbiol.* **2016**, *6* (1559), 1–13. https://doi.org/10.3389/fmicb.2015.01559.

Bhardwaj, D.; Ansari, M. W.; Sahoo, R. K.; Tuteja, N. Biofertilizers function as key player in sustainable agriculture by improving soil fertility, plant tolerance and crop productivity. *Microb. Cell. Fact.* **2014**, *13*(1), 1–10. https://doi.org/10.1186/1475–2859–13–66

Bhat, M.; Yadav, S.; Ali, T.; Bangroo, S. Combined effects of *Rhizobium* and vesicular arbuscular fungi on green gram (*Vigna radiata* L.) under temperate conditions. *Indian. J. Ecol.* **2010**, *37*, 157–161.

Boyer, L. R.; Brain, P.; Xu, X. M.; Jeffries, P. Inoculation of drought-stressed strawberry with a mixed inoculum of two arbuscular mycorrhizal fungi: Effects on population dynamics of fungal species in roots and consequential plant tolerance to water. *Mycorrhiza* **2014**, *25* (3), 215–227.

Cabral, C.; Sabine, R.; Ivanka, T.; Bernd, W. Arbuscular mycorrhizal fungi modify nutrient allocation and composition in wheat (*Triticum aestivum* L.) subjected to heat-stress. *Plant Soil* **2016**, *408*(1–2), 385–399.

Calvo-Polanco, M.; Sanchez-Romera, B.; Aroca, R. Mild salt stress conditions induce different responses in root hydraulic conductivity of *Phaseolus vulgaris* over-time. *PLoS One* **2014**, *9*(3), 1–13.

Calvo-Polanco, M.; Sanchez-Romera, B.; Aroca, R.; Asins, M. J.; Declerck, S.; Dodd, I. C. et al. Exploring the use of recombinant inbred lines in combination with beneficial microbial inoculants (AM fungus and PGPR) to improve drought stress tolerance in tomato. *Environ. Exp. Bot.* **2016**, *131*, 47–57.

Castellanos-Morales, V.; Villegas, J.; Wendelin, S.; Vierheiling, H.; Eder, R.; Cardenas-Navarro, R. Root colonization by the arbuscular mycorrhizal fungus *Glomus intraradices* alters the quality of strawberry fruit (*Fragaria ananassa* Duch.) at different nitrogen levels. *J. Sci. Food Agric.* **2010**, *90*, 1774–1782.

Chintala, R.; Gelderman, R.; Thomas, H.; Schumacher, E.; Douglas, D. M. Vegetative corn growth and nutrient uptake in Biochar amendends soil from an eroded landscape. *Proceedings of the Joint Annual Meeting of the Association for the Advancement of Industrial Crops and the USDA National Institute of Food and Agriculture* **2015**, 200–216.

Christie, P.; Li, X.; Chen, B. Arbuscular mycorrhiza can depress translocation of zinc to shoots of host plants in soils moderately polluted with zinc. *Plant Soil* **2004**, *261*, 209–217.

Citterio, S.; Prato, N.; Fumagalli, P.; Aina, R.; Massa, N.; Santagostino, A.; Sgorbati, S.; Berta, G. The arbuscular mycorrhizal fungus *Glomus mosseae* induces growth and metal accumulation changes in *Cannabis sativa* L. *J. Chemosphere* **2005**, *59*(1), 21–29.

Daei, G.; Ardekani, M.; Rejali, F.; Teimuri, S.; Miransari, M. Alleviation of salinity stress on wheat yield, yield components, and nutrient uptake using arbuscular mycorrhizal fungi under field conditions. *J. Plant. Physiol.* **2009**, *166*, 617–625.

De-Carvalho, K.; de Campos, M. K.; Domingues, D. S.; Pereira, L. F.; Vieira, L. G. The accumulation of endogenous proline induces changes in gene expression of several antioxidant enzymes in leaves of transgenic *Swingle citrumelo*. *Mol. Biol. Rep.* **2013**, *40*, 3269–3279.

Ehteshami, M. R.; Tehrani, A.; Samadi, B. Effect of planting date on yield physiology of different rapeseed (*Brassica napus* L.) cultivars in Varamin region, Iran. *J. Plant Process Funct.* **2012**, *1*(1), 71–87.

Eidi-Zadeh, K. H.; Mahdavi D. A.; Sabahi, H.; Soofizadeh, S. Effects of application of biological fertilizers in corporation of chemical fertilizers on growth of Shooshtar *Zea mays* cultivar. *J. Agro.* **2012**, 292–301.

Garg, N.; Bhandari, P. Cadmium toxicity in crop plants and its alleviation by arbuscular mycorrhizal (AM) fungi: An overview. *Plant Biosyst.* **2014**, *148*, 609–621.

Garg, N.; Singh, S. Arbuscular mycorrhiza *Rhizophagus irregularis*, and silicon modulate growth, proline biosynthesis and yield in *Cajanus cajan* L. Millsp. (Pigeon pea) genotypes under cadmium and zinc stress. *J. Plant Growth Regul.* **2017**, *37*, 1–18.

Gilbert, L.; Johnson, D. Plant-mediated 'apparent effects' between mycorrhiza and insect herbivores. *Curr. Opin. Plant. Biol.* **2015**, *26*, 100–105.

Giri, B.; Kapoor, R.; Mukerji, K. G. Improved tolerance of *Acacia nilotica*, to salt stress by arbuscular mycorrhiza, *Glomus fasciculatum*, may be partly related to elevated K/Na ratios in root and shoot tissues. *Microbiol. Ecol.* **2007**, *54*, 753–760.

Goicoechea, N.; Antol, M. C. Increased nutritional value in food crops. *Microbiol. Biotechnol.* **2017**, *10*, 1004–1007.

Grümberg, B. C.; María, U. C.; Shroeder, A.; Vargas-Gil, S.; Luna, C. M. The role of inoculum identity in drought stress mitigation by arbuscular mycorrhizal fungi in soybean. *Biol. Fert. Soils* **2015**, *51*, 1–10.

Hajiboland, R.; Aliasgharzadeh, N.; Laiegh, S. F.; Poschenrieder, C. Colonization with arbuscular mycorrhizal fungi improves salinity tolerance of tomato *Solanum lycopersicum* L. plants. *Plant Soil* **2010**, *331*, 313–327.

Hamzei, J.; Sadeghi, M. F. Study of root colonization percentage of grain sorghum cultivars by two species of mycorrhizal fungi and its effect on some morphological and agronomic traits. *J. Agron. Sci.* **2013**, *9*, 25–36.

Hart, M. M.; Forsythe, J. A. Using arbuscular mycorrhizal fungi to improve the nutrient quality of crops; nutritional benefits in addition to phosphorus. *Sci. Hort.* **2012**, *148*, 206–214.

Hasanuzzaman, M.; Bhuyan, M. B.; Zulfiqar, F.; Raza, A.; Mohsin, S. M.; Mahmud, J. A.; Fujita, M.; Fotopoulos, V. Reactive oxygen species and antioxidant defense in plants under abiotic stress: Revisiting the crucial role of a universal defense regulator. *Antioxidants* **2020**, *9* (8), 681. https://doi.org/10.3390/antiox9080681.

Hashem, A.; Alqarawi, A. A.; Radhakrishnan, R.; Al-Arjani, A. F.; Aldehaish, H. A.; Egamberdieva, D. et al. Arbuscular mycorrhizal fungi regulate the oxidative system, hormones and ionic equilibrium to trigger salt stress tolerance in *Cucumis sativus* L. *Saudi J. Biol. Sci.* **2018**, *25*(6), 1102–1114.

Heike, B.; Elliot, L.; Prashant, A. The role of the mycorrhizal symbiosis in nutrient uptake of plants and the regulatory mechanisms underlying these transport processes. In: *Plant Science*; Nabin, K. D., Sudam, C. S., Eds.; IntechOpen: London, 2011; pp 108–138.

Hetrick, B. A. D.; Wilson, G. W. T.; Cox, T. S. Mycorrhizal dependence of modern wheat cultivars and ancestors: A synthesis. *Can. J. Bot.* **1993**, *71*, 512–517.

Hildebrandt, U.; Regvar, M.; Bothe, H. Arbuscular mycorrhiza and heavy metal tolerance. *Phytochemistry* **2007**, *68*, 139–146.

Hu, Z. H.; Zhuo, F.; Jing, S. H.; Li, X.; Yan, T. X.; Lei, L. L. et al. Combined application of arbuscular mycorrhizal fungi and steel slag improves plant growth and reduces Cd, Pb accumulation in *Zea mays*. *Int. J. Phytoremed.* **2019**, *21*, 857–865.

Igiehon, N. O.; Babalola, O. O. Biofertilizers and sustainable agriculture: Exploring arbuscular mycorrhizal fungi. *Appl. Microbiol. Biotechnol.* **2017**, *101*(12), 4871–4881. https://doi.org/10.1007/s00253–017–8344-z.

Janos, D. P. Plant responsiveness to mycorrhizas differs from dependence up on mycorrhizas. *Mycorrhiza* **2007**, *17*, 75–91.

Janoušková, M.; Pavlíková, D.; Vosátka, M. Potential contribution of arbuscular mycorrhiza to cadmium immobilisation in soil. *Chemosphere* **2006**, *65*, 1959–1965.

Jatav, K. S.; Agarwal, R. M.; Singh, R. P.; Shrivastava, M. Growth and yield responses of wheat (*Triticum aestivum* L.) to suboptimal water supply and different potassium doses. *Environ. Exp. Bot.* **2012**, *2*(1), 39–51.

Jeffries, P.; Gianinazzi, S.; Perotto, S.; Turnau, K.; Barea, J. M. The contribution of arbuscular mycorrhizal fungi in sustainable maintenance of plant health and soil fertility. *Biol. Fert. Soils* **2003**, *37*, 1–16.

Jianfeng, H.; Xiangui, L.; Rui, Y.; Yufang, S. Effects of arbuscular mycorrhizal fungi inoculation on arsenic accumulation by tobacco (*Nicotiana tabacum* L.). *J. Environ. Sci.* **2009**, *21*, 1214–1220.

Jixiang, L.; Yingnan, W.; Shengnan, S.; Chunsheng, M.; Xiufeng, Y. Effects of arbuscular mycorrhizal fungi on the growth, photosynthesis and photosynthetic pigments of *Leymus chinensis* seedlings under salt-alkali stress and nitrogen deposition. *Sci. Total Environ.* **2017**, *576*, 234–241.

Ju, H. W.; Min, J. H.; Chung, M. S.; Kim, C. S. The atrzf1 mutation of the novel RING-type E3 ubiquitin ligase increases proline contents and enhances drought tolerance in *Arabidopsis*. *Plant Sci.* **2013**, *1*(7), 203–204.

Khalloufi, M.; Martínez-Andújar, C.; Lachaâl, M.; Karray-Bouraoui, N.; Pérez-Alfocea, F.; Albacete, A. The interaction between foliar GA3 application and arbuscular mycorrhizal fungi inoculation improves growth in salinized tomato *Solanum lycopersicum* L. plants by modifying the hormonal balance. *J. Plant Physiol.* **2017**, *214*, 134–144.

Khan, A.; Singh, J.; Upadhayay, V. K.; Singh, A.V.; Shah, S. Microbial biofortification: A green technology through plant growth promoting microorganisms. In: *Sustainable green technologies for environmental management*; Shah, S., Venkatramanan, V., Prasad, R., Eds.; Springer: Singapore, 2019; pp 255–269.

Khan, M. A.; Mahmood, R.; Ramzani, P. M. A.; Zubair, M.; Rasool, B.; Khan, M. K. et al. Associative effects of lignin derived biochar and arbuscular mycorrhizal fungi applied to soil polluted from Pb acid batteries effluents on barley grain safety. *Sci. Total Environ.* **2020,** *710*, 136294. https://doi. org/10.1016/j.scitotenv.2019.136294.

Koide, R. T. Nutrient supply, nutrient demand and plant response to mycorrhizal infection. *New Phytol.* **1991**, *117*, 365–386.

Kramer, P. J. *Water relations of plants*; Academic Press: Cambridge, 1983; pp 312–110.

Lanfranco, L.; Fiorilli, V.; Gutjahr, C. Partner communication and role of nutrients in the arbuscular mycorrhizal symbiosis. *New Phytol.* **2018**, *220*, 1031–1046.

Larry P. R.; Massicotte, B.; Melville, L. H. *Mycorrhiza: Anatomy and cell biology*; NRC Research Press: Canada, 2006, p 173.

Li, J.; Meng, B.; Chai, H.; Yang, X.; Song, W.; Li. S.; Lu, A.; Zhang, T.; Sun, W. Arbuscular mycorrhizal fungi alleviate drought stress in C3 (*Leymus chinensis*) and C4 (*Hemarthria altissima*) grasses via altering antioxidant enzyme activities and photosynthesis. *Front. Plant Sci.* **2019**, *10*(499), 1–12.

Lin, A. J.; Zhang, X. H.; Wong, M. H.; Ye, Z. H.; Lou, L. Q.; Wang, Y. S. Increase of multi-metal tolerance of three leguminous plants by arbuscular mycorrhizal fungi colonization. *Environ. Geochem. Health.* **2007**, *29*, 473–481.

Liu, L.; Li, J.; Yue, F.; Yan, X.; Wang, F.; Bloszies, S.; Wang, Y. Effects of arbuscular mycorrhizal inoculation and biochar amendment on maize growth, cadmium uptake and soil cadmium speciation in Cd-contaminated soil. *Chemosphere* **2018**, *194*, 495–503.

Liu, M.; Che, Y.; Wang, L.; Zhao, Z.; Zhang, Y.; Wei, L.; Xiao, Y. Rice straw biochar and phosphorus inputs have more positive effects on the yield and nutrient uptake of *Lolium multiflorum* than arbuscular mycorrhizal fungi in acidic Cd-contaminated soils. *Chemosphere*. **2019**, *235*, 32–39.

Liu, X.; Li, Y.; Han, B.; Zhang, Q.; Zhou, K.; Zhang, X.; Hashemi, M. Yield response of continuous soybean to one-season crop disturbance in a previous continuous soybean field in Northeast China. *Field. Crops. Res.* **2012**, *138*, 52–56.

Mardukhi, B.; Rejali, F.; Daei, G.; Ardakani, M. R.; Malakouti, M. J.; Miransari, M. Arbuscular mycorrhizas enhance nutrient uptake in different wheat genotypes at high salinity levels under field and greenhouse conditions. *Comptes Rendus Biol.* **2011**, *334*, 564–571.

Mathur, S.; Sharma, M. P.; Jajoo, A. Improved photosynthetic efficacy of maize *Zea mays* plants with arbuscular mycorrhizal fungi (AMF) under high temperature stress. *J. Photochem. Photobiol B.* **2016**, *180*, 149–154.

Mirshad, P. P.; Puthur, J. T. Arbuscular mycorrhizal association enhances drought tolerance potential of promising bioenergy grass *Saccharum arundinaceum* Retz. *Environ. Monit. Assess.* **2016**, *188*, 425.

Muleta, D.; Assefa, F.; Nemomissa, S.; Granhall, U. Composition of coffee shade tree species and density of indigenous arbuscular mycorrhizal fungi (AMF) spores in Bonga Natural Coffee, Forest Southwestern Ethiopia. *Forest. Ecol. Manag.* **2007**, *241*, 145–154.

Munns, R. Physiological responses limiting plant growth in saline soils: some dogmas and hypotheses. *Plant Cell Environ.* **1993**, *16*, 15–24.

Naheeda, B.; Cheng, Q.; Muhammad, A. A.; Sajjad, R.; Muhammad, K. Role of arbuscular mycorrhizal fungi in plant growth regulation: Implications in abiotic stress tolerance. *Front. Plant Sci.* **2019**, *10*, 1068. doi: 10.3389/fpls.2019.01068.

Nelson, D. M.; Sheng, H. F.; Jian, T.; Ivanka, S.; Brown, T. A. Response of C3 and C4 plants to middle-Holocene climatic variation near the prairie-forest ecotone of Minnesota. *Proc. Natl. Acad. Sci. USA* **2004**, *101*, 562–567.

Niether, W.; Glawe, A.; Pfohl, K.; Adamtey, N.; Schneider, M.; Karlovsky, P.; Parvelzik, E. The effect of short-term vs. long-term soil moisture stress on the physiological response of three cocoa (*Theobroma cacao* L.) cultivars. *Plant Growth Regul.* **2020**, *92*, 295–306.

Pal, A.; Pandey, S. Role of arbuscular mycorrhizal fungi on plant growth and reclamation of barren soil with wheat (*Triticum aestivum* L.) crop. *Int. J. Soil Sci.* **2016**, *12*, 25–31.

Palumbo, G.; Carfagna, S.; Stoleru, V.; Torino, V.; Romano, P. M.; Letizia, F.; Di Martino, C. Environmental sustainability fruit quality and production in mycorrhizal tomato plants without P fertilizing. *Agron. Res.* **2020**, *18*(4), 2535–2549.

Pandey, R.; Agarwal, R. M.; Jeevaratnam, K.; Sharma, G. L. Osmotic stress-induced alterations in rice (*Oryza sativa* L.) and recovery on stress release. *Plant Growth Regul.* **2004**, *42*, 79–87.

Pavithra, D.; Yapa, N. Arbuscular mycorrhizal fungi inoculation enhances drought stress tolerance of plants. *Ground Water Sust. Dev.* **2018**, *7*, 490–494.

Pearcy, R. W.; Ehleringer, J. Comparative ecophysiology of C3 and C4 plants. *Plant Cell Environ.* **1984**, *7*, 1–13.

Pedranzani, H.; Rodrãguez-Rivera, M.; Gutiarrez, M.; Porcel, R.; Hause, B.; Ruiz-Lozano, J. M. Arbuscular mycorrhizal symbiosis regulates physiology and performance of *Digitaria eriantha* plants subjected to abiotic stresses by modulating antioxidant and jasmonate levels. *Mycorrhiza* **2016**, *26*, 141–152.

Peterson, R. L.; Massicotte, H. B.; Melville, L. H. *Mycorrhizas: Anatomy and cell biology*; NRC Research Press: Canada, 2004; p 173.

Porcel, R.; Redondogómez, S.; Mateosnaranjo, E.; Aroca, R.; Garcia, R.; Ruizlozano, J. M. Arbuscular mycorrhizal symbiosis ameliorates the optimum quantum yield of photosystem II and reduces non-photochemical quenching in rice plants subjected to salt stress. *J. Plant Physiol.* **2015**, *185*, 75–83.

Qiao, Y. H.; Crowley, D.; Wang, K.; Zhang, H. Q.; Li, H. F. Effects of biochar and arbuscular mycorrhizae on bioavailability of potentially toxic elements in an aged contaminated soil. *Environ. Pollut.* **2015**, *206*, 636–643.

Rafique, M.; Ortas, I.; Rizwan, M.; Sultan, T.; Chaudhary, H. J.; Isik, M.; Aydin, O. Effects of *Rhizophagus clarus* and biochar on growth, photosynthesis, nutrients, and cadmium (Cd) concentration of maize (*Zea mays*) grown in Cd-spiked soil. *Environ. Sci. Pollut. Res.* **2019**, *26*, 20689–20700.

Raju, P. S.; Clark, R. B.; Ellis, J. R.; Maranville, J. W. Effects of species of VA mycorrhizal fungi on growth and mineral uptake of Sorghum at different temperatures. *Plant and Soil* **1990,** *121*, 165–170. https://doi.org/10.1007/BF00012308

Rangan, P.; Furtado, A.; Henry, R. J. Commentary: New evidence for grain specific C4 photosynthesis in wheat. *Front. Plant Sci.* **2016**, *7*(1537), 1–2.

Rani, B. Effect of arbuscular mycorrhiza fungi on biochemical parameters in wheat *Triticum aestivum* L. under drought conditions. *Doctoral dissertation*, CCSHAU, Hisar, 2016.

Ravnskov, S.; Wu, Y.; Graham, J. H. Arbuscular mycorrhizal fungi differentially affect expression of genes coding for sucrose synthases in maize roots. *New Phytol.* **2003**, *157*, 539–545.

Riaz, M.; Kamran, M.; Fang, Y.; Wang, Q.; Huayuan, C.; Huayuan, C. et al. Arbuscular mycorrhizal fungi-induced mitigation of heavy metal phytotoxicity in metal contaminated soils: A critical review. *J. Hazard. Mater.* **2021**, *15*(*402*), 123919.

Ruiz-Lozano, J. M.; Aroca, R.; Zamarreño, Á. M.; Molina, S.; Andreo-Jiménez, B.; Porcel, R. et al. Arbuscular mycorrhizal symbiosis induces *strigolactone* biosynthesis under drought and improves drought tolerance in lettuce and tomato. *Plant Cell Environ.* **2015**, *39*(2), 441–452.

Ruiz-Lozano, J.; Porcel, R.; Bárzana, G.; Azcón, R.; Aroca, R. Contribution of arbuscular mycorrhizal symbiosis to plant drought tolerance: State of the art. In: *Plant responses to drought stress*; Aroca, R., Ed.; Springer: Berlin, Heidelberg, 2012; pp 335–362. https://doi.org/10.1007/978–3-642–32653–0_13

Sabannavar, S. J.; Lakshman, H. C. Interactions between *Azotobacter*, *Pseudomonas* and arbuscular mycorrhizal fungi on two varieties of *Sesamum indicum* L. *J. Agron. Crop Sci.* **2008**, *194*(6) 470–478.

Sadhana, B. Arbuscular mycorrhizal fungi (AMF) as a biofertilizer. A review. *Int. J. Curr. Microbiol. Appl. Sci.* **2014**, *3*, 384–400.

Saia, S.; Amato, G.; Frenda, A. S.; Giambalvo, D.; Ruisi, P. Influence of arbuscular mycorrhizae on biomass production and nitrogen fixation of berseem clover plants subjected to water stress. *PLoS One* **2014**, *9*(3), 1–7.

Saleh Rastin, N. Biological fertilizers and their role in sustainable agriculture. *Collection Res. Necessity Ind. Product. Bio-fert.* **2001**, *34*, 1–54 (In Persian).

Sánchez-Romera, B.; Ruiz-Lozano, J. M.; Zamarreño, Á. M.; García-Mina, J. M.; Aroca, R. Arbuscular mycorrhizal symbiosis and methyl jasmonate avoid the inhibition of root hydraulic conductivity caused by drought. *Mycorrhiza* **2016**, *26*(2), 111–122. https://doi.org/10.1007/s00572–015–0650–7.

Santaniello, A.; Scartazza, A.; Gresta, F.; Loreti, E.; Biasone, A.; Di Tommaso, D.; Piaggesi, A.; Perata, P. *Ascophyllum nodosum* seaweed extract alleviates drought stress in *Arabidopsis* by affecting photosynthetic performance and related gene expression. *Front. Plant Sci.* **2017**, *8*(1362), 1–15. https://doi.org/10.3389/fpls.2017.01362.

Sara, O.; Ennajeh, M.; Zrig, A.; Gianinazzi, S.; Khemira, H. Estimating the contribution of arbuscular mycorrhizal fungi to drought tolerance of potted olive trees (*Olea europaea*). *Acta Physiol. Plant.* **2018**, *40*, 1–81.

Sarmadnia, G. H. The importance of environmental stresses in agriculture. In: *The first congress of agriculture and plant breeding of Iran*; Tehran University: Iran, 1993; pp 157–169.

Siddiqui, Z. A.; Pichtel, J. Mycorrhizae: An overview. In: *Mycorrhizae: Sustainable agriculture and forestry*; Siddiqui Z. A., Akhtar, M. S., Futai, K., Eds.; Springer: Dordrecht, 2008; pp 1–35.

Singh, R. P.; Choudhary, A.; Gulati, A.; Dahiya, H. C.; Jaiwal, P. K.; Sengar, R. S. Response of plants to salinity in interaction with other abiotic and factors. In: *Strategies for improving salt tolerance in higher plants*; Jaiwal, P. K., Singh, R. P., Gulati, A., Eds.; Science Publishers: Enfield, 1999; pp 239–241. https://doi.org/10.1017/S0014479799232103.

Slama, I.; Tayachi, S.; Jdey, A.; Rouached, A.; Abdelly, C. Differential response to water deficit stress in alfalfa (*Medicago sativa*) cultivars: Growth, water relations, osmolyte accumulation and lipid peroxidation. *Afr. J. Biotechnol.* **2011**, *10*(72), 16250–16259.

Smith, S. E.; Read, D. J. *Mycorrhizal symbiosis*; Elsevier Science, 2010; p 800.

Song, H. Effects of VAM on host plant in the condition of drought stress and its mechanisms. *Electron. J. Biol.* **2005**, *1*(3), 44–48.

Teotia, P.; Kumar, M.; Prasad, R.; Kumar, V.; Tuteja, N.; Varma, A. Mobilization of micronutrients by mycorrhizal fungi. In: *Mycorrhiza—Function, Diversity, State of the Art*; Varma, A.; Prasad, R.; Tuteja N.; Eds.; Springer, Cham, 2017; pp 121–129.

Tian, C. Y.; Feng, G.; Li, X. L. Different effects of arbuscular mycorrhizal fungal isolates from saline or non-saline soil on salinity tolerance of plants. *Applied Soil Ecology* **2004**, *26*, 143–148.

Tian, H.; Drijber, R.; Zhang, J.; Li, X. (2013) Impact of long-term nitrogen fertilization and rotation with soybean on the diversity and phosphorus metabolism of indigenous arbuscular mycorrhizal fungi within the roots of maize (*Zea mays* L.). *Agric. Ecosyst Environ.* **2013**, *164*, 53–61.

Tsoata, E.; Njock, S. R.; Youmbi, E.; Nwaga, D. Early effects of water stress on some biochemical and mineral parameters of mycorrhizal *Vigna subterranea* (L.) Verdc. (Fabaceae) cultivated in Cameroon. *Int. J. Agron. Agric. Res.* **2015**, *7*, 21–35.

Watts-Williams, S. J.; Cavagnaro, T. R. Arbuscular mycorrhizas modify tomato responses to soil zinc and phosphorus addition. *Biol. Fert. Soil.* **2012**, *48*, 285–294.

Watts-Williams, S. J.; Smith, F. A.; Mc Laughlin, M. J.; Patti, A. F.; Cavagnaro, T. R. How important is the mycorrhizal pathway for plant Zn uptake?. *Plant Soil* **2015**, *390*, 157–166. https://doi.org/10.1007/s11104–014–2374–4.

Whitfield, L.; Richards, A. J.; Rimmer, D. L. Relationships between soil heavy metal concentration and mycorrhizal colonisation in *Thymus polytrichus* in northern England. *Mycorrhiza J.* **2004**, *14*(1), 55–62. https://doi.org/10.1007/s00572–003–0268-z.

Wu, H.; Zou, Y.; Rahman, M.; Ni, Q. D.; Wu, Q. S. Mycorrhizas alter sucrose and proline metabolism in trifoliate orange exposed to drought stress. *Sci. Rep.* **2017**, *7*(1), 1–10. https://doi.org/10.1038/srep42389

Xiao, Y.; Liu, M.; Chen, L.; Ji, L.; Zhao, Z.; Wang, L.; Wei, L.; Zhang, Y. Growth and elemental uptake of *Trifolium repens* in response to biochar addition, arbuscular mycorrhizal fungi and phosphorus fertilizer applications in low-Cd-polluted soils. *Environ. Pollut.* **2020**, *260*, 113761.

Xiao, Y.; Zhao, Z.; Chen, L.; Li, Y. Arbuscular mycorrhizal fungi and organic manure have synergistic effects on *Trifolium repens* in Cd-contaminated sterilized soil but not in natural soil. *Appl. Soil Ecol.* **2020**, *149*, 103485.

Yang, Y.; Tang, M.; Sulpice, R.; Chen, H.; Tian, S.; Ban, Y. Arbuscular mycorrhizal fungi alter fractal dimension characteristics of *Robinia pseudoacacia*, L. seedlings through regulating plant growth, leaf water status, photosynthesis, and nutrient concentration under drought stress. *J. Plant Growth Regul.* **2014**, *33*, 612–625.

Yooyongwech, S.; Samphumphuang, T.; Tisarum, R.; Theerawitaya, C.; Chaum S. Arbuscular mycorrhizal fungi (AMF) improved water deficit tolerance in two different sweet potato genotypes involves osmotic adjustments *via* soluble sugar and free proline. *Sci. Hort.* **2016**, *198*, 107–117.

Zeng, L.; Li, J.; Liu, J.; Wang, M. Effects of arbuscular mycorrhizal (AM) fungi on citrus fruit quality under nature conditions. *Southwest China J. Agric. Sci.* **2014**, *27*(5), 2101–2105.

Zhang, F.; Jia-Dong, H. E.; Qiu-Dan, N. I.; Qiang-Sheng, W. U.; Zou, Y. N. Enhancement of drought tolerance in trifoliate orange by mycorrhiza: changes in root sucrose and proline metabolisms. *Not. Bot. Hort. Agrobot. Cluj-Napoca* **2018**, *46*, 270.

Zhang, F.; Liu, M.; Li, Y.; Che, Y.; Xiao, Y. Effects of arbuscular mycorrhizal fungi, biochar and cadmium on the yield and element uptake of *Medicago sativa*. *Sci. Total Environ.* **2019**, *655*, 1150–1158.

Zhang, H.; Wu, X.; Li, G.; Qin, P. Interactions between arbuscular mycorrhizal fungi and phosphate-solubilizing fungus (*Mortierella* sp.) and their effects on *Kostelelzkya virginica* growth and enzyme activities of rhizosphere and bulk soils at different salinities. *Biol. Fert. Soils.* **2011**, *47*, 543–554.

Zhao, R.; Guo, W.; Bi, N.; Guo, J.; Wang, L.; Zhao J. et al. Arbuscular mycorrhizal fungi affect the growth, nutrient uptake and water status of maize (*Zea mays*, L.) grown in two types of coal mine spoils under drought stress. *Appl. Soil Ecol.* **2015**, *88*, 41–49.

Zhuo, F.; Zhang, X. -F.; Lei, L. -L.; Yan, T. -X.; Lu, R. -R.; Hu, Z. -H.; Jing, Y. -X. The effect of arbuscular mycorrhizal fungi and biochar on the growth and Cd/Pb accumulation in *Zea mays*. *Int. J. Phytoremed.* **2020**, 1–10.

CHAPTER 9

ARBUSCULAR MYCORRHIZAL FUNGI IN SALINITY TOLERANCE AND GROWTH RESPONSE IN PLANTS UNDER SALT STRESS CONDITIONS

ROHAN SACHAN[1], ARUN KARNWAL[1], ABDEL RAHMAN MOHAMMAD AL-TAWAHA[2], DEVARAJAN THANGADURAI[3], and JEYABALAN SANGEETHA[4]

[1]*Department of Microbiology, School of Bioengineering and Biosciences, Lovely Professional University, Phagwara, Punjab, India*

[2]*Department of Biological Sciences, Al Hussein Bin Talal University, Maan, Jordan*

[3]*Department of Botany, Karnatak University, Dharwad, Karnataka, India*

[4]*Department of Environmental Science, Central University of Kerala, Kasaragod, Kerala, India*

ABSTRACT

Abiotic stresses have a detrimental effect on plant growth and productivity. The effects of abiotic stresses on crop productivity and on ecosystem deterioration have been exacerbated by climate change and agricultural malpractices such as the excessive use of fertilizers and pesticides. Salinity is one of the most severe environmental stresses because it reduces crop yield on more than 20% of irrigated land throughout the world, making it one of the most severe environmental stresses. To combat this, salt-tolerant

Mycorrhizal Technology: Managing Plant Stress and Mitigating Climate Change using Mycorrhizae.
Jeyabalan Sangeetha, Abdel Rahman Mohammad Al-Tawaha, & Devarajan Thangadurai (Eds.)

crops must be developed. Therefore, it is necessary to understand the mechanisms that allow plants to grow in salty environments. Salinization of plants involves activating molecular networks involved in stress sensing, signaling, and the production of stress-related genes and metabolites. On the other hand, arbuscular mycorrhizal fungi (AMF) are frequently referred to as bio-fertilizers. Additionally, it is widely believed that inoculating host plants with AMF confers tolerance on them against a variety of stressful conditions such as heat, salinity, drought, metals, and extreme temperatures. This chapter provides a comprehensive up-to-date knowledge on the role of AMF in salinity tolerance and growth response in plants under salt stress conditions.

9.1 INTRODUCTION

Many plants are affected by numerous environmental factors, both physiological and physical changes that affect the plant's growth (Hayyawi et al., 2018; Amanullah et al., 2019; Hani et al., 2019; Alu'datt et al., 2019; Turk et al., 2003a; Turk et al., 2003b; Turk et al., 2003c; Nikus et al., 2004a; Nikus et al., 2004b; Tawaha and Turk, 2004; Turk and Tawaha, 2004). Soil health and fertility are currently being addressed in order to make our agricultural system more sustainable (Al-Tawaha et al., 2005; Lee et al., 2005; Sulpanjan et al., 2005; Tawaha et al., 2005; Assaf et al., 2006; Turk et al., 2006; Al-Kiyyam et al., 2008; Al-Ajlouni et al., 2009; Al-Tawaha et al., 2010 Al-Tawaha et al., 2018a; Al-Tawaha et al., 2018b). On the other hand, appropriate plant nutrition is essential for the production of commercial crops (Hameed et al., 2008; Al-Tawaha et al., 2010; Ananthi et al., 2017; Al-Tawaha et al., 2017; Abu Obaid et al., 2018; Amanullah et al., 2019a, 2019b; Hani et al., 2019; Rajendran et al., 2019). On the other hand, sustainable production in an organic farming system is dependent on managing organic matter to improve soil health and supply nutrients to growing plants (Turk and Tawaha, 2001; Tawaha and Turk, 2002a; Tawaha and Turk, 2002b; Turk and Tawaha, 2002a; Turk and Tawaha, 2002b; Tawaha et al., 2003; Abebe et al., 2005; Abera et al., 2005; Al-Tawaha et al., 2005; Elser et al., 2007; Al-Kiyyam et al., 2008; Hameed et al., 2008; Al-Tawaha et al., 2010; Ananthi et al., 2017; Al-Tawaha et al., 2017; Abu Obaid et al., 2018). Soil-salinity in many areas of the earth is a serious concern in arid and semi-arid areas (Evelin et al., 2009; Simpson et al., 2011; Bezza et al., 2015; Akram et al., 2016). In fact, 7% of the earth's surface seems to have become saline susceptible. The main reasons for growing the salinization of farm fields are an improper use of chemical fertilizers, irrigation by a groundwater reservoir, flood irrigation, and no crop rotation (Sharma et al., 2009; Eddouaouda et al., 2012; Upadhyay et al., 2012; Zhang et al., 2016). Agricultural performance shall be influenced by the transformation of fertile soil into saline soil, with reduced soil fertility and decreasing soil efficiency to 20% only (Nadeem et al., 2016). Deeper salinization of farmland contributes to depletion of 30% of agriculture over the next 25 years as well as of more to 50% over the next 40 years. Elevated soil salt concentration reduces soil permeability, soil ventilation, and water conductivity that induce the plant to be water-deficient and physiological drought (Carvalho et al., 2004; Hammer et al., 2011; Chen et al., 2012). Salt exposures may reduce the photosynthetic intensity, minimize the function of antioxidant enzymes, reduce stomatal conductance, trigger ion losses, impact the membrane stabilization index, and restrict the residual amount of water (Miransari, 2010; Hammer et al., 2011).

Salt content impacts standard crop growth by damaging the leaf, nutritional deficiencies, reducing soil fertility, nitrate content, and reducing the activity of nitrate reductase (Ahmad et al., 2011; Bezza et al., 2015). Salinity stress destructs the PSII reaction center and prevents the transportation of electrons in photosynthetic structures that decrease the net photosynthetic efficiency (Zhang et al., 2016). Soil salinity is often damaged growing plants by adverse effects on Na and Cl ions that contribute to denaturing of enzymes, cell organelles damage, reduced photosynthesis, reduced respiration, confounding physiological drought imbalances, and plant nutrient imbalance (Eddouaouda et al., 2012; Egamberdieva et al., 2016). Many of these consequences jointly resulting from salt stress gradually result in reduced plant growth and lowered agricultural output. Plant cells take up many dissolved salts from soil, resulting in a plasmolytic-synthetic collapse of plant cells, which have impact on the anatomical variables of plants, such as the expansion of leaves and reducing fresh and dry matter and leaf and root tissue content (Ahmad et al., 2006; Paul and Nair, 2008; Kuźmiuk-Glembin et al., 2018). There are certain claims that a higher salinity amount doesn't really impact the productivity of legumes but greatly reduces the weight, fresh, and dried content of plants. Various studies on various plants have been performed to explain how osmotic stress caused by soil salinity is harmful (Heydarian et al., 2016). The degradation of the plant by producing reactive oxygen species (ROS) is also accountable for salt content. Plants generate various types of salt stress proteins and osmolytes, including prolines, glycine betaine, and malonaldehydes, as a reaction to salt stress to shield the plant from an osmotic attack (Chang et al., 2014; Heydarian et al., 2016). Such biochemical components build up after exposure to salt stress in higher concentrations.

9.2 EFFECT OF SOIL SALINITY ON GROWTH OF PLANTS

Agricultural plants display a variety of actions during salt distress. Salt content not only reduces agriculture crop yields but also impacts the physical and chemical characteristics of the soil and the areas of ecological balance (Sharma et al., 2016; Zhao et al., 2016). The salinity effects comprise low crop productivity, low economic profits, and desertification. The effects of salinity are the consequence of interrelationships between anatomical, physiological, and biochemical functions, such as sprouting, crop production, and water and nutrient consumption (Paul and Nair, 2008; Islam et al., 2016). Salinity influences almost all parts of plant production like germination, flowering, and reproduction (Berg et al., 2013; Shrivastava and Kumar, 2015; Sáenz-Mata et al., 2016). The salinity in soils causes ion resistance and osmotic tension on plants, and thus reduces water intake through soil, as well as nutrient deficit (N, Ca, K, P, Fe, and Zn) (Feng et al., 2002; Sheng et al., 2008; Porcel et al., 2012; Liu et al., 2014).

Soil salinity minimizes crop phosphorus (P) absorption substantially because Ca ions are precipitated by phosphate ions. Many chemicals, such as sodium, chlorine, and boron, are particularly toxic to plants (Kuźmiuk-Glembin et al., 2018). Excessive sodium deposition in plant cell walls will contribute quickly to osmotic stress and cell mortality. Plants susceptible to these elements can be damaged at a fairly lower sodium level if ample toxic components are found in soil (Ali et al., 2014; Panwar et al., 2016; Singh and Jha, 2016; Ungureanu et al., 2017). Since many of the chemicals are essential to plant nutrients, elevated amounts of salt in the soil can interrupt the plant's nutritional balance. Salinity often influences photosynthesis primarily through a decline of the leaf volume, chlorophyll and stomatal permeability, and a lower level of output of the photosystem II (Sheng et al., 2008). Salinity negatively impacts reproductive growth by infesting microsporogenesis and stamen tract elongation, ovule abortions, and fertilized embryo senescence (Ahmad et al., 2011; Maheshwari, 2012; Singh and Jha, 2016). The saline soil matrix has a variety of detrimental effects on the development of plants owing to its weak osmotic ability (osmotic stress), direct stress ion impact, nutrient imbalances, or a mixture of those causes. All these influences negatively affect the metabolic, biological, and genetic development of plants (Ahmad et al., 2011; Liu et al., 2014; Singh and Jha, 2016) (Fig. 9.1).

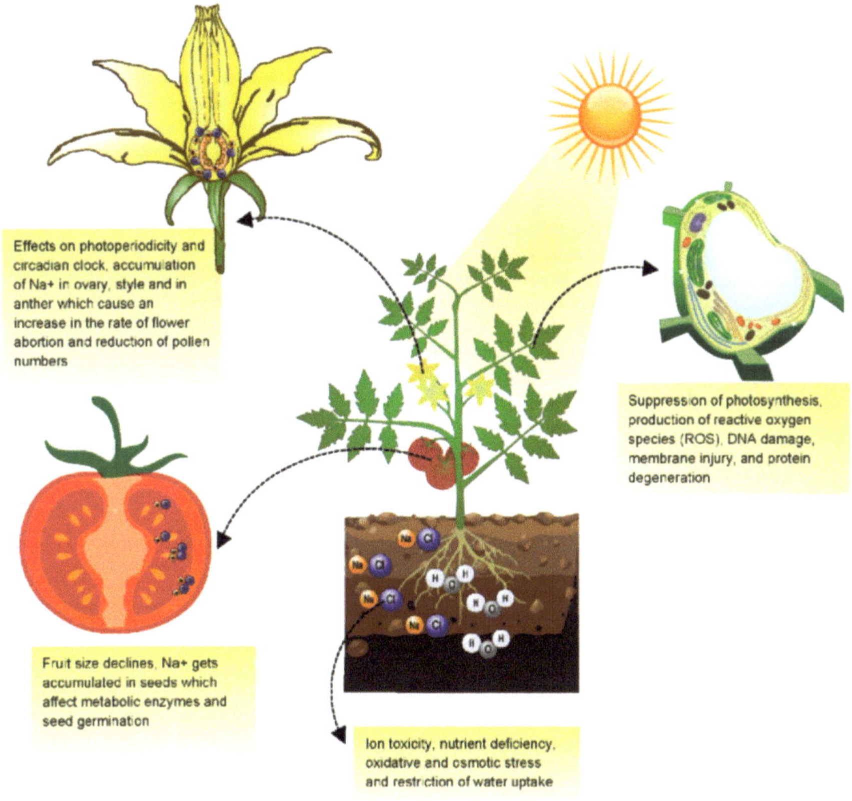

FIGURE 9.1 Effect of salinity stress on plant development.

Source: Reprinted from Egamberdieva et al. (2019), https://creativecommons.org/licenses/by/4.0/

The assessment of plants' tolerance to salinity stress, growth, or plant survival is managed by the integration of the forward or backward control of several physiological processes within the plant. With plants that grow in the saline environment, osmotic equilibrium is important. If this equilibrium is not achieved, it will lead to loss of turgidity, cell starvation, and eventually cell death (Ahmad et al., 2011). Nevertheless, the negative impact of salinity on the growth of the plant may also be induced by a decrease in the availability of photosynthetic assimilates or hormones to the developing tissue (Sheng et al., 2008; Liu et al., 2014). Ion toxicity is the outcome of biochemical processes replacing K^+ with Na^+ and Na^+ and Cl^- causing confirmative modifications in protein. K^+ serves as a co-factor for many enzymes and can't be replaced with Na^+ (Abouseoud et al., 2010; George et al., 2013; Marti et al., 2014). The attachment of tRNA to ribosomes and, therefore, the production of proteins often needs high K^+ concentration. The ion toxicity and osmotic tension triggers the metabolic imbalance resulting in oxidative stress in addition (De Oliveira et al., 2016). Throughout the developmental process, the deleterious consequences of salinity become greater on plant growth. The stresses on wheat plants in 100–175 mM NaCl show a significant decrease in spikelet's development, a delayed spike appearance, and lower fertility, leading to poor grain outputs (De Oliveira et al., 2016). Nevertheless, in the shooting peak of these wheat plants, the Na^+ and Cl^- concentrations of both were below 50 and 30 mM, which is too small to inhibit metabolism reactions. Consequently, the detrimental effects of salt content on the cell cycle and differentiation can be attributed to the salt stress effect (Sharma et al., 2009; De Oliveira et al., 2016).

Salinity briefly halts the cell cycle by raising the production and behavior of cyclines and cycline-dependent kinases, thereby reducing development (Almeida, 1997). In fact, post-translation suppression throughout salt stress reduces the function of cycline-dependent kinases (Pierce et al., 1998; Hussain et al., 2013). Previous studies (Bezza et al., 2015; Panwar et al., 2016; Vaishnav et al., 2016; Kapoor et al., 2017) have also shown that salinity has a negative impact on plant development, preventing germination, reducing growth of seeds, lower enzyme function, DNA, RNA, protein synthesis, and mitosis.

9.3 SALINITY OPTIMIZATION

Salinization can be reduced by leaching root salt, modified agricultural methods, and the usage of resistant salt plants. Improved irrigation strategies like adopting a partial drying method for root zone drying and dripping or micro-jet irrigation to optimize water use can sustain irrigated agriculture (Abouseoud et al., 2008; Gandhimathi et al., 2009). By limiting the volume of water that goes beyond the rooted areas, the expansion of dry land salt content can be managed. This can be achieved by planting deeply rooted perennial trees/plants that keep growing and access groundwater in periods that do not sustain annual plant crops (Tapia et al., 2008).

This can restore the balance of precipitation and water consumption, avoiding increasing levels of water and the flow of salt onto the land. Farming systems are subject to change to include perennials in annual crop rotation (phase cultivation), mixed crop plantation (alley cultivation, intercropping), or site-specific plants (precision agriculture) (Hettwer et al., 1995; Reddy et al., 2016; Vaishnav et al., 2016). While these environmental management strategies may minimize yields in salinity tension, the price and accessibility of decent water supply or water resources often hinder their application. It is a major challenge to create effective, inexpensive, and readily adaptable abiotic stress management methods (Berg et al., 2013; Vaishnav et al., 2016). Comprehensive research is underway worldwide to develop strategies to tackle abiotic stress, develop tolerant varieties of salt and droughts, shift crop calendars, management of resources, etc.

9.4 ARBUSCULAR MYCORRHIZAL FUNGI

Arbuscular mycorrhizal fungi (AMF) are soil microorganisms establishing a beneficial symbiotic relationship to most plants (Chandran and Das, 2011; Satyanarayana et al., 2012). Being obligatory biotrophs such fungi in the absence of the parent plant cannot achieve their life cycle. AMF facilitate the vigorous growth of host crops mostly under tense situations by facilitating a complex series incidents between the host plant and the fungus that result in higher photosynthetic frequency, gas exchange, and an enhanced absorption of water (Pfeffer et al., 1999; Quilambo, 2003; Mensah et al., 2015). Multiple sources mentioned increased abiotic and biotic stress tolerance, like drought, salt content, humidity, metals, and mycorrhizal fungi-related diseases. Almost 90% of plant species, bryophytes, and ferns, may establish interrelated links to AMF (Clark and Zeto, 2000; Govindarajulu et al., 2005; Armada et al., 2016).

Arbuscular mycorrhizal fungi form arbuscules, vesicles, and hyphae, in the plant rhizosphere. The development of the fungal hyphal web by the AMF with the plant roots substantially increases root access to a wider zone of the soil, thereby improving plant development (Cairney, 2000; Jansa et al., 2002). AMF improve plant health by improving accessibility and translocation of abundant nutrients. AMF increase soil quality through influencing soil structure, texture, and plant health. The degradation cycle in the soil organic matter can be intensified by fungal hyphae. Moreover, fungi can influence the atmospheric fixation of carbon dioxide by host crops, as the "sink effect" increases, and photo-assimilates migrate from air to the roots (Jung et al., 2012). Consistent with the importance of the AMFs, the current chapter focuses on the role of the AMF as bio-fertilizer and abiotic stress tolerance trait in regulating plant growth and nutrient-enhancing growth in stressful environments (Rillig and Mummey, 2006). This chapter highlights the importance of AMF in the field of agriculture applications.

Arbuscular mycorrhizal fungi are soil-borne microorganisms, which would enhance the absorption and resistance of a range of abiotic stress considerably. A significant proportion of AMF species are of the Mucoromycota phylum and Glomeromycotina sub-phylum (Spatafora et al., 2016). In that subphylum, four AMF orders have been identified, which also contain 25 genera, namely Glomerales, Archaeosporales, Paraglomerales, and Diversisporales (Spatafora et al., 2016).

These are obligatory biotrophic organisms that consume lipids and other plant photosynthesis nutrients to complete their lifespan. The plant development that AMF mediate is not only improved by the absorption of water and mineral nutrients from the neighboring soil but is also protect plants from fungal pathogens (Parniske, 2008; Miransari, 2012; Lugtenberg, 2014). Therefore, AMF are vital endosymbionts performing an important role in plant development, production, and the flourishing of the ecosystem. They are essential in order to boost crops sustainably.

9.5 DEVELOPMENT AND ORIGIN OF ARBUSCULAR MYCORRHIZAL FUNGI

Previous research suggests the emergence of earlier plants through non-photosynthetic eukaryotes by engulfing and domesticating the photo-cyanobacterium in the freshwater climate (Sengupta and Chaudhuri, 2002). Plants are, therefore, attributed to the development of an endosymbiosis, which allows them to sublimate *via* the majority of marine ecosystems. What developments permitted plants to overcome the issue of drylands? Many of the apparent modifications needed to colonize in the existing climate include elevated levels of radiation defense, a waterproof cuticle, and water-conductive vascular systems (Bianciotto et al., 2001). Moreover, in the lack of advanced absorptive structures like roots, which only developed later, an equally significant adaptation was needed to allow plants to absorb nutrients and water from the surfaces. Fungal symbiosis may probably have been useful for the colonization of land by aquatic algae representatives (Parniske, 2008).

Even though AMF interactions are not the only fungal symbiotic relationship carried out with ancient soil plants, recent evidence indicates that

Arbuscular mycorrhiza (AM) in early land plants might have been a major incident. Thus, AM appears as a breakthrough that has permitted the rapid colonization of AM with vascular plants (Parniske, 2008). It is, therefore, possible for the early root-less plants to be involved in different sorts of fungal interactions since they are still seen in the early diverging plantations and for which the roots in the vascular plants have co-evolved with AM. Arbuscular mycorrhizal fungal interactions have been so productive that somehow the majority of soil plants often involve in this symbiotic relationship in most habitats (except aquatic environments) (Fortin et al., 2002; Parniske, 2008).

9.6 ARBUSCULAR MYCORRHIZAL FUNGI INTERACTION CHARACTERISTICS

The symbiotic relationship with plants of the AMF was recorded 400 million years ago (Spinosa, 2008). These ways of connections are developed as a sequence of ecological operations that generate a spectrum of beneficial consequences in both natural and agricultural flora environments. AMF's symbiotic arrangement is a typical illustration of reciprocal relationships that can influence crop production and growth (Bianciotto et al., 2001). The mycelial fungal matrix spreads below plant roots and facilitates the absorption of nutrients that are not usually accessible for plants.

Although they adhere to several plant species, the fungal mycelium surrounds the roots of several land plants and results in a common mycorrhizal network (CMN) (Selosse et al., 2006; Igiehon and Babalola, 2017). This CMN has a noticeable impact in several plant community groups, especially invasive crops and fungal-mediated phosphorus (P) and nitrogen (N) transfer to the plants. This CMN is recognized as a critical element of the agricultural environment. Furthermore, population resources transfer from fungi to plants together with the additional consequences includes increasing plant resistance for biotic and abiotic factors due to AMF (Pieterse et al., 2014; Vimal et al., 2017). They can optimize soil conditions and, therefore, improve the performance of plants in both usual and adverse environments. Mycorrhizal colonization increases the resistance of the plants to adverse conditions by modifying certain morpho-physiological characteristics (Oehl et al., 2011). AMF are known to be the regulators of the natural growth of most land vegetation. AMF were employed and investigators promote its use in agricultural production as popular bio-fertilizers (Smith and Read, 2008).

In contrast, the AMF-inoculated soil seems to be more stable than that of the non-AMF-treated soil. The soil protein, Glomalin-related soil protein (GRSP) associated with glomalin is assumed to retain a water content in soil subjected to multiple abiotic stresses that later controls water levels between soil and plants, stimulating plant growth spontaneously (Rillig and Mummey, 2006; Miransari, 2012). Glomalin comprises 30–40% C and its metabolites that protect soil against desiccation by boosting soil water retaining efficiency. The impact of arbuscular mycorrhizal fungal inoculation is correlated with growth rate, for example, atomic conductance, leaf water capacity, RWC, PSII output, and the integration of the fungi (Rillig and Mummey, 2006; Treseder et al., 2004). AMF can also enhance water tension resistance by modifying tissue and organs physiologically.

In addition, inoculation of AMF increases dried content retention and boosts water humidity, which improves immunity for plants to stresses such as dehydration and salt content. The use of AMF in different ecological environments to grow plants may make a major contribution to organic crops for promoting growth and maximizing yields (Spinosa, 2008).

9.7 ROOT COLONIZATION OF ARBUSCULAR MYCORRHIZAL FUNGI

It was documented that AMF can boost the absorption of nutrients through the surface of the host root by 10 times. It was reported that ions like P, Zn, and Cu are not easily diffused *via* the soil. Subject to this slow diffusion, the roots directly around the root deplete these immobile soil nutrients from the field (Miransari, 2012; Averill et al., 2014). The improvement in plant productivity due to AM symbiosis typically involves a rise in the absorption of nutrients from the soil by the hyphae. A fungal hyphal web connected with the host plant's roots is commonly accepted for being allowed to invade roots of many other plants growing near the host plant (Averill et al., 2014). Fungal hyphae spread to the soil, penetrate the mineral deficiency area, and improve the viability of immovable resources up to 60-fold. In certain environments, however, Zn, Cu, and NH_4 are also essential in order to facilitate the absorption of phosphates by the plant (Averill et al., 2014). Indirect advantages might comprise an enhanced accumulation of soil nutrients or soil stabilization coupled with fungal hyphae in soil. It is reported that mycorrhizae enhance the root growth of affected plants, and it also raises or generates no modifications in xylem pressure (Jung et al., 2012) (Fig. 9.2).

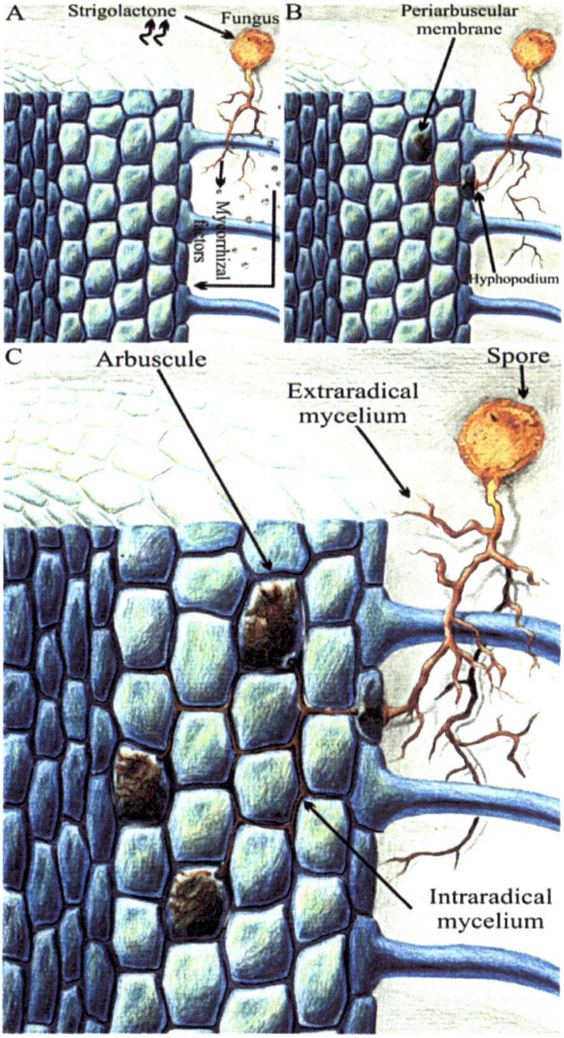

FIGURE 9.2 Hypothetical root colonized by an arbuscular mycorrhizal fungus (AMF). (A) The root of the host plant signals that it is apt to establish arbuscular mycorrhizal symbiosis, by releasing strigolactone, which promotes the germination of the fungal spores and stimulates the branching of the extraradical mycelium of the AMF, and the fungus responds with the production of the mycorrhizal factors (lipo-chitooligosaccharides and chitooligosaccharides) that are recognized by the host plant. (B and C) The physical contact between the root surface and the AMF allows of the formation of the hyphopodium, which leads to the intercellular proliferation of the intraradical mycelium in the cortex cells and to the intracellular development of the arbuscules surrounded by the periarbuscular membrane derived from the plasma membrane of the plant cell.

Source: Reprinted from Vergara et al. (2019), https://creativecommons.org/licenses/by/4.0/

AM fungal community may whether to improve crop development explicitly or implicitly by soil enhancement (Satyanarayana et al., 2012). AMF root colonization may have a protective effect against parasites and nematodes. Increased output, an abundance of nutrients, and reproductive performance can be the advantages of mycorrhizal interactions. The hyphae network, assisted by predominant plants, can help plants to grow or can lead to shaded plant growth (Kloppholz et al., 2011; Zuccaro et al., 2014). Mycorrhizas' key benefit was its improved exploration of the soils and the increased supply of nutrients to host roots, that is, N, P, K, Zn, Cu, S, Fe, Ca, Mg, and Mn. *Glomus etunicatum* stored higher phosphorous than non-mycorrhizal plants in sandy loam soils in water-stressed environments (Al-Karaki et al., 2004; Pawlowska and Taylor, 2004). The relevance of AMF to the stress resistance in *Acacia nilotica* was found to be limited. Fungal symbiotics have a fertilizing and seed defensin function, as colonized plants are best suited to obtain food and to withstand environmental stresses (Khan et al., 2012). In agriculture, it is appropriate to allow significant reducing fertilizers and pesticide inputs, thus achieving similar or even greater agricultural productivity through enhanced soil mineral intake by colonized plants. While maintaining mycorrhizae in farming, the consistency and fertility of the soil can also be protected, thus preserving the ecosystem as well as reducing production costs over the long run (Spinosa, 2008). The benefits and disadvantages of using mycorrhizal fungi in farming helps users to imagine the extent of this phenomenon at the crop level and the effect on the quality nutritional improvements from its long-term acceptance, water tension response, low-temperature resistance, root morphology transformation, the variety of microbes in the field of soil resistance growth (Ahmad et al., 2011).

Mycorrhizal fungi make a significant contribution to soil carbon accumulation by modifying soil organic matter content and abundance. AMF change the mechanical attributes of the root and enhance its nutrients absorption potential. It is, therefore, evident that the nutrient processing and sustainability of crops play a vital part in mycorrhizal fungi (Cairney, 2000; Miransari, 2012). AMF colonize a variety of tropical plants like vegetables. The AM organization encourages the production of different crops, including vegetable crops such as carrots or tomatoes (Hildebrandt et al., 2002; Engelmoer and Kiers, 2015) (Fig. 9.3).

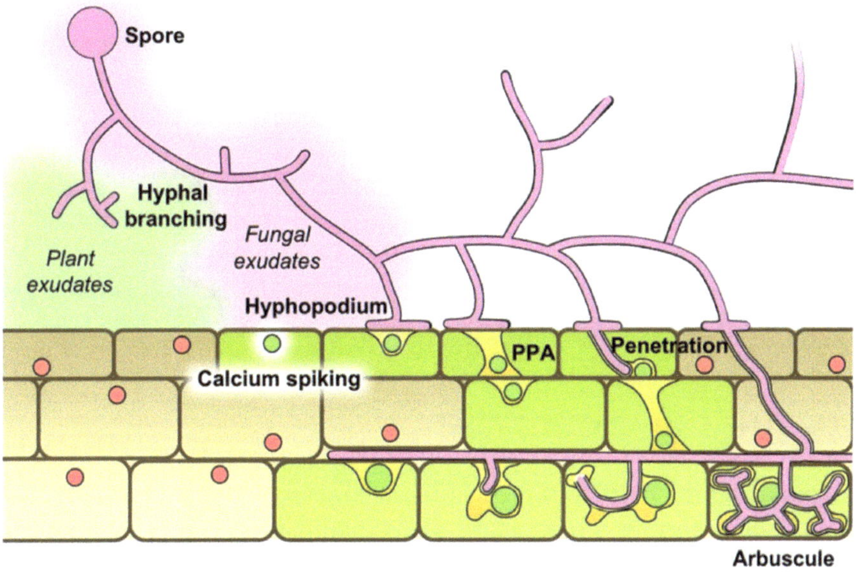

FIGURE 9.3 Schematic summary of the root colonization process by AM fungi. The germination of a resting spore is followed by the production of a short explorative mycelium. The perception of plant exudates, released by the host root, induces recursive hyphal branching, increasing the probability of a direct contact between the symbionts. In the meantime, fungal exudates are perceived by the root, where they trigger calcium spiking through the activation of the common SYM pathway (Box 1). Signal transduction leads to the activation of cellular and transcriptional responses (green cells and nuclei). The contact between the plant and fungus is followed by the adhesion of a hyphopodium to the root surface. This triggers the assembly of a broad aggregation of cytoplasm (yellow), named the prepenetration apparatus (PPA) in the contacted epidermal cell and underlying outer cortical cell. Subsequent intracellular fungal colonization strictly follows the route of PPAs from the epidermis to the inner cortex. Here, intercellular hyphae can develop along the root axis. The PPA mechanism is then replicated in the contacted inner cortical cells, both before fungal entry and—on a smaller scale—branching. Eventually, a highly branched arbuscule occupies most of the cell volume, forming an extensive surface for nutrient exchange.

Source: Reproduced with permission from Bonfante and Genre (2010), Copyright © Springer Nature (2010).

9.8 ARBUSCULAR MYCORRHIZAL FUNGI AS BIO-FERTILIZER

Bio-fertilizers are a combination of natural compounds with beneficial microbes to boost the quality of the soil (Islam et al., 2013). Such fertilizers are highly beneficial for crop growth and development, along with soil quality. Various studies performed over the past two decades in AMF have demonstrated its abundant nutrition and crop yield benefits (Srivastava et al., 2007). AMF may also probably be considered a substitute for synthetic chemical fertilizers in the future, since any usage of mycorrhizal fungi will significantly minimize the use of the input of chemicals in agriculture, particularly phosphorus, in measurable applications. Due to adverse effects on the foodstuffs, soil, air, and water supplies, persistent use of inorganic synthetic chemical fertilizers, herbicides, and fungicides have created extensive soil, plant, and human health issues (Hussain et al., 2018). AMF are thought to minimize the use of chemicals in agriculture by up to 50%. This number relies on both the plant species and the predominant stressful regime (Siddiqui et al., 2008) (Table 9.1).

9.9 AMF AND ABIOTIC STRESS

Maintaining the fertility of the soil for maximum productivity is very much important in today's world, where agricultural products have become basic necessities for one's life. Various chemical fertilizers, pesticides have led the soil quality to compromise such chemicals either affect crop quality or soil quality. There are certain abiotic and biotic factors that affect the soil quality and hence plant growth and crop productivity. These factors are regarded as abiotic stress and biotic stress. Salinity, drought, temperature, and pollutants are some of the abiotic stress encountered by various plants and crops and hence affect plant growth and crop productivity.

Among the listed abiotic stresses, salinity is the severe environmental stress. The increased salt content in the soil affects the uptake of water by plants, and hence osmotic balance is disturbed, and the condition of osmotic and ionic stress is encountered (Yu et al., 2016; Elhindi et al., 2017). Such effects on plants affect the biochemical processes, physiological processes, and photosynthetic rate as salinity in the soil creates dehydration conditions (Ebrahim and Saleem, 2017). Although plants have various means to tackle such stress by various signal transduction pathways like Salt Overly Sensitive (SOS), Mitogen-Activated Protein Kinase (MAPK) cascades, Phosphatidylinositol (PI), Abscisic Acid (ABA) mediated signal transduction (Fernando and Schroeder, 2016). Plants also have molecular signals to adapt such stress through activating transcription factors by molecular signaling. These signals amplify and regulate genes and also re-establish homeostasis of cellular ions, osmotic balance, and reaction oxygen species (ROS) (Yang and Guo, 2018). AMF are the endomycorrhizal type that protects plant species

TABLE 9.1 Manufacturers of Mycorrhizal Product Across the Globe.

Country	Location	Company	Product
USA	Colorado	AgBio, Inc., Westminster	MycoApply
	Ohio	Accelerator Horticulture Products	MycorT AM
	California	Bio-Organic Supply, Camarillo	Mycorrhizal Landscape Inoculant (LA), Mycorrhizal Root Dip Inoculant (RD)
	Iowa	Becker Underwood, Ames	Myke®
	Arizona	BioScientific, Inc., Avondale	Ascend ST
	California	EcoLife Corporation, Moorpark	Bio Terra Plus
	MN	Sustane Corporate Headquarters	Sustane®
	Westminster, Colombia	AgBio, Inc.	AgBio-Endos™, AgBio-Ectos™
	Carpinteria, CA	Albright Seed Co./S & S Seeds	TurboStart™
	Ames, Iowa	Becker Underwood, Biologicals	Rhizanova™
	Warrenton, Oregon	Bio-Oregon	BioVita™
	La Pine, Oregon	Bio-Organics	Bio-Organics™ Endomycorrhizal Inoculant (BEI), Bio-Organics™ Mycorrhizal Landscape Inoculant (LA), Bio-Organics™ Mycorrhizal Root Dip Inoculant (RD)
	Triadelphia, WV	First Fruits, LLC	EaethRoots™
	Olympia, WA	Fungi Perfecti	MycoGrow™, Planr Success™
	Chino, CA	Gro-Power	GroLife™
	Parkland, FL	Hoodridge International	Mycoroot™
	Sarasota, FL	Horticultural Alliance, Inc.	DIEHARD™
	Independence, MO	ROOTS, Inc.	MycorrhizaROOTS™ endoROOTS™, M-ROOTS™

TABLE 9.1 *(Continued)*

Country	Location	Company	Product
	Ontario, Canada	Mikko-Tek Labs	
	Pittsburg, PA	Plant Health Care	Mycor™, Mycor™, Plant Sever™, Mycor™, Tree Sever™, Mycor™, Flower Sever, MycorTree™, Root Dip, MycorTree™, Ecto Spore Spray, PHC™, Colonize™, VAM Stimulant
	Red Hill, PA	Premier Enterprises Ltd.	MYKE™, MYKE™ Pro, MycorRise
	Lakeland, FL	Poulenger USA, Inc.	RUTOPIA+M™
	Salinas, CA	Reforestration Technologies International	Silva Dip™, AM120™, MycoPaks™
	Knoxville TN	The Tree Doctor	DieHard™ Injectable, DieHard™ Root Reviver
	West Lafayette, IN	Tree Pro	MycorTree™
Canada	Ontario	Mikro-Tek, Inc., Timmins	MIKRO-CONE®, MIKRO-VAM®
	Quebec	Premier Tech	Myke®
Chile	Sodegaura	Idemitsu Kosan Co.	R-10
France	Dijon	Biorize	Endorize
UK	Kent	Biological Crop Protection Ltd.	Mycorrhizal inoculants
	Royston, Herts	MicroBio Ltd.	Vaminoc™
	Sittingbourne	PlantWorks Ltd.	Root Grow
	Lincoln	Crop Intellect Ltd.	Formulations of arbuscular mycorrhiza
	Mells	Zander Corporation	Arbuscular mycorrhiza
Spain	Zaragoza	Arvensis Agro	Arbuscular mycorrhiza
Belgium	Watou	Clemens Consult	Arbuscular mycorrhiza
Germany	Bitterfeld	Triton Umweltschutz GmbH	Myke Pro Endo AM

TABLE 9.1 *(Continued)*

Country	Location	Company	Product
Poland	Koeskowola	Mykoflor Wáodzimierz SzaáaĚski	Mykoflor
Spain	Pedreguer	Odd Distributions	Root WebStar
Netherlands	Lelystad	Global Horticare	BioRoot
Czech Republic	Sázava, Lanškroun	Symbiom	Rhodovit, Symbivit®
Italy	Larino	Sacom	Seme
	Lansing, MI	VAMTech, Inc.	Mycoform® and Myconate®
India	Chennai, TN	Biotrack Technology Pvt. Ltd.	RHIZAgold
	Andhra Pradesh	KCP Sugar and Industries Corporation Ltd.	Mycorrhiza VAM
	Ahmedabad	Candila Pharmaceuticals, Ltd.	Josh
	Goa	Cosme Biotech	Shubhodaya
	Chennai, Tamil Nadu	ManiDharma Biotech Pvt. Ltd.	ManiDharma VAM
	New Delhi	TERI	Ecorhiza-VAM/Nurserrhiza-VAM
	Bhopal, MP	Ambika Biotech	Root Care
	TN	Dr. Rajan Laboratories	Mycorrhizas
	Thanjavur, TN	TARI Biotech	Vesicular Arbascular Mycorrhiza (VAM)
	Surat, Gujarat	Sundaram Overseas Operation	Vesicular Arbascular Mycorrhiza (VAM)
	Coimbatore, Tamil Nadu	T. Stanes and Company Limited	Solid formulations of arbuscular mycorrhiza
	Coimbatore, Tamil Nadu	Neesa Agritech Private Limited	Arbuscular mycorrhiza (powder and tablets)
	Ahmedabad, Gujarat	Neesa Agritech Private Limited	

TABLE 9.1 *(Continued)*

Country	Location	Company	Product
	Una, HP	Majestic Agronomics Pvt. Ltd.	Arbuscular mycorrhiza
	Dwarka, Delhi	Krishidhan Seeds Pvt. Ltd.	Arbuscular mycorrhiza (powder)
	Coimbatore, TN	GreenMax AgroTech	Gmax VAM
Japan	Tokyo	Central Glass Co., Chemicals Section	Cerakinkong
	Tokyo	Idemitsu Kosan Co. Ltd.	R-10
Malaysia	Bangi, Selangor	Agri Hi-Tech Sdn	MycoGold, Myco-V
	Klang, Selangor	N-Viron Sdn Bhd	Mycorrhiza VAM
South Africa	Cape Town	Biocult	Biocult™

Source: Reproduced with permission from Pal et al. (2016), Copyright © Springer Nature (2016).

from the deleterious effects of stresses. Such AMF helps plant by intensifying the ionic homeostasis, antioxidant enzyme activity, osmotic balance (Porcel et al., 2016; Hashem et al., 2018), and even improve the photosynthetic rate, nutrients uptake, and water uptake by plants (Fig. 9.4).

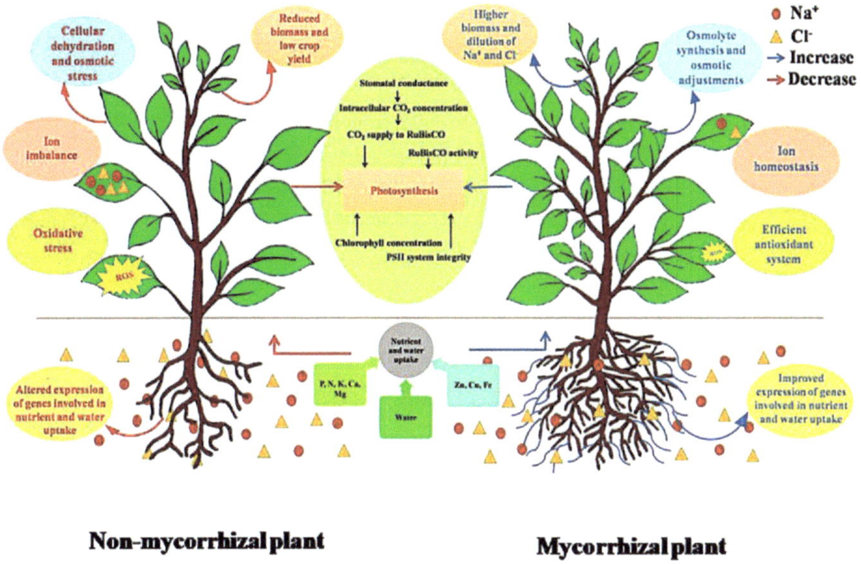

FIGURE 9.4 Differential response of non-mycorrhizal and mycorrhizal plants under salt stress. Accumulation of salt in soil creates competition for nutrient uptake and transport. This leads to imbalance of the ionic composition of plant, thereby affecting plant's physiological traits. AMF increase the volume of soil explored by plant roots, upregulate several cation transporters, leading to improved nutrient uptake, and also maintains ionic homeostasis. Salinity lowers soil water potential causing cellular dehydration due to decrease in water uptake. AM negates this effect by mediating accumulation of osmolytes and also improves plant's water status by improving root hydraulic conductivity. Salinity induces oxidative stress due to imbalance in ROS (reactive oxygen species) generation and the quenching activities of antioxidants. AMF are known to improve both enzymatic and non-enzymatic antioxidant systems of plants. Photosynthesis is also negatively affected by salinity. AM has a positive effect on photosynthesis under salt stress. Overall, AMF improve the performance of plant under salt stress.

Source: Reprinted from Evelin et al. (2019), https://creativecommons.org/licenses/by/4.0/

Drought stress is due to the scarcity of underground water. Underground water is the most valuable source for the uptake of nutrients for the productive growth of the plants. But the stress has become the very ominous abiotic stress for the plants to utilize the underground water due to less availability of the water just affecting the total crop productivity in every region of the world as it has been increasing significantly in the dimension of every part of the world (Shukla et al., 2012). The negative impact on the plants is seen in terms of productivity, biochemical and physiological parameters, respiration processes, translocation of the photoassimilates, and reduction in plant overall growth (Kahil et al., 2015). AMF helps and protects plants from such stress conditions and also helps in reducing groundwater pollution. AMF activity in water stress or drought conditions is complex mechanisms and is seen in accumulative physical, nutritional, physiological, and cellular effects of the plants. Colonization of AMF in watermelon plants has given a positive impact against water stress or drought conditions (Mo et al., 2016), whereas findings of Pavithra and Yapa (2018) showed that AMF colonization in soybeans did not give positive impact against the water stress or drought conditions.

Another abiotic stress like temperature stress, which of two types, high-temperature stress and low-temperature stress. Both stresses have a negative impact on plant growth, depending on the severity of the temperature. Chilling effects low-temperature and intensity of the high-temperature affect the crop and plant productivity worldwide. The plants and crop growth depend on the range of the temperature, and there are three ranges of temperature minimum, optimum, and maximum range (Mathur et al., 2018). The chilling effect of low temperature causes many disruption changes in plants like a mechanical constraint, macromolecular changes, alteration of the permeability of the plasma membrane, reduced osmotic potential (Janicka-Russak et al., 2012). At the high intensity of the temperature, photosynthetic activity is disturbed. Both light and dark reactions are heat-labile and are disturbed at high temperatures (Haque et al., 2014). According to Chen et al. (2013), less colonization of AMF was observed in cucumber than in control under low-temperature stress, and an increase in total fresh mass and total biomass of seedling growth inoculated with AMF of 19.1 and 18.8% was reported, respectively, than control under normal conditions. There was no significant increment in plant growth with AMF colonization under low temperature (Chen et al., 2013). The AMF colonization in maize plants under 45–55°C temperature showed less growth of plants but significantly positive than the control one without AMF (Mathur et al., 2018). The plant grows faster when colonized with AMF than plants without AMF under a high-temperature range as the root system is large in the case of AMF-aided plants (Gong et al., 2013; Zhu et al., 2014).

Yet another abiotic stress which hinders plant growth is the pollutant stress. Mining waste, heavy metals, and toxic metals from various anthropogenic activities are mainly responsible for contaminating the soil, just affecting the growth of plants and crops. The phytotoxicity is seen in heavy metals and toxic metals at a low concentration, which acts as a potentially detrimental effect on plants and crops (Mondal et al., 2015). Phytotoxicity symptoms include withering, chlorosis, necrosis, and poor development of meristem and roots of the plants (Ghosh and Sethy, 2013). Such pollutants are biomagnified during human food chain and causes detrimental health issues. AMF colonization in plant root acts as phytoremediation agent against pollutants by precipitating the metal phosphates or neutralizing the effect of the heavy metals and chelates the metals or heavy metals.

9.10 SALINITY STRESS AND TOLERANCE

Salinity stress is found to be a severe form of stress that accounts for the majority of crop and plant destruction and hence decreases the overall productivity. There are several mechanisms in which plants adopt to tolerate salinity stress like compartmentalization, excretion, and secretion of Na^+ and H^+ into vacuoles (Arzani, 2016), homeostasis of Na^+ and K^+ by Na^+/H^+ exchangers proteins which is localized in the intracellular membrane. The vacuoles of the various plants exchange Na^+/H^+ through the tonoplast, a membrane of the vacuole, to reduce Na^+ concentration in the cytoplasm of the plant cells (Arzani, 2016). The V-type ATPases maintains the acidic environment in the vacuolar lumen through the uptake of Na^+ during salt stress (Yarra, 2019).

There are plants that have an additional way to tolerate salt stress; wheat *NHX* (TαNHX1, TαNHX2, TαNHX3) genes play an important role in protecting themself from salt stress. The expression proteins from TαNHX2 and TαNHX3 genes are higher in leaves and roots and high expression of TαNHX1 gene transcript only in roots of the wheat plant under the salt stress (Lu et al., 2014). According to Yarra (2019), such *NHX* genes from wheat plants isolated and used to transform yeast mutant cells were able to tolerate the salt stress when the transformed yeast cell with NHX genes grown under 400 mM and 200 mM NaCl concentration (Yarra, 2019). The report of Brini et al. (2005) successfully generated transgenic *Arabidopsis thaliana* plants using *NHX* genes to tolerate salt stress. Lu et al. (2014) successfully generated transgenic tobacco plants by overexpressing TαNHX3 genes to improve the tolerance rate during salt stress.

Another way plants tolerate salt stress by various signal transduction pathways (Yang et al., 2010, 2012), an important enzyme governing such a signal pathway is Protein Kinase, which provides tolerance power against drought, saline, and temperature stress. MAPK (Mitogen-activated protein kinase), a serine/threonine-protein kinase also plays an important role in tolerance against stress problems (Bergmann et al., 2004). In soybeans, tolerance against salt stress has been regulated by ABA-dependent or independent signaling pathways (Fujita et al., 2011), and GsMAPK4 plays an important signaling role in soybeans against high salt tolerance (You-wen et al., 2019).

A facultative halophytic plant Quinoa (*Chenopodium quinoa* Wild.) can tolerate high-salt concentration and it belongs to the family of Amaranthaceae s.l. A study was done by Shabala et al. (2012), where quinoa plants were grown under gradually increased salt concentration for 7 days, and it was observed that plants developed morphological and physiological mechanisms to tolerate salt stress. Yet another factor, Plant Natriuretic Peptides (PNP) stimulate tolerance against salt stress in plants (Wang et al., 2011). Liu et al. (2020) reported that the overexpression of PNP from *Stellera chamaejasme*, a toxic plant confers tolerance against salt stress in *Arabidopsis thaliana* plants.

9.11 SALINITY TOLERANCE BY AMF

Arbuscular Mycorrhizae fungi protect plants against salinity stress and confers tolerance in many ways. Zhang et al. (2019) worked on the effect of AMF (*Rhizobium irregularis*) on the growth of garden asparagus, that is, *Asparagus officinalis* L. under salinity stress by an experimental setup. He designed four treatment methods to grow *A. officinalis:* (1) no AMF treatment and no salt stress; (2) AMF treatment but no salt stress; (3) no AMF treatment with salt stress; (4) AMF treatment with salt stress. The leaves were analyzed after 6 days of growth and roots were subjected to AMF colonization at 0, 5, and 10 days under salt stress. The result observed was the colonization of AMF in the plant without salt stress, where hyphae, arbuscules, and vesicles were found in plant roots at the rate of 54.17, 52.3, and 38.97%, respectively, and there was no growth of AMF in (2) no AMF-treated plants without salt stress. In (3) case, the growth of the plant was slow due to salt stress as compared to (1) case where plant growth was normal without AMF under no salt stress and (2) case where plants were inoculated with AMF. There was an increment in the plant growth inoculated with AMF under salt stress in (4) case as compared to the slow growth of plants with salt stress in (3) case (Zhang et al., 2019).

Hashem and co-workers studied the effect of salt stress on seeds of soybeans (*Glycine max*) with two different genotypes, genotype Clark (salt-tolerant), and genotype Kint (salt-sensitive) with or without the inoculation of 2 g of inoculant of AMFs *Funneliformis mosseae* (syn. *Glomus mosseae*), *Rhizophagus intraradices* (syn. *Glomus intraradices*), and *Claroideoglomus etunicatum* (syn. *Glomus etunicatum*) per 200 seed of the soybeans in different pots, respectively. The seedlings were grown under greenhouse conditions like 11 h of sunlight and 6 h of the dark cycle, with 70% humidity at 27°C for 60 days. It was observed that there was a decrease in the rate of the germination in Clark genotype by 6.8% and in Kint by 1.4% under 10, 200, and 300 mM NaCl salt concentration, respectively (Hashem et al., 2016).

To study the effect of salinity stress in seeds of sebania (*S. sesban* [L.] Merr; syn. *Sesban aegyptiaca* Poiret) with or without AMF, the AMF selected were *Funneliformis mosseae* (syn. *Glomus mosseae*), *Rhizophagus intraradices* (syn. *Glomus intraradices*), and *Claroideoglomus etunicatum* (syn. *Glomus etunicatum*) and the salt concentration used were 75 and 150 mM NaCl solution. The result observed was the percent reduction in shoot and root length was 21.17 and 40.85% under 75 and 150 mM NaCl concentration, respectively, without AMF inoculation. Approximately 23.8 and 47.5% increments in shoot and root were observed when AMF were inoculated as compared to control under 75 and 150 mM NaCl concentration, respectively. There was a reduction in fresh shoot and root weight of 25.64 and 48.9% under 75 mM NaCl concentration, respectively, and 57 and 80.1% reduction in fresh weight of shoot and root under 150 mM NaCl concentration, respectively. The shoot and fresh root weight showed 37.6 and 29% increments when inoculated with AMF under 75 and 150 mM NaCl concentration, respectively (Abdallah et al., 2015).

9.12 TOLERANCE THROUGH NUTRIENT UPTAKE AND IONIC BALANCE

AMF are beneficial to plants in stress conditions, as AMF share a positive symbiotic relationship with plants. For example, *Ocimum basilicum* tolerates salt stress when colonizing with AMF and increases the uptake of minerals, improves chlorophyll synthesis, and water use efficiency (Shekoofeh et al., 2012). AMF colonization in the tomato plant showed positive growth tolerance against salt stress as there was an increase in leaf area and certain minerals like nitrogen, potassium, calcium, and phosphorus content (Balliu et al., 2015). Even high water content cucumber showed enhanced antioxidant, mineral uptake, and ionic balance under salinity stress when AMF colonized the plant. *Glomus mosseae*, an AMF conferred tolerance against salt stress in *Solanum lycopersicum* by increasing the nutrient uptake of phosphorus and potassium from the soil (Abdel Latef and Chaoxing, 2011). AMF ameliorated salt stress by improving the uptake of potassium, calcium, magnesium, manganese, zinc, iron, copper in the roots, and leaves of the *Cucumis sativus*. The AMF also decreases the amount of Na^+ content in the leaves (Hashem et al., 2018). *G. mosseae* colonized in roots of *Lycopersicon esculentum* cv Zhongzha105 (tomato) showed increased uptake of potassium and phosphorus but there was a decrease in sodium uptake under salt stress (Latef and Chaoxing, 2011).

Alleviation against salt stress is achieved by the colonization of AMF as they increase the uptake of calcium and potassium by plants and decrease the uptake of sodium (Hammer et al., 2011). The reports of Amanifar et al. (2019) showed that inoculation of *F. mosseae* mycorrhizal into the roots of liquorice plants increases the uptake of potassium and phosphorus in roots and shoots of the plants under stress conditions. The absorption of sodium by AMF in plants was significantly reduced than plants without AMF colonization maintaining the ionic balance in the cell and there was an increase K^+/Na^+ ratio in AMF inoculated liquorice plants in the shoot portion as K^+/Na^+ of non-inoculated plants showed decreased ratio (Amanifar et al., 2019).

The sweet basil plant is regarded as an important medicinal plant. But due to salinity stress, these plants differ in their medicinal properties and hardiness. Inoculation of AMF, *Glomus deserticola* improves the nutrient uptake of the plant species like it enhances the uptake of potassium, calcium, and phosphorus. The AMF also maintains the ionic balance by decreasing the absorption of sodium and chloride in the plant species. The K^+/Na^+ and Ca^{2+}/Na^+ ratio significantly increased in plant leaves when inoculated with AMF under salt stress conditions (Elhindi et al., 2017) (Table 9.2).

9.13 TOLERANCE THROUGH PHYSIOLOGICAL RESPONSE

The effect of salinity is reduced photosynthetic process and hence reduces the overall productivity. The salinity stress affects plants by lowering the leaves area, and less availability of CO_2 content to plants so assimilation of the photosynthetic product is less, the conductivity of stomata is also

TABLE 9.2 Effect of Salinity and AMF on Nutrient Concentration and Ionic Ratios in Plants.

Plant (family)	Fungus	Parameters	Effects	
			salinity	AMF on salt stressed plants
Citrus tangerine (Rutaceae)	*Glomus mosseae* and *Paraglomus occultum*	Na, K, Ca, Mg content and K^+:Na^+, Ca^{2+}:Na^+, Mg^{2+}:Na^+ ratios	Increased- Na, Ca, and Mg content, decreased- K content and ionic ratios	Increased-K, Mg content and ionic ratios, Ca^{2+} content increased by *P. occultum* Decreased-Na and Ca content, Ca^{2+}:Na^+ ratio by *G. mosseae* only
Trigonella foenum-graecum (Fabaceae)	*Glomus intraradices*	Shoot and root nutrient status and K^+:Na^+, Ca^{2+}:Na^+, Ca^{2+}:Mg^{2+} ratios	Increased-shoot Na, Mn and root Na, Mg content Decreased-shoot and root N, P, K, Ca, Cu, Fe, Zn content, and ionic ratios	Increased-root and shoot N, P, K, Ca, Mg, Cu, Fe, Mn, Zn content and K^+:Na^+ ratio Decreased-root and shoot Na content
Medicago sativa (Fabaceae)	*Glomus viscosum*	Shoot and root Na, K, Ca, Mg, Cl content.	Increased-Na and Cl content Decreased-K, Ca, and Mg content	Increased-K and Mg content Decreased-Na, Ca, and Cl content
Kosteletzkya virginica (Malvaceae)	*Glomus mosseae*	Root and shoot N, P, K content and ionic ratios of different root tissues	Increased- K^+:Na^+ and Ca^{2+}:Mg^{2+}ratios Decreased-root and shoot N, P, K content, Na^+:Ca^{2+} ratio	Increased-root and shoot N, P, K content, Na^+:Ca^{2+} ratio Decreased-K^+:Na^+ and Ca^{2+}:Mg^{2+} ratios
Oryza sativa (Poaceae)	*Claroideoglomus etunicatum*	Root and shoot P, Na and K content and expression of genes with a role in the uptake, transport or compartmentation of Na^+ and/or K^+	Increased-root and shoot Na content, Na root-to-shoot distribution Decreased-root and shoot K^+:Na^+ ratio, expression of *OsNHX3, OsSOS1, OsHKT2;1*, and *OsHKT1;5* genes	Increased-root Na content, root and shoot P, upregulation of *OsNHX3, OsSOS1, OsHKT2;1*, and *OsHKT1;5* genes Decreased-shoot Na content, root and shoot K^+:Na^+ ratio, root-to-shoot distribution of Na

TABLE 9.2 *(Continued)*

Plant (family)	Fungus	Parameters	Effects	
			salinity	**AMF on salt stressed plants**
Cicer arietinum (Fabaceae)	*Funneliformis mosseae*	Nutrient status and $K^+:Na^+$ ratio	Increased-Na content Decreased-N, P, K, Mg content and leaf and root $K^+:Na^+$ ratio	Increased-N, P, K, Mg content, leaf and root $K^+:Na^+$ ratio Decreased-Na content
Cucumis sativus (Cucurbitaceae)	*Claroideoglomus etunicatum, Rhizoglomus intraradices*, and *Glomus mosseae*	Ionic status and electrolytic leakage	Increased-Na content and electrolytic leakage Decreased-K, Ca, Mg, Fe, Zn, Mn, and Cu content	Increased-K, Ca, Mg, Fe, Zn, Mn, and Cu content Decreased-Na content and electrolytic leakage

Source: Adapted from Evelin et al. (2019), Creative Commons Attribution 4.0 International.

affected. AMF help the plant to tolerate salinity stress by minimizing the light harvest and dissipating energy through non-photochemical quenching and also increases the leaves area, thus making maximum CO_2 available for photosynthesis.

The report of Galmero et al. (2010) states the effect of 75 mM of salt solution on *Cucumis sativus* when colonizing with AMF *Gigaspora rosea* BEG9. They reported an overall enhanced photosynthetic index in AMF plants than non-AMF plants under salt stress conditions. According to Wu et al. (2010), when *Glomus mosseae* and *Paraglomus occultum* inoculated in *Citrus tangerina*, respectively, increased the rate of the photosynthesis under 100 mM of salt concentration.

The chlorophyll content was enhanced when *Glomus intraradices* were inoculated in *Capsicum annum* kept in 50, 100, and 200 mM salt concentrations (Beltrano et al., 2013). Campanelli et al. (2013) reported an overall increment in chlorophyll content and stomatal density in *Medicago sativa* inoculated with *Glomus viscosum* under 0, 50, 100, 150, and 200 mM salt concentrations.

The high salinity in soil creates drought condition in the soil as water is not available for the plants. The AMF-inoculated plants enhance the uptake of the water even under salt stress. The AMF plants express more aquaporin which are present on leaves and roots and enhance the uptake of the water when subjected to saline condition. There are some reports in which different aquaporins have different effects depending upon the salt stress.

The colonization of the AMF in plants with salicylic acid has shown tolerance against salt stress. This response enhanced the proline accumulation, certain protein synthesis, increased K^+ content, and also limited the peroxidation of the lipids. The report of Hashem et al. (2018) showed the increase concentration of JA (jasmonic acid) under salt stress in *Cucumis sativus*. The JA is a phytohormone of octadecanoid family and has an important function in salinity stress.

9.14 TOLERANCE THROUGH ANTIOXIDANT RESPONSE

The antioxidant response seen in plants against salt stress helps plants to clear the excess of ROS generated in plants during the stress conditions. There are two types of antioxidant responses, that is, enzymes-based and non-enzyme-based. The plants inoculated with AMF also enhance the concentration of antioxidant enzymes in the plant system. These AMF enhance the activities of the antioxidant enzymes by increasing the production of the antioxidant molecules. These molecules are α-tocopherol, glutathione, ascorbic acid, and carotenoids which are commonly found in plants and common antioxidant enzymes are found in plant cells, such as SOD (superoxide dismutase), catalase, peroxidase, glutathione reductase (Velarde-Buendía et al., 2012).

The high content of antioxidant molecules like α-tocopherol in the AMF plants reduces the peroxidation of the lipids. The high levels of the glutathione reductase and glutathione in the AMF plants increase the production of the ascorbic acids. Similarly, the high level of the α-tocopherol increases the tocopheroxyl radicals which are facilitated by a high level of ascorbic acids on the AMF plants. The carotenoids present in the AMF plants remove the oxygen radicals from the cells. The effect of such antioxidant molecules depends upon the level of salinity, species of AMF inoculated, plant species used, and duration of the stress-induced.

Ait-El-Mokhtar and his associates worked on the effect of AMF on the date plant. They used a consortium of AMF *Glomus* sp., *Sclerocystis* sp., and *Acaulospora* sp. to inoculate in *Phoenix dactylifera* commonly known as date plant. The antioxidant property was increased even under the salt stress condition. The antioxidant enzymes particularly SOD, catalase, ascorbate peroxidase, and peroxidase activity were in high amount to tackle the stress. The underlying mechanism is the SOD scavenges the superoxide into water and hydrogen peroxide as SOD acts as a first line of defense. The hydrogen peroxide produced and then converted to water and oxygen by catalase or ascorbate peroxidase. The hydrogen peroxide can also be removed by peroxidase which is present in lignified tissues of the plant (Ait-El-Mokhtar et al., 2019).

Hashem et al. (2018) worked on the plant *Cucumis sativus* which was inoculated with (*Claroideoglomus etunicatum* [syn. *Glomus etunicatum*], *Rhizophagus intraradices* [syn. *Glomus intraradices*], and *Funneliformis mosseae* [syn. *Glomus mosseae*], respectively, showed increased antioxidant properties under salt stress. SOD, ascorbate peroxidase, glutathione reductase, catalase, and ascorbic acid concentration were elevated in AMF-inoculated plants under salt stress condition.

Solanum lycopersicon L. (tomato), when inoculated with the spores of *Glomus fasciculatum* showed an elevated concentration of SOD and peroxidase antioxidant responses in plants in salt stress condition. But the catalase activity was limited even AMF when inoculated in the plants (Ebrahim et al., 2017). The report published by Abdel Latef and Chaoxing (2011) showed that AMF *Glomus mosseae* inoculated tomato plant under salt stress condition, the leaves of the plant showed an increase in SOD, peroxidase, catalase, and ascorbate peroxidase activity.

9.15 THROUGH PROLINE ACCUMULATION

Proline is a compatible solute synthesized during stress conditions like salinity stress to enhance water uptake even in such stressful conditions. The proline is regarded as osmoregulating against oxidative damage through the maintenance of water uptake and its balance. There are many reports that gave an idea that proline accumulation can be increased in AMF plants than non-AMF plants or it can be reduced in AMF plants also. If proline content and accumulation are increased in AMF plants under salt stress condition, the proline synthesis gene, that is, P5CS gene is highly expressed in the cell and increases the enzyme activity of glutamate dehydrogenase enzyme for the synthesis of the glutamate as it acts as a precursor for proline synthesis and lastly, the inhibition of proline dehydrogenase as it catabolizes the proline. Wu et al. (2016) reported that proline accumulated in AMF plants and increases the efficiency of water uptake in plants under salt stress conditions. The percent of 83.01 of proline accumulated in Cucumber plant when AMF was inoculated under salt stress condition than percent of 24.54 of proline accumulated in non-AMF plants under the same salt stress conditions (Hashem et al., 2018).

The work on the effect of AMF colonization in fenugreek plants under salt stress condition, observed that there was an increase in proline content in AMF colonized plants than in non-AMF plants under 75 mM salt conditions. It gave a possible explanation that this increases the proline content in AMF plants due to an increase in proline synthesizing enzymes rather than catabolizing the proline content (Metwally and Abdelhameed, 2018).

Elhindi and his workers reported that there were lower production and accumulation of proline in the AMF-inoculated sweet basil plants than in non-AMF plants under salt stress and also reported that sweet basil plants are less stressed in salinity conditions. They also compared their data with other findings of Garg and Manchanda (2009) that proline accumulated enhanced the sweet basil plants by improving the osmotic balance and reducing the radicals when inoculated with AMF (Elhindi et al., 2017).

9.16 CONCLUSION AND FUTURE SCOPE

The stress conditions created in plants lead to reduced productivity and poor plant growth. The different stresses encountered are salinity, drought, temperature, and pollutant (metals, heavy metals, and pesticides). However, there are certain plants that have adapted to such stress which we called as plant-tolerant system. The plants accumulate proline as a compatible solute, decrease the uptake of sodium and chloride ions to maintain the ionic balance inside the cells, and even increase the phytohormones, increases the antioxidant molecules like ascorbic acid, α-tocopherol, glutathione, and activity of antioxidant enzymes is enhanced. However, AMF offer many more benefits to plants to withstand salinity stress. AMF have a fungal association with the plants and many AMF are the member of the family Glomerulomycota. AMF-assisted salinity stress tolerance shows a great potential to increase the plant in terms of productivity. The detailed AMF protection and tolerance rate toward plants against the salinity stress have been discussed such that AMF tolerance by enhancing the proline accumulation, maintaining the ionic balance, increasing antioxidant molecules to provide defensin against the ROS generated in salinity stress condition, and increasing the uptake of the nutrients like potassium, phosphorus, zinc, iron, and calcium. There are many approaches been mentioned in the above text which have been reported by many authors for growing plants with AMF inoculated under the salinity stress conditions. In the future, more research concerning the exploration of AMF species and its gene tolerance against stress conditions should be done, isolating genes that possess tolerance properties in AMF and creating transgenic plants.

In the future, there will be more such salinity soil with increasing soil pollution. There are more AMF left to be discovered and need to exploit in enhancing productivity and making plants tolerant against any stress that will be encountered in the future. The AMF-based biofertilizers will be more efficiently used in the near future due to their non-toxicity composition and easy way of the production. The biofertilizers will replace conventional chemical fertilizers. The AMF-based phytoremediation shows great potential in cleaning pollutants present in the soil. However, more detailed research needs to be done for more efficiency of the AMF to kill a broad range of the soil organisms which includes bacteria, fungi, nematodes, and insects.

KEYWORDS

- **arbuscular mycorrhizal fungi**
- **salinity**
- **stress conditions**
- **tolerance**
- **abiotic stress**

REFERENCES

Abdallah, E. F.; Hashem, A.; Alqarawi, A. A.; Bahkali, A. H.; Alwhibi, M. S. Enhancing growth performance and systemic acquired resistance of medicinal plant *Sesbania sesban* (L.) Merr using arbuscular mycorrhizal fungi under salt stress. *Saudi J. Biol. Sci.* **2015**, *22*(3), 274–283.

Abebe, G.; Hattar, B.; Al-Tawah, A. M. Nutrient availability as affected by manure application to cowpea (*Vigna unguiculata* L. Walp.) on calacarious soils. *J. Agric. Soc. Sci.* **2005**, *1*(1), 1–6.

Abera, T.; Feyisa, D.; Yusuf, H.; Nikus, O.; Al-Tawaha, A. R. Grain yield of maize as affected by biogas slurry and NP fertilizer rate at Bako, Western Oromiya, Ethiopia. *Biosci. Res.* **2005**, *2*, 31–38.

Abouseoud, M.; Maachi, R.; Amrane, A.; Boudergua, S.; Nabi, A. Evaluation of different carbon and nitrogen sources in production of biosurfactant by *Pseudomonas fluorescens*. *Desalination* **2008**, *1–3*, 143–151.

Abouseoud, M.; Yataghene, A.; Amrane, A.; Maachi, R. Effect of pH and salinity on the emulsifying capacity and naphthalene solubility of a biosurfactant produced by *Pseudomonas fluorescens*. *J. Hazard. Mater.* **2010**, *180*(1–3), 131–136.

Abu Obaid, A. M.; Melnyk, A.V.; Onychko, V. I.; Usmael, F. M.; Abdullah, M. J.; Rifaee, M. K.; Tawaha, A. M. Evaluation of six sunflower cultivar for forage productivity under salinity condition. *Adv. Environ. Biol.* **2018**, *12*, 13–15.

Ahmad, I.; Ahmad, F.; Pichtel, J. *Microbes and microbial technology: Agricultural and environmental applications*; Springer Science + Business Media, 2011.

Ahmad, M.; Yousuf, S.; Khan, M. B.; Hoda, M. N.; Ahmad, A. S.; Ansari, M. A.; Islam, F. Attenuation by *Nardostachys jatamansi* of 6-hydroxydopamine-induced parkinsonism in rats: behavioral, neurochemical, and immunohistochemical studies. *Pharmacol. Biochem. Behav.* **2011**, *83*(1), 150–160.

Ahmad, M.; Zahir, Z. A.; Asghar, H. N.; Asghar, M. Inducing salt tolerance in mung bean through coinoculation with rhizobia and plant-growth-promoting rhizobacteria containing 1-aminocyclopropane-1-carboxylate deaminase. *Can. J. Microbiol.* **2011**, *57*(7), 578–589.

Ait-El-Mokhtar, M.; Laouane, R. B.; Anli, M.; Boutasknit, A.; Wahbi, S.; Meddich, A. Use of mycorrhizal fungi in improving tolerance of the date palm (*Phoenix dactylifera* L.) seedlings to salt stress. *Sci. Hort.* **2019**, *253*, 429–438.

Akram, M. S.; Shahid, M.; Tariq, M.; Azeem, M.; Javed, M. T.; Saleem, S.; Riaz, S. Deciphering *Staphylococcus sciuri* SAT-17 mediated anti-oxidative defensin mechanisms and growth modulations in salt stressed maize (*Zea mays* L.). *Front. Microbiol.* **2016**, *7*, 867.

Al Zoubi, O. M.; Al-Tawaha, A. R. Allelopathic effect of Beetroot (*Beta vulgaris* L.) on germination and growth of *Zea mays* and *Vigna umbellata*. *Int. J. Bot. Stud.* **2019**, *4*(4), 47–51.

Al-Ajlouni, M. M.; Al-Ghzawi, A. L. A.; Al-Tawaha, A. R. Crop rotation and fertilization effect on barley yield grown in arid conditions. *J. Food Agric. Environ.* **2010**, *88*(3), 869–872.

Ali, S.; Charles, T. C.; Glick, B. R. Amelioration of high salinity stress damage by plant growth-promoting bacterial endophytes that contain ACC deaminase. *Plant Physiol. Biochem.* **2014**, *80*, 160–167.

Al-Juthery, H. W.; Habeeb, K. H.; Jawad, F.; Altaee, K.; AL-Taey, D. K. A.; Al-Tawaha, A. R. M. Effect of foliar application of different sources of nano-fertilizers on growth and yield of wheat. *Biosci. Res.* **2018**, *15*(4), 3988–3997.

Al-Karaki, G.; McMichael, B.; Zak, J. Field response of wheat to arbuscular mycorrhizal fungi and drought stress. *Mycorrhiza* **2004**, *14*(4), 263–269.

Al-Kiyyam, M. A.; Turk, M.; Al-Mahmoud, M.; Al-Tawaha, A. R. Effect of plant density and nitrogen rate on herbage yields of marjoram under Mediterranean conditions. *J. Am.-Eurasian J. Agric. Environ. Sci.* **2008**, *3*, 153–158.

Almeida, O. P. D. Biologia molecular da doença de Alzheimer: uma luz no fim do túnel? *Revista da Associação Médica Brasileira* **1997**, *43*(1), 77–81.

Al-Tawaha, A. M.; Turk, M.; Hameed, K.; Assaf, T.; Tahhan, R.; Al-Jamali, A. F. Effects of application of olive mill by-products on lentil yield and their symbiosis with mycorrhizal fungi under arid conditions. *Crop Res.* **2010**, *40*(3), 66–75.

AL-Tawaha, A. R. M.; Nidal, O. Use of sorghum and maize allelopathic properties to inhibit germination and growth of wild barley (*Hordeum spontaneum*). *Notulae Botanicae Horti Agrobotanici Cluj-Napoca* **2010**, *38*(3), 124–127.

Al-Tawaha, A. R.; Al-Karaki, G.; Al-Tawaha, A. R.; Sirajuddin, S. N.; Makhadmeh, I.; Wahab, P. E. M.; Youssef, R. A.; Al Sultan, W.; Massadeh, A. Effect of water flow rate on quantity and quality of lettuce (*Lactuca sativa* L.) in nutrient film technique (NFT) under hydroponics conditions. *Bulgarian J. Agric. Sci.* **2018a**, *245*,791–798.

Al-Tawaha, A. R.; Al-Tawaha, A. R.; Aludatt, M.; Al-Ghzawi, A.L.; Wedyan, M.; Al-Obaidy, S. D. A.; Al-Ramamneh, E. A. D. Effects of soil type and rainwater harvesting treatments in the growth, productivity and morphological trains of barley plants cultivated in semi-arid environment. *Aust. J. Crop Sci.* **2018b**, *12*(6), 975–979.

Al-Tawaha, A. R.; McNeil, D. L.; Yadav, S. S.; Turk, M.; Ajlouni, M.; Abu-Darwish, M. S.; Al-Ghzawi, A. L. A.; Aludatt, M.; Aladaileh, S. Integrated legume crops production and management technology. In: *Climate change and management of cool season grain legume crops*; Yadav, S., McNeil, D. L., Redden, R., Patil, S. A., Eds.; Springer: Dordrecht, 2010; pp 325–349.

Al-Tawaha, A. R.; Turk, M. A.; Abu-Zaitoon, Y. M.; Aladaileh, S. H.; Al-Rawashdeh, I. M.; Alnaimat, S.; Al-Tawaha, A. R. M.; Alu'datt, M. H.; Wedyan, M. Plants adaptation to drought environment. *Bulgarian J. Agric. Sci.* **2017**, *23*(3), 381–388.

Alu'datt, M. H.; Rababah, T.; Alhamad, M. N.; Al-Tawaha, A.; Al-Tawaha, A. R.; Gammoh, S.; Ereifej, K. I.; Al-Karaki, G.; Hamasha, H. R.; Tranchant, C. C.; Kubow, S. Herbal yield, nutritive composition, phenolic contents and antioxidant activity of purslane (*Portulaca oleracea* L.) grown in different soilless media in a closed system. *Industrial Crops and Products* **2019**, *141*, 111746.

Amanifar, S.; Khodabandeloo, M.; Fard, E. M.; Askari, M. S.; Ashrafi, M. Alleviation of salt stress and changes in glycyrrhizin accumulation by arbuscular mycorrhiza in liquorice (*Glycyrrhiza glabra*) grown under salinity stress. *Environmental and Experimental Botany* **2019**, *160*, 25–34.

Ananthi, T.; Amanullah, M. M.; Al-Tawaha, A. R. M. S. A review on maize-legume intercropping for enhancing the productivity and soil fertility for sustainable agriculture in India. *Adv. Environ. Biol.* **2017**, *11*(5), 49–63.

Armada, E.; Probanza, A.; Roldán, A.; Azcón, R. Native plant growth promoting bacteria *Bacillus thuringiensis* and mixed or individual mycorrhizal species improved drought tolerance and oxidative metabolism in *Lavandula dentata* plants. *J. Plant Physiol.* **2016**, *192*, 1–12.

Arzani, A.; Ashraf, M. Smart engineering of genetic resources for enhanced salinity tolerance in crop plants. *Crit. Rev. Plant Sci.* **2016**, *35*(3), 146–189.

Averill, C.; Turner, B. L.; Finzi, A. C. Mycorrhiza-mediated competition between plants and decomposers drives soil carbon storage. *Nature* **2014**, *505*(7484), 543–545.

Beltrano, J.; Ruscitti, M.; Arango, M. C.; Ronco, M. Effects of Arbuscular mycorrhiza inoculation on plant growth, biological and physiological parameters and mineral nutrition in pepper grown under different salinity and P levels. *J. Soil Sci. Plant Nutr.* **2013**, *13*(1), 123–141.

Berg, G.; Alavi, M.; Schmidt, C. S.; Zachow, C.; Egamberdieva, D.; Kamilova, F.; Lugtenberg, B. J. Biocontrol and osmoprotection for plants under salinated conditions. In *Molecular Microbial Ecology of the Rhizosphere*, de Bruijn Frans, J., Ed., Wiley-Blackwell, USA 2013, pp 561–573.

Bergmann, D. C.; Lukowitz, W.; Somerville, C. R. Stomatal development and pattern controlled by a MAPKK kinase. *Science* **2004**, *304*(5676), 1494–1497.

Bezza, F. A.; Beukes, M.; Chirwa, E. M. N. Application of biosurfactant produced by *Ochrobactrum intermedium* CN3 for enhancing petroleum sludge bioremediation. *Process Biochem.* **2015**, *50*(11), 1911–1922.

Bianciotto, V.; Andreotti, S.; Balestrini, R.; Bonfante, P.; Perotto, S. Extracellular polysaccharides are involved in the attachment of *Azospirillum brasilense* and *Rhizobium leguminosarum* to arbuscular mycorrhizal structures. *Eur. J. Histochem.* **2001**, 39–50.

Bonfante, P.; Genre, A. Mechanisms underlying beneficial plant-fungus interactions in mycorrhizal symbiosis. *Nat. Commun.* **2010**, *1*, 48. https://doi.org/10.1038/ncomms1046.

Brini, F.; Gaxiola, R. A.; Berkowitz, G. A.; Masmoudi, K. Cloning and characterization of a wheat vacuolar cation/proton antiporter and pyrophosphatase proton pump. *Plant Physiol. Biochem.* **2005**, *43*(4), 347–354.

Cairney, J. W. G. Evolution of mycorrhiza systems. *Naturwissenschaften* **2000**, *87*(11), 467–475.

Campanelli, A.; Ruta, C.; De Mastro, G.; Morone-Fortunato, I. The role of arbuscular mycorrhizal fungi in alleviating salt stress in *Medicago sativa* L. var. icon. *Symbiosis* **2013**, *59*(2), 65–76.

Carvalho, L. M.; Correia, P. M.; Martins-Loução, M. A. Arbuscular mycorrhizal fungal propagules in a salt marsh. *Mycorrhiza* **2004**, *14*(3), 165–170.

Chandran, P.; Das, N. Degradation of diesel oil by immobilized *Candida tropicalis* and biofilm formed on gravels. *Biodegradation* **2011**, *22*(6), 1181–1189.

Chang, P.; Gerhardt, K. E.; Huang, X. D.; Yu, X. M.; Glick, B. R.; Gerwing, P. D.; Greenberg, B. M. Plant growth-promoting bacteria facilitate the growth of barley and oats in salt-impacted soil: implications for phytoremediation of saline soils. *Int. J. Phytoremed.* **2014**, *16*(11), 1133–1147.

Chen, J.; Huang, P. T.; Zhang, K. Y.; Ding, F. R. Isolation of biosurfactant producers, optimization and properties of biosurfactant produced by *Acinetobacter* sp. from petroleum-contaminated soil. *J. Appl. Microbiol.* **2012**, *112*(4), 660–671.

Clark, R. Á.; Zeto, S. K. Mineral acquisition by arbuscular mycorrhizal plants. *J. Plant Nutr.* **2000**, *23* (7), 867–902.

De Oliveira, M. V. V.; Intorne, A. C.; Vespoli, L. D. S.; Madureira, H. C.; Leandro, M. R.; Pereira, T. N.; Olivares, F. L.; Berbet-Molina, M. A.; De Souza Filho, G. A. Differential effects of salinity and osmotic stress on the plant growth-promoting bacterium *Gluconacetobacter diazotrophicus* PAL5. *Arch. Microbiol.* **2016**, *198*(3), 287–294.

Dorchin, N.; Freidberg, A.; Aloni, R. Morphogenesis of stem gall tissues induced by larvae of two Cecidomyiid species (Diptera: Cecidomyiidae) on *Suaeda monoica* (Chenopodiaceae). *Can. J. Bot.* **2002**, *80*(11), 1141–1150.

Ebrahim, M. K.; Saleem, A. R. Alleviating salt stress in tomato inoculated with mycorrhizae: photosynthetic performance and enzymatic antioxidants. *J. Taibah Univ. Sci.* **2017**, *11*(6), 850–860.

Eddouaouda, K.; Mnif, S.; Badis, A.; Younes, S. B.; Cherif, S.; Ferhat, S.; Mhiri, N.; Chamkha, M.; Sayadi, S. Characterization of a novel biosurfactant produced by *Staphylococcus* sp. strain 1E with potential application on hydrocarbon bioremediation. *J. Basic Microbiol.* **2012**, *52*(4), 408–418.

Egamberdieva, D.; Lugtenberg, B. Use of plant growth-promoting rhizobacteria to alleviate salinity stress in plants. In: *Use of microbes for the alleviation of soil stresses*, Vol. 1; Miransari, M., Ed.; Springer: New York, 2014; pp 73–96.

Egamberdieva, D.; Li, L. I.; Lindström, K.; Räsänen, L. A. A synergistic interaction between salt-tolerant *Pseudomonas* and *Mesorhizobium* strains improves growth and symbiotic performance of liquorice (*Glycyrrhiza uralensis* Fish.) under salt stress. *Appl. Microbiol. Biotechnol.* **2016**, *100* (6), 2829–2841.

Egamberdieva, D.; Wirth, S. J.; Bellingrath-Kimura, S. D.; Mishra, J.; Arora, N. K. Salt tolerant plant growth promoting rhizobacteria for enhancing crop productivity of saline soils. *Front. Microbiol.* **2019**, *10*, 2791. doi:10.3389/fmicb.2019.02791.

Elhindi, K. M.; El-Din, A. S.; Elgorban, A. M. The impact of arbuscular mycorrhizal fungi in mitigating salt-induced adverse effects in sweet basil (*Ocimum basilicum* L.). *Saudi J. Biol. Sci.* **2017**, *24*(1), 170–179.

Engelmoer, D. J.; Kiers, E. T. Host diversity affects the abundance of the extraradical arbuscular mycorrhizal network. *New Phytol.* **2015**, *205*(4), 1485–1491.

Evelin, H.; Devi, T. S.; Gupta, S.; Kapoor, R. Mitigation of salinity stress in plants by arbuscular mycorrhizal symbiosis: Current understanding and new challenges. *Front. Plant Sci.* **2019**, *10*, 470. https://doi.org/10.3389/fpls.2019.00470.

Evelin, H.; Kapoor, R.; Giri, B. Arbuscular mycorrhizal fungi in alleviation of salt stress: A review. *Ann. Bot.* **2009**, *104* (7), 1263–1280.

Feng, G.; Zhang, F.; Li, X.; Tian, C.; Tang, C.; Rengel, Z. Improved tolerance of maize plants to salt stress by arbuscular mycorrhiza is related to higher accumulation of soluble sugars in roots. *Mycorrhiza* **2002**, *12*(4), 185–190.

Fernando, V. D.; Schroeder, D. F. Role of ABA in Arabidopsis salt, drought, and desiccation tolerance. In *Abiotic and biotic stress in plants - recent advances and future perspectives*; Intech Open: London, 2016; pp 508–524.

Fujita, Y.; Fujita, M.; Shinozaki, K.; Yamaguchi-Shinozaki, K. ABA-mediated transcriptional regulation in response to osmotic stress in plants. *J. Plant Res.* **2011**, *124*(4), 509–525.

Gamalero, E.; Berta, G.; Massa, N.; Glick, B. R.; Lingua, G. Interactions between *Pseudomonas putida* UW4 and iBEG9 and their consequences for the growth of cucumber under salt-stress conditions. *J. Appl. Microbiol.* **2010**, *108*(1), 236–245.

Gandhimathi, R.; Kiran, G. S.; Hema, T. A.; Selvin, J.; Raviji, T. R.; Shanmughapriya, S. Production and characterization of lipopeptide biosurfactant by a sponge-associated marine actinomycetes *Nocardiopsis alba* MSA10. *Bioprocess Biosyst. Eng.* **2009**, *32*(6), 825–835.

Garg, N.; Manchanda, G. Role of arbuscular mycorrhizae in the alleviation of ionic, osmotic and oxidative stresses induced by salinity in *Cajanus cajan* (L.) Millsp. (pigeonpea). *J. Agron. Crop Sci.* **2009**, *195*(2), 110–123.

George, P.; Gupta, A.; Gopal, M.; Thomas, L.; Thomas, G. V. Multifarious beneficial traits and plant growth promoting potential of *Serratia marcescens* KiSII and *Enterobacter* sp. RNF 267 isolated from the rhizosphere of coconut palms (*Cocos nucifera* L.). *World J. Microbiol. Biotechnol.* **2013**, *29*(1), 109–117.

Gong, M.; Tang, M.; Chen, H.; Zhang, Q.; Feng, X. Effects of two *Glomus* species on the growth and physiological performance of *Sophora davidii* seedlings under water stress. *New Forests* **2013**, *44*(3), 399–408.

Govindarajulu, M.; Pfeffer, P. E.; Jin, H.; Abubaker, J.; Douds, D. D.; Allen, J. W.; Bucking, H.; Lammers, P.; Shachar-Hill, Y. Nitrogen transfer in the arbuscular mycorrhizal symbiosis. *Nature* **2005**, *435*(7043), 819–823.

Hammer, E. C.; Nasr, H.; Pallon, J.; Olsson, P. A.; Wallander, H. Elemental composition of arbuscular mycorrhizal fungi at high salinity. *Mycorrhiza* **2005**, *21*(2), 117–129.

Hani, N. B.; Al-Ramamneh, E. A. D.; Haddad, M.; Al-Tawaha, A. R. and Al-Satari, Y. The impact of cattle manure on the content of major minerals and nitrogen uptake from ^{15}N isotope-labeled ammonium sulphate fertilizer in maize (*Zea mays* L.) plants. *Pak. J. Bot.* **2019**, *51*(1), 185–189.

Haque, M. S.; Kjaer, K. H.; Rosenqvist, E.; Sharma, D. K.; Ottosen, C. O. Heat stress and recovery of photosystem II efficiency in wheat (*Triticum aestivum* L.) cultivars acclimated to different growth temperatures. *Environ. Exp. Bot.* **2014**, *99*, 1–8.

Hashem, A.; Alqarawi, A. A.; Radhakrishnan, R.; Al-Arjani, A. B. F.; Aldehaish, H. A.; Egamberdieva, D.; Abdallah, E. F. Arbuscular mycorrhizal fungi regulate the oxidative system, hormones and ionic equilibrium to trigger salt stress tolerance in *Cucumis sativus* L. *Saudi J. Biol. Sci.* **2018**, *25*(6), 1102–1114.

Hettwer, U.; Gross, M.; Rudolph, K. Purification and characterization of an extracellular levansucrase from *Pseudomonas syringae* pv. phaseolicola. *J. Bacteriol.* **1995**, *177*(10), 2834–2839.

Heydarian, Z.; Yu, M.; Gruber, M.; Glick, B. R.; Zhou, R.; Hegedus, D. D. Inoculation of soil with plant growth promoting bacteria producing 1-aminocyclopropane-1-carboxylate deaminase or expression of the corresponding acdS gene in transgenic plants increases salinity tolerance in *Camelina sativa*. *Front. Microbiol.* **2016**, *7*, 1966.

Hildebrandt, U.; Janetta, K.; Bothe, H. Towards growth of arbuscular mycorrhizal fungi independent of a plant host. *Appl. Environ. Microbiol.* **2002**, *68*(4), 1919–1924.

Hussain, M.; Ahmad, S.; Hussain, S.; Lal, R.; Ul-Allah, S.; Nawaz, A. Rice in saline soils: physiology, biochemistry, genetics, and management. *Adv. Agron.* **2018**, *148*, 231–287.

Hussain, Z.; Khattak, R. A.; Irshad, M.; Eneji, A. E. Ameliorative effect of potassium sulphate on the growth and chemical composition of wheat (*Triticum aestivum* L.) in salt-affected soils. *J. Soil Sci. Plant Nutr.* **2013**, *13*(2), 401–415.

Igiehon, N. O.; Babalola, O. O. Biofertilizers and sustainable agriculture: exploring arbuscular mycorrhizal fungi. *Appl. Microbiol. Biotechnol.* **2017**, *101*(12), 4871–4881.

Islam, F.; Ali, B.; Wang, J.; Farooq, M. A.; Gill, R. A.; Ali, S.; Wang, D.; Zhou, W. Combined herbicide and saline stress differentially modulates hormonal regulation and antioxidant defensin system in *Oryza sativa* cultivars. *Plant Physiol. Biochem.* **2016**, *107*, 82–95.

Islam, M. R.; Sultana, T.; Joe, M. M.; Yim, W.; Cho, J. C.; Sa, T. Nitrogen-fixing bacteria with multiple plant growth-promoting activities enhance growth of tomato and red pepper. *J. Basic Microbiol.* **2013**, *53*(12), 1004–1015.

Janicka-Russak, M.; Kabała, K.; Wdowikowska, A.; Kłobus, G. Response of plasma membrane H^+-ATPase to low temperature in cucumber roots. *J. Plant Res.* **2012**, *125*(2), 291–300.

Jansa, J.; Mozafar, A.; Anken, T.; Ruh, R.; Sanders, I.; Frossard, E. Diversity and structure of AMF communities as affected by tillage in a temperate soil. *Mycorrhiza* **2002**, *12*(5), 225–234.

Jung, S. C.; Martinez-Medina, A.; Lopez-Raez, J. A.; Pozo, M. J. Mycorrhiza-induced resistance and priming of plant defenses. *J. Chem. Ecol.* **2012**, *38*(6), 651–664.

Kahil, M. T.; Dinar, A.; Albiac, J. Modeling water scarcity and droughts for policy adaptation to climate change in arid and semiarid regions. *J. Hydrol.* **2015**, *522*, 95–109.

Kapoor, R.; Gupta, M. K.; Kumar, N.; Kanwar, S. S. Analysis of nhaA gene from salt tolerant and plant growth promoting *Enterobacter ludwigii*. *Rhizosphere* **2017**, *4*, 62–69.

Khan, N. A.; Ajmal, M.; Haq, M. I. U.; Javed, N.; Ali, M. A.; Binyamin, R.; Khan, S. A. Impact of sawdust using various woods for effective cultivation of oyster mushroom. *Pak. J. Bot.* **2012**, *44* (1), 399–402.

Kloppholz, S.; Kuhn, H.; Requena, N. A secreted fungal effector of *Glomus intraradices* promotes symbiotic biotrophy. *Curr. Biol.* **2011**, *21* (14), 1204–1209.

Latef, A. A. H. A.; Chaoxing, H. Effect of arbuscular mycorrhizal fungi on growth, mineral nutrition, antioxidant enzymes activity and fruit yield of tomato grown under salinity stress. *Sci. Hort.* **2011**, *127* (3), 228–233.

Liu, W.; Hou, J.; Wang, Q.; Ding, L.; Luo, Y. Isolation and characterization of plant growth-promoting rhizobacteria and their effects on phytoremediation of petroleum-contaminated saline-alkali soil. *Chemosphere* **2014**, *117*, 303–308.

Liu, X.; Guan, H.; Wang, T.; Meng, D.; Yang, Y.; Dai, J.; Fan, N.; Guo, B.; Fu, Y.; He, W.; Wei, Y. 2020. ScPNP-A, a plant natriuretic peptide from *Stellera chamaejasme*, confers multiple stress tolerances in Arabidopsis. *Plant Physiol. Biochem.* **2020**, *149*, 132–143.

Lu, W.; Guo, C.; Li, X.; Duan, W.; Ma, C.; Zhao, M.; Gu, J.; Du, X.; Liu, Z.; Xiao, K. Overexpression of TaNHX3, a vacuolar Na^+/H^+ antiporter gene in wheat, enhances salt stress tolerance in tobacco by improving related physiological processes. *Plant Physiol. Biochem.* **2014**, *76*, 17–28.

Maheshwari, D. K. *Bacteria in agrobiology: Plant probiotics*; Springer Science + Business Media, 2012.

Marti, M. E.; Colonna, W. J.; Reznik, G.; Pynn, M.; Jarrell, K.; Lamsal, B.; Glatz, C. E. Production of fatty-acyl-glutamate biosurfactant by *Bacillus subtilis* on soybean co-products. *Biochem. Eng. J.* **2015**, *95*, 48–55.

Mathur, S.; Sharma, M. P.; Jajoo, A. Improved photosynthetic efficacy of maize (*Zea mays*) plants with arbuscular mycorrhizal fungi (AMF) under high temperature stress. *J. Photochem. Photobiol. B* **2018**, *180*, 149–154.

Mensah, J. A.; Koch, A. M.; Antunes, P. M.; Kiers, E. T.; Hart, M.; Bücking, H. High functional diversity within species of arbuscular mycorrhizal fungi is associated with differences in phosphate and nitrogen uptake and fungal phosphate metabolism. *Mycorrhiza* **2015**, *25*(7), 533–546.

Metwally, R. A.; Abdelhameed, R. E. Synergistic effect of arbuscular mycorrhizal fungi on growth and physiology of salt-stressed *Trigonella foenum-graecum* plants. *Biocatal. Agric. Biotechnol.* **2018**, *16*, 538–544.

Miransari, M. Contribution of arbuscular mycorrhizal symbiosis to plant growth under different types of soil stress. *Plant Biol.* **2010**, *12*(4), 563–569.

Miransari, M. Microbial products and soil stresses. In *Bacteria in agrobiology: Stress management*; Springer: Berlin, Heidelberg, 2012; pp 65–75.

Mo, Y.; Wang, Y.; Yang, R.; Zheng, J.; Liu, C.; Li, H.; Ma, J.; Zhang, Y.; Wei, Ch.; Zhang, X. Regulation of plant growth, photosynthesis, antioxidation and osmosis by an arbuscular mycorrhizal fungus in watermelon seedlings under well-watered and drought conditions. *Front. Plant Sci.* **2016**, 7, 644.

Mondal, N. K.; Das, C.; Datta, J. K. Effect of mercury on seedling growth, nodulation and ultrastructural deformation of *Vigna radiata* (L.) Wilczek. *Environ. Monitor. Assess.* **2015**, *187*(5), 241.

Moreno, F. T.; Miranda, C. L.; Moreno, M. G.; Royo, E. R. Bayesian model for optimization adaptive e-learning process. *Int. J. Emerg. Technol. Learn.* **2008**, *3*(2), 38–52.

Nadeem, S. M.; Ahmad, M.; Naveed, M.; Imran, M.; Zahir, Z. A.; Crowley, D. E. Relationship between in vitro characterization and comparative efficacy of plant growth-promoting rhizobacteria for improving cucumber salt tolerance. *Arch. Microbiol.* **2016**, *198*(4), 379–387.

Nikus, O.; Al-Tawaha, A. M.; Turk, M. A. (2004a). Effect of manure supplemented with phosphate fertilizer on the fodder yield and quality of two sorghum cultivars (*Sorghum bicolor* L.). *Biosci. Res.* **2004a**, *1*, 1–7.

Nikus, O.; Al-Tawaha, A. M.; Turk, M. A. (2004b). Yield response of Sorghum (*Sorghum bicolor* L.) to manufacture supplemented with phosphate fertilizer under semi-arid Mediterranean conditions. *Int. J. Agric. Biol.* **2004b**, *6*, 889–893.

Oehl, F.; Sieverding, E.; Palenzuela, J.; Ineichen, K.; da Silva, G. A. Advances in Glomeromycota taxonomy and classification. *IMA Fungus* **2011**, *2*(2), 191–199.

Pal, S.; Singh, H. B.; Farooqui, A.; Rakshit, A. Commercialization of arbuscular mycorrhizal technology in agriculture and forestry. In: *Agriculturally Important Microorganisms*; Singh, H., Sarma, B., Keswani, C., Eds.; Springer: Singapore, 2016; pp 97–105. https://doi.org/10.1007/978–981–10–2576–1_6

Panwar, M.; Tewari, R.; Nayyar, H. Native halo-tolerant plant growth promoting rhizobacteria *Enterococcus* and *Pantoea* sp. improve seed yield of Mungbean (*Vigna radiata* L.) under soil salinity by reducing sodium uptake and stress injury. *Physiol. Mol. Biol. Plants* **2016**, *22*(4), 445–459.

Parniske, M. Arbuscular mycorrhiza: the mother of plant root endosymbioses. *Nat. Rev. Microbiol.* **2008**, *6*(10), 763–775.

Paul, D.; Nair, S. Stress adaptations in a plant growth promoting rhizobacterium (PGPR) with increasing salinity in the coastal agricultural soils. *J. Basic Microbiol.* **2008**, *48*(5), 378–384.

Pawlowska, T. E.; Taylor, J. W. Organization of genetic variation in individuals of arbuscular mycorrhizal fungi. *Nature* **2004**, *427*(6976), 733–737.

Pfeffer, P. E.; Douds, D. D.; Bécard, G.; Shachar-Hill, Y. Carbon uptake and the metabolism and transport of lipids in an arbuscular mycorrhiza. *Plant Physiol.* **1999**, *120*(2), 587–598.

Pierce, R. C.; Quick, E. A.; Reeder, D. C.; Morgan, Z. R.; Kalivas, P. W. Calcium-mediated second messengers modulate the expression of behavioral sensitization to cocaine. *J. Pharmacol. Exp. Therap.* **1998**, *286*(3), 1171–1176.

Pieterse, C. M.; Zamioudis, C.; Berendsen, R. L.; Weller, D. M.; Van Wees, S. C.; Bakker, P. A. Induced systemic resistance by beneficial microbes. *Annu. Rev. Phytopathol.* **2014**, *52*, 347–375.

Porcel, R.; Aroca, R.; Ruiz-Lozano, J. M. Salinity stress alleviation using arbuscular mycorrhizal fungi. A review. *Agron. Sustain. Dev.* **2012**, *32*(1), 181–200.

Porcel, R.; Aroca, R.; Azcon, R.; Ruiz-Lozano, J. M. Regulation of cation transporter genes by the arbuscular mycorrhizal symbiosis in rice plants subjected to salinity suggests improved salt tolerance due to reduced Na^+ root-to-shoot distribution. *Mycorrhiza* **2016**, *26*(7), 673–684.

Qiu, Y. W.; Zhe, F.; Fu, Y.; Luo, C. C.; Yu, Y. B.; Yan-Zhong, F.; Qi, W.; Li, F. L. GsMAPK4, a positive regulator of soybean tolerance to salinity stress. *J. Integr. Agric.* **2019**, *18*(2), 372–380.

Quilambo, O. A. The vesicular-arbuscular mycorrhizal symbiosis. *Afr. J. Biotechnol.* **2003**, *2*(12), 539–546.

Reddy, K. S.; Khan, M. Y.; Archana, K.; Reddy, M. G.; Hameeda, B. Utilization of mango kernel oil for the rhamnolipid production by *Pseudomonas aeruginosa* DR1 towards its application as biocontrol agent. *Bioresour. Technol.* **2016**, 221, 291–299.

Rillig, M. C.; Mummey, D. L. Mycorrhizas and soil structure. *New Phytol.* **2006**, *171*(1), 41–53.

Sáenz-Mata, J.; Palacio-Rodríguez, R.; Sánchez-Galván, H.; Balagurusamy, N. Plant growth promoting rhizobacteria associated to halophytes: Potential applications in agriculture. In: *Sabkha ecosystems*; Khan, M. A., Ed.; Springer International Publishing, Switzerland; 2016; pp 411–425.

Satyanarayana, T.; Johri, B. N.; Prakash, A. *Microorganisms in sustainable agriculture and biotechnology*; Springer Science + Business Media, 2012.

Selosse, M. A.; Richard, F.; He, X.; Simard, S. W. Mycorrhizal networks. *Trends Ecol. Evol.* **2006**, *21* (11), 621–628.

Sengupta, A.; Chaudhuri, S. Arbuscular mycorrhizal relations of mangrove plant community at the Ganges river estuary in India. *Mycorrhiza* **2002**, *12* (4), 169–174.

Sethy, S. K.; Ghosh, S. Effect of heavy metals on germination of seeds. *J. Natural Sci. Biol. Med.* **2013**, *4* (2), 272.

Shabala, L.; Mackay, A.; Tian, Y.; Jacobsen, S. E.; Zhou, D.; Shabala, S. Oxidative stress protection and stomatal patterning as components of salinity tolerance mechanism in quinoa (*Chenopodium quinoa*). *Physiologia Plantarum* **2012**, *146* (1), 26–38.

Sharma, S.; Kulkarni, J.; Jha, B. Halotolerant rhizobacteria promote growth and enhance salinity tolerance in peanut. *Front. Microbiol.* **2016**, 7, 1600.

Sharma, S.; Singh, P.; Raj, M.; Chadha, B. S.; Saini, H. S. Aqueous phase partitioning of hexachlorocyclohexane (HCH) isomers by biosurfactant produced by *Pseudomonas aeruginosa* WH-2. *J. Hazard. Mater.* **2009**, *171*(1–3), 1178–1182.

Sheng, M.; Tang, M.; Chen, H.; Yang, B.; Zhang, F.; Huang, Y. Influence of arbuscular mycorrhizae on photosynthesis and water status of maize plants under salt stress. *Mycorrhiza* **2008**, *18*(6–7), 287–296.

Shrivastava, P.; Kumar, R. Soil salinity: A serious environmental issue and plant growth promoting bacteria as one of the tools for its alleviation. *Saudi J. Biol. Sci.* **2015**, *22*(2), 123–131.

Shukla, N.; Awasthi, R. P.; Rawat, L.; Kumar, J. Biochemical and physiological responses of rice (*Oryza sativa* L.) as influenced by *Trichoderma harzianum* under drought stress. *Plant Physiol. Biochem.* **2012**, *54*, 78–88.

Siddiqui, Z. A.; Futai, K. *Mycorrhizae: Sustainable agriculture and forestry*; Springer: New Delhi, 2011; p 359.

Simpson, D. R.; Natraj, N. R.; McInerney, M. J.; Duncan, K. E. Biosurfactant-producing *Bacillus* are present in produced brines from Oklahoma oil reservoirs with a wide range of salinities. *Appl. Microbiol. Biotechnol.* **2011**, *91*(4), 1083.

Singh, R. P.; Jha, P. N. The multifarious PGPR *Serratia marcescens* CDP-13 augments induced systemic resistance and enhanced salinity tolerance of wheat (*Triticum aestivum* L.). *PLoS One* **2016**, 11(6).

Smith, S. E.; Read, D. J. The symbionts forming arbuscular mycorrhizas. *Mycorrhizal Symbiosis* **2008**, *2*, 13–41.

Spatafora, J. W.; Chang, Y.; Benny, G. L.; Lazarus, K.; Smith, M. E.; Berbee, M. L.; Bonito, G.; Corradi, N.; Grigoriev, I.; Gryganskyi, A.; James, T. Y.; O'Donnell, K.; Roberson, R. W. Taylor, T. N.; Uehling, J.; Vilgalys, R.; White, M. M.; Stajich, J. E. A phylum-level phylogenetic classification of zygomycete fungi based on genome-scale data. *Mycologia* **2016**, *108*(5), 1028–1046.

Spinosa, R.; Stamets, P.; Running, M. Fungi and sustainability. *Fungi* **2008,** *1* (1), 38–43.

Srivastava, R.; Roseti, D.; Sharma, A. K. The evaluation of microbial diversity in a vegetable based cropping system under organic farming practices. *Appl. Soil Ecol.* **2007,** *36* (2–3), 116–123.

Tawaha, A. M.; Turk, M. A. Lentil (*Lens culinaris* Medic.) productivity as influenced by rate and method of phosphate placement in a Mediterranean environment. *Acta Agronomica Hungarica* **2002**, *50*(2), 197–201.

Tawaha, A. M.; Turk, M. A. Field pea seeding management for semi-arid mediterranean conditions. *J. Agron. Crop Sci.* **2004**, *190*, 86–92.

Tawaha, A. M.; Singh, V. P.; Turk, M. A. A review on growth, yield components and yield of barley as influenced by genotypes, herbicides and fertilizer application. *Res. Crop* **2003**, *4*(1), 1–9.

Tawaha, A. R. M.; Turk, M. A.; Lee, K. D.; Zheng, W. J.; Ababneh, M.; Abebe, G.; Musallam, I. W. Impact of fertilizer and herbicide application on performance of ten barley genotypes grown in northeastern part of Jordan. *Int. J. Agric. Biol.* **2005**, *7*(2), 162–166.

Treseder, K. K.; Mack, M. C.; Cross, A. Relationships among fires, fungi, and soil dynamics in Alaskan boreal forests. *Ecol. Appl.* **2016**, *14*(6), 1826–1838.

Turk, M. A.; Tawaha, A. M. Common vetch (*Vicia sativa* L.) productivity as influenced by rate and method of phosphate fertilization in a Mediterranean environment. *Agricultura Mediterranea* **2001**, *131*, 108–111.
Turk, M. A.; Tawaha, A. M. Onion (*Allium cepa* L.) as influenced by rate and method of phosphorus placement. *Crop Res.* **2002a**, *23*(1), 105–107.
Turk, M. A.; Tawaha, A. M. Impact of seeding rate, seeding date and method of phosphorous application in faba bean (*Vicia faba* L. minor) in the absence of moisture stress. *Biotechnol. Agron. Soc. Environ.* **2002b**, *6*(3), 171–178.
Turk, M. A.; Tawaha, A. M. (2003a). Allelopathic effect of black mustard (*Brassica nigra* L.) on germination and growth of wild oat (*Avena fatua* L.). *Crop Protect.* **2003a**, *22*(4), 673.
Turk, M. A.; Tawaha, A. M. (2003b). Allelopathic effect of black mustard (*Brassica nigra* L.) on germination and growth of wild barley (*Hordeum spontaneum*). *J. Agron. Crop Sci.* **2003b**, *189*(5), 298–303.
Turk, M. A.; Tawaha, A. M. Effect of variable sowing ratios and sowing rates of bitter vetch on the herbage yield of oat-bitter vetch mixed cropping. In *Proceedings 7th International Oat Conference*; Peltonen-Sainio, P., Topi-Hulmi, M., Eds.; MTT, 2004.
Turk, M. A.; Hmaeed, K. M.; Aqeel, A. M.; Tawaha, A. M. Nutritional status of durum wheat grown in soil supplemented with olive mill by-products. *Agrochimica* **2003a**, *47*(5–6), 209–219.
Turk, M. A.; Tawaha, A. M.; Shatnawi, M. Lentil (*Lens culinaris* Medik) response to plant density, sowing date, phosporous fertilization and Ethephon application in the absence of moisture stress. *J. Agron. Crop Sci.* **2003b**, *189*(1), 1–6.
Turk, M.A.; Tawaha, A.M.; Samara, N. Effects of seedling rate and date, and phosphorous application on growth and yield of carbon vetch (*Vicia narbonensis*). *Agronomie* **2003c**, *23*, 1–4.
Turk, M. A.; Lee, K. D.; Tawaha, A. M. Inhibitory effects of aqueous extracts of black mustard on germination and growth of radish. *Res. J. Agric. Biol. Sci.* **2005**, *1*(3), 227–231.
Ungureanu, G.; Ignat, G.; Vintu, C. R.; Diaconu, C. D.; Sandu, I. G. Study of utilization of agricultural waste as environmental issue in Romania. *Rev Chim.* **2017**, *68*(3), 570–575.
Upadhyay, S. K.; Singh, G.; Singh, D. Mechanism and understanding of PGPR: An approach for sustainable agriculture under abiotic stresses. *J. Plant Sci.* **2016**, *134*(2), 24–75.
Vaishnav, A.; Kumari, S.; Jain, S.; Varma, A.; Tuteja, N.; Choudhary, D. K. PGPR-mediated expression of salt tolerance gene in soybean through volatiles under sodium nitroprusside. *J. Basic Microbiol.* **2016**, *56*(11), 1274–1288.
Velarde-Buendía, A. M.; Shabala, S.; Cvikrova, M.; Dobrovinskaya, O.; Pottosin, I. Salt-sensitive and salt-tolerant barley varieties differ in the extent of potentiation of the ROS-induced K^+ efflux by polyamines. *Plant Physiol. Biochem.* **2012**, *61*, 18–23.
Verbruggen, E.; Toby Kiers, E. Evolutionary ecology of mycorrhizal functional diversity in agricultural systems. *Evol. App.* **2010**, *3*(5–6), 547–560.
Vergara, C.; Araujo, K. E. C.; de Souza, S. R.; Schultz, N.; Júnior, O. J. S.; Sperandio, M.V. L.; Zilli, J. É. Plant-mycorrhizal fungi interaction and response to inoculation with different growth-promoting fungi. *Pesquisa Agropecuária Brasileira* **2019**, *54*, e25140. https://doi.org/10.1590/S1678-3921.pab2019.v54.25140.
Vimal, S. R.; Singh, J. S.; Arora, N. K.; Singh, S. Soil-plant-microbe interactions in stressed agriculture management: A review. *Pedosphere* **2017**, *27*(2), 177–192.
Wang, Y. H.; Gehring, C.; Irving, H. R. Plant natriuretic peptides are apoplastic and paracrine stress response molecules. *Plant Cell Physiol.* **2011**, *52*(5), 837–850.

Wu, N.; Li, Z.; Wu, F.; Tang, M. Comparative photochemistry activity and antioxidant responses in male and female *Populus cathayana* cuttings inoculated with arbuscular mycorrhizal fungi under salt. *Sci. Rep.* **2016**, *6*(1), 1–15.

Wu, Q. S.; Zou, Y. N.; Liu, W.; Ye, X. F.; Zai, H. F.; Zhao, L. J. Alleviation of salt stress in citrus seedlings inoculated with mycorrhiza: changes in leaf antioxidant defensin systems. *Plant Soil Environ.* **2010**, *56*(10), 470–475.

Yang, L.; Ji, W.; Gao, P.; Li, Y.; Cai, H.; Bai, X.; Chen, Q.; Zhu, Y. GsAPK, an ABA-activated and calcium-independent SnRK2-type kinase from *G. soja*, mediates the regulation of plant tolerance to salinity and ABA stress. *PLoS One* **2012**, *7*(3).

Yang, Y.; Guo, Y. Unraveling salt stress signalling in plants. *J. Integr. Plant Biol.* **2018**, *60*(9), 796–804.

Yarra, R. The wheat NHX gene family: Potential role in improving salinity stress tolerance of plants. *Plant Gene* **2016**, 100178.

Yu, C.; Xu, S.; Yin, Y. Transcriptome analysis of the Taxodium 'Zhongshanshan 405' roots in response to salinity stress. *Plant Physiol. Biochem.* **2016**, *100*, 156–165.

Zhang, S.; Gan, Y.; Xu, B. Application of plant-growth-promoting fungi *Trichoderma longibrachiatum* T6 enhances tolerance of wheat to salt stress through improvement of antioxidative defensin system and gene expression. *Front. Plant Sci.* **2016**, *7*, 1405.

Zhang, X.; Han, C.; Gao, H.; Cao, Y. Comparative transcriptome analysis of the garden asparagus (*Asparagus officinalis* L.) reveals the molecular mechanism for growth with arbuscular mycorrhizal fungi under salinity stress. *Plant Physiol. Biochem.* **2019**, *141*, 20–29.

Zhao, S.; Zhou, N.; Zhao, Z. Y.; Zhang, K.; Wu, G. H.; Tian, C. Y. Isolation of endophytic plant growth-promoting bacteria associated with the halophyte *Salicornia europaea* and evaluation of their promoting activity under salt stress. *Curr. Microbiol.* **2016**, *73*(4), 574–581.

Zhu, X. Q.; Wang, C. Y.; Chen, H.; Tang, M. Effects of arbuscular mycorrhizal fungi on photosynthesis, carbon content, and calorific value of black locust seedlings. *Photosynthetica* **2014**, *52*(2), 247–252.

Zuccaro, A.; Lahrmann, U.; Langen, G. Broad compatibility in fungal root symbioses. *Curr. Opin. Plant Biol.* **2014**, *20*, 135–145.

CHAPTER 10

ARBUSCULAR MYCORRHIZAL FUNGI IN ALLEVIATION OF COLD STRESS IN PLANTS

SAMIA KHANUM[1], ABDEL RAHMAN MOHAMMAD AL-TAWAHA[2], ABDEL RAZZAQ AL-TAWAHA[3], HIBA ALATRASH[4], ABDUR RAUF[5], ARUN KARNWAL[6], ABHIJIT DEY[7], NUJOUD ALIMAD[8], DEVARAJAN THANGADURAI[9], JEYABALAN SANGEETHA[10], SAHER ISLAM[11], IMRAN[12], AMANULLAH[12], SHAH KHALID[12], PALANI SARANRAJ[13], and ELIF GUNAL[14]

[1]*Department of Botany, University of the Punjab, Lahore, Pakistan*

[2]*Department of Biological Sciences, Al Hussein Bin Talal University, Maan, Jordan*

[3]*Department of Crop Science, Faculty of Agriculture, Universitiy Putra Malaysia, Serdang, Selangor, Malaysia*

[4]*General Commission for Scientific Agricultural Research, Syria*

[5]*Department of Chemistry, University of Swabi, Anbar, Khyber Pakhtunkhwa, Pakistan*

[6]*Department of Microbiology, School of Bioengineering and BioSciences, Lovely Professional University, Phagwara, Punjab, India*

[7]*Department of Life Sciences, Presidency University, Kolkata, India*

[8]*Damascus University, Faculty of Agriculture, Damascus, Syria*

[9]*Department of Botany, Karnatak University, Dharwad, Karnataka, India*

Mycorrhizal Technology: Managing Plant Stress and Mitigating Climate Change using Mycorrhizae.
Jeyabalan Sangeetha, Abdel Rahman Mohammad Al-Tawaha, & Devarajan Thangadurai (Eds.)

[10]*Department of Environmental Science, Central University of Kerala, Kasaragod, Kerala, India*

[11]*Institute of Biochemistry and Biotechnology, University of Veterinary and Animal Sciences, Lahore, Pakistan*

[12]*Department of Agronomy, The University of Agriculture, Peshawar, Pakistan*

[13]*Department of Microbiology, Sacred Heart College (Autonomous), Tirupattur, Tamil Nadu, India*

[14]*Department of Soil Science and Plant Nutrition, Faculty of Agriculture, Harran University, Sanliurfa, Turkey*

ABSTRACT

Plants are facing many biotic and abiotic stresses while interacting with their adjacent environment. Sudden environmental changes are the cause of many stresses faced by plants. Some stresses are very dangerous for the proper growth and development of plant, for example, cold stress especially in their flowering stage. The best way to alleviate cold stress, plant adapts in many ways. To increase the efficiency of nutrient uptake, tree plants can make use of novel resources, such as modification of the rhizosphere environment and symbiotic association with mycorrhizal fungi. Mycorrhizal association plays a very important role in helping plants to develop tolerance and can fight against cold stress. Arbuscular mycorrhizal fungi (AMF) involve in many transcriptional modifications and also involve in the regulation of various useful genes which can tolerate the cold stress in plants. Mycorrhiza plays a significant role in developing tolerance in the plant against the low temperature which includes ROS (reactive oxygen species) system, osmotic adjustment, regulating hydraulic conductance in roots, improving respiration and photosynthesis rate and nutrients uptake.

10.1 INTRODUCTION

Plants are sessile organisms and interact with their immediate environment constantly and that's why they have to face the abrupt environmental changes. The common ecological stress that plants experience is the exposing to extreme temperature which is variable in nature. The chilling temperature ranges from 0 to 15°C and the freezing temperature for plants is below 0°C (Jan and Andrabi, 2009), and both of these temperatures contributed in cold stress. The productivity and growth of plants have been limited by cold stress (Xin and Browse, 2001; Sanghera et al., 2011). Cold stress directly affects the plants on cellular levels like proteins, DNA, RNA, and membranes which results into the decrease of plant metabolism, rigidified cell membrane, and disturbed the cell membrane function (Jewell et al., 2010). The plants belonging to temperate areas have tolerance against chilling temperature (Levitt, 1980). The plants from tropical and subtropical regions are affected by chilling temperatures because they don't have any mechanism for cold acclimatization. Because developing resistance against the chilling temperatures, plant required many complex mechanisms and metabolic paths in cell organelles (Hannah et al., 2005).

Conventional breeding is not very effective in order to overcome the cold tolerance in crop plants because of the absence of selection criteria, complex traits of cold resistance, and also with little genetic variations (Sanghera et al., 2011). Novel approaches for the development of transgenic plants with the improved ability of cold resistance have been described (Wani et al., 2008; Byun et al., 2015; Wang et al., 2016). It is appreciated and the need of the hour to build new strategies for developing cold tolerance in plants. Arbuscular mycorrhizal association helps the plant to develop resistance against many stresses (Evelin et al., 2009; Smith and Read, 2010; Miransari, 2011; Sharma et al., 2017). Around 90% of terrestrial plants are in this symbiotic association (Smith and Read, 2008; Zhu et al., 2010a). The fungal part of mutualistic association depends on the host plants for carbon sources and in contrast plants get benefit with improved nutrient levels (Sawers et al., 2008; Smith and Read, 2008).

Fungi formed intracellular structures known as arbuscules which play the role for interchange of nutrients and carbon among partners. The Arbuscular Mycorrhizal Fungi (AMF) hyphae play a key role in the absorption of minerals and water (Candido et al., 2015). AM association helps in the alleviation of the effects of abiotic and biotic stresses. The cold stress impacts the plant in two major phases. At the first stage, stress signals were received by plants and as the result, membrane rigidification occurred (Beck et al., 2004) and in the second stage, many variations occurred in the morphology and metabolism of the cell (Xiong et al., 2002). The mycorrhizal association controls the cold stress by manipulating the second phase by changing many physiological and gene expression activities (Winfield et al., 2010; Wu and Zou, 2010).

This chapter discusses how AMF alleviate the cold stress in plants. This is followed by what are the major cellular responses induced by cold stress, rigidification of membrane, disturbance in the relationship of plant and water because of reduced hydraulic conductance of root, dehydration at cellular level because of disturbance in osmotic potential, respiration, and photosynthesis also effected by low temperature and also some cold-inducible genes activated ultimately affect the plant efficiency overall.

10.2 COLD STRESS EFFECT ON AMF

The symbiotic association between plants and arbuscular fungi depends on the soil environmental condition, nutrients, light, and cropping system (Liu et al., 2004). The main factor influences the function of the mutualistic association is the temperature (Gavito et al., 2005). The extraradical mycelium of the fungi more exposed to the rhizosphere and that's why more affected by soil parameters just like temperature (Smith and Read, 1997). The establishment of AMF can be affected by temperature directly or indirectly. The extraradical mycelium mostly affected by temperature because of its direct contact with the soil. It is studied that the effect of temperature at 10 and 15°C both on extra- and intraradical mycelium in *Pisum sativum* (Gavito et al., 2002). It is noted that the intraradical colonization happened at both temperatures but the growth of extraradical mycelium was only observed at 15°C so it's proved that low temperature inhibited the extraradical mycelium. The average development of mycorrhiza was reported at 20°C and 25°C when growing in temperate regions (Matsubara et al., 2000) and the growth decreased at temperature below 15°C and at 10°C, the growth of AMF was completely inhibited (Liu et al., 2004). The relationship between the increase in temperature and growth of AMF has been reported through various researches. The increase in temperature improved the mycorrhizal association, but it is also reported that low temperature has no influence on the AMF association and this is most probably because of low exposure of low temperature. The reason behind the reduction in AMF colonization at low temperature is the decrease in carbohydrates supply (Gavito et al., 2005) and as it is already discussed that this mutualistic association is based on the two-way exchange of nutrients among two partners so when the plants are exposed to cold temperature it will affect their growth and development because of that they are unable to supply carbohydrate to their fungal partner which leads to inhibit the symbiotic relationship efficiency (Ma et al., 2015) (Table 10.1).

TABLE 10.1 Contribution of AMF in Helping Plants to Cope with Extreme Temperatures Stress.

Host plants	AMF strains	Responses related to AMF inoculation	References
Cucumis sativus (L.)	*R. irregularis*	Increases the photosynthetic efficiency of cold-stressed cucumber seedlings by protecting their photosynthetic apparatus against light-induced damage and increasing their carbon sink	Ma et al. (2015, 2018, 2019)
Zea mays (L.)	*Funneliformis* (*Glomus*) species	Regulated photosystem (PS) II heterogeneity	Mathur and Jajoo (2020)
Hordeum vulgare (L.)	*G. versiforme* *R. irregularis*	Increasing the survival rate, alleviation of low-temperature stress	Hajiboland et al. (2019)
Cyclamen persicum (Mill.)	*R. fasciculatum*	Enhanced biomass production and heat stress response; increased activity of antioxidative enzymes such as superoxide dismutase and ascorbate peroxidase	Maya and Matsubara (2013)
Elymus nutans (Griseb.)	*F. mosseae*	Less oxidative damage; promoted plant growth and enhanced the level of chlorophyll and antioxidant compounds such as glutathione and soluble sugars	Chu et al. (2016)

Source: Adapted from Diagne et al. (2020).

10.3 AM AND COLD TOLERANCE IN PLANTS

Using mycorrhiza against cold stress for plants is reported very effective in different plant species. Mycorrhiza plays a significant role in developing tolerance in the plant against the low temperature which includes ROS (reactive oxygen species) system, osmotic adjustment, regulating hydraulic conductance in roots, improving respiration and photosynthesis rate and nutrient's uptake (Ma et al., 2018). The mycorrhizal association removed the osmotic stress in plants because of low temperature by adjusting osmotic pressure by production of sugar, controlling proline, and other osmolytes. The plant–water connection was sustained by the regulation of hydraulic conductance and the function of aquaporin. Low-temperature induced oxidative stress also controlled by AMF with the production of oxidant compounds. At the molecular level, AM produced some functional proteins like aquaporin, sugars, and iron transporters. All of the mechanisms works together to help plants to develop tolerance against cold stress.

10.4 EFFECTIVE ROS SCAVENGE SYSTEM

Low temperature-mediated oxidative stress is the reason of cold injury in the plants (Liu et al., 2016). Oxidative damage induced by cold temperature favored by ROS like superoxide, hydroxyl ion, and hydrogen peroxide. When the condition is under control and without stress, ROS production and quenching are in equilibrium but this equilibrium is disturbed under cold stress and the production of ROS is increased (Bowler et al., 1992). Plants also have many mechanisms to lessen these damages (Lázaro et al., 2013). These antioxidant systems in plants comprise of three types of antioxidants: (1) enzymatic, (2) water-soluble reductases, and (3) lipid-soluble antioxidants like sugars.

It is reported in recent studies that AM improves the antioxidant systems and efficiently alleviates the cold-mediated ROS. AMF also have effect on enzymatic activity of antioxidants. AM improved antioxidant systems in plants by improving SOD. Plants colonized with arbuscular mycorrhiza have high rates of SOD activities in comparison with plants which are not in symbiotic association (Yang et al., 2015; Chu et al., 2016). The cold tolerance mediated by AM also neutralizes H_2O_2 efficiently by the regulation of recycling of ascorbate and glutathione (Liu et al., 2016). In this mechanism, APX eliminates the H_2O_2 by using AsA as a donor of the electron and oxidation of AsA into dehydroascorbate. The oxidized ascorbates produced by the APX activity are decreased by MDHAR and GHAR, respectively (Ordoñez et al., 2014). The dehydroascorbate was reduced by DHAR which involves the oxidation of GSH into ascorbate and glutathione disulfide leads to the regulation of AsA into its reduced cellular form (its oxidized form), thus shaping the responsiveness and environmental ROS tolerance in the cell (Chen and Gallie, 2006). When the ROS increased it caused the depletion of the production site of AsA so the activity of DHAR increased meant to recover AsA regeneration from DHA *via* feedback regulated mechanism (Locato et al., 2009). The reduced level of glutathione maintained by another GR enzyme activity which regulates the ascorbate and glutathione cycle

(Rao and Reddy, 2008). The normal redox status of the cell was maintained by glutathione and thus reduced the ROS-mediated oxidative loss to plants (Noctor et al., 2012). The plants colonized with arbuscular mycorrhizal fungi have improved ascorbate and glutathione redox ration when grows under cold stress because they show maximum expression of these genes.

10.5 OSMOTIC ADJUSTMENT

Low temperature damage the cell membrane by saturating the phospholipids present in the tails of fatty acids and hence rigid the cell membrane. It decreases the fluidity of the cell membrane leads to a reduction in permeability and influence the transport of biomolecules in the cells and disturbed the osmotic potential ultimately affects the function of the cell (Chen et al., 2013). To tolerate the cold stress plants adjusted their osmotic potential. Plants having tolerance against cold can adjust their osmotic potential by producing compatible osmolytes (organic molecules with low molecular weight) including soluble sugar (Chu et al., 2016). Production of these compatible osmolytes helps in reducing the osmotic potential of cells and thus maintains turgor pressure. It is also reported that AM also helps in the adjustment of osmotic potential and also helps in alleviating the cold stress (Chu et al., 2016). These soluble sugars act as osmo and cryoprotectants in order to protect the cell. Soluble sugars replaced the water molecules due to the formation of hydrogen bonds with the lipid molecules of cell membrane to protect the plant from physiological dehydration induced by chilling. The soluble sugars also contribute in stress responses mediated by hormones by acting as signaling molecules (Theocharis et al., 2012). The high amount of these sugar molecules has been produced in mycorrhizal plants (Chu et al., 2016). Mycorrhizal plants produced soluble sugars by hydrolyzing the starch molecules present in the leaf to cope up with cold stress (Zhu et al., 2015).

Proline reaction to abiotic stresses and reported as osmo and cryoprotectant as well. The relation between increased production of proline and increase tolerance against chilling temperature has been reported in plants which are sensitive to chilling, but these plants are not able to tolerate chilling stress until the high concentration of proline was not applied before the stress hence its proved that the proline has the ability to withstand the chilling stress (Kaur et al., 2011). Proline acts as the scavenger for hydroxyl radical, protein stabilizer, regulate the reduction, cytoplasmic acidosis ameliorator (Theocharis et al., 2012). In mycorrhizal plants, they stored the high concentration of proline in their roots and leaves than in non-mycorrhizal (Chen et al., 2014). It is also found that in mycorrhizal plants, the amount of proline is less in leaves than in non-mycorrhizal and thus it is proved that chilling stress has less effect on mycorrhizal than in non-mycorrhizal plants, that's why they have no need to produce the extra amount of proline in their leaves. Cold stress results in the formation of ice in the apoplast and causes dehydration of the tissues. The gap in the uptake and loss of water also results in dehydration. The acclimatization capability of plants to cope up the cold stress is depending upon the availability and uptake of water. The water uptake capacity of roots regulated by the hydraulic conductance with the assistance of the aquaporin (Luu and Maurel, 2005). The aquaporin is acting as the channel for water and other solutes across the membrane in plants, animals, and bacteria (Takata et al., 2004). The plants in AM association can uptake water at the higher rate under water-deficient conditions. There are very few researches explained that how mycorrhizal association affects the hydraulic properties of roots when subjected to cold stress. However, it is reported that the conductance capacity decreases in non-mycorrhizal plants but unchanged in mycorrhizal plants.

10.6 EFFECT ON PHOTOSYNTHESIS RATE

Low temperature also affects the photosynthesis rate of plants (Paredes and Quiles, 2015), and so for tolerating chilling stress the efficient photosynthesis rate is very important. Photosynthesis gives energy in the form of photosynthates which help in initiating the cold acclimation responses. Chlorophyll is an important molecule in photosynthesis and also adversely affected by the cold stress as low temperature alter the biosynthesis of chlorophyll. It is reported that when exposed to low temperature, the chlorophyll content reduced upto 66% (Liu et al., 2013b). The gene expression profiling and *de novo* transcriptomics studies show that the expression of cold, tolerant and cold sensitive genes which help in the biosynthesis of chlorophyll *a* significantly reduced (Fu et al., 2016). The decline in the level of chlorophyll leads to the variations in antenna complex and then alters the function of photosystem II and ultimately affects photosynthesis. It is reported that the AMF plays an important role to decrease the damage caused by chilling temperature to chlorophyll content. Chilling temperature reduced the photosynthetic pigments and if plants colonized with AMF the concentration of these pigments improved under chilling stress. The AMF improved photosynthesis by fortifying the carbon sink as AMF improved the chlorophyll level and gaseous exchange through leaves so the increase in photosynthesis rate by AMF involves both stomatal and non-stomatal factors (Porcel et al., 2015). The photosynthetic rate controlled by two key mechanisms Rubisco carboxylation and RuBP regeneration and both of these mechanisms improved under mycorrhizal association. When plants are inoculated with AMF, it increased the photosynthesis rate, stomatal conductance, the intercellular concentration of CO_2, boosts Rubisco carboxylation, and RuBP regeneration and also improves the efficiency of photosystem II and thus improves the photosynthesis rate in plants (Yang et al., 2015).

10.7 EFFECT ON RESPIRATION RATE

Cold stress decreases the transport of leaf carbohydrates and sinks activity which changes the carbon assimilation level and ultimately affects the respiration rate in plants (Bagnall et al., 1988). Chilling temperature decreases the ATP synthesis as well as the respiration rate, which is considering as cold-induced changes in respiration (Karasawa et al., 2012). There are two pathways, cytochrome C oxidase and alternative oxidase, of the electron transport chain in respiratory systems present in mitochondria of the cell. When the conditions are normal the main pathway is COX because it is directly involved in ATP production. AOX pathway activated when plants exposed to the stresses like drought, chilling, heat, and salinity (Li et al., 2013) which reduced the ATP production per consumption of oxygen. The high concentration of pyruvate reported when the AOX activity increased. The AMF improved the respiration rate in plants grow under cold stress by increasing the substrate level like sugars and starch and also by optimizing the pyruvate. The AMF activates the COX pathway instead of using AOX pathway when exposed to cold stress and thus increase the production of ATP. It is also reported the increased rate of pyruvate because of the AOX pathway. The AMF increases the expression of genes which are used in the metabolism of pyruvate (Liu et al., 2015). To maintain the level of pyruvate the pyruvate dehydrogenase degraded the pyruvate to maintain the COX pathway activated because of the high expression level of the COX5c gene in mycorrhizal plants. For the efficient TCA cycle and respiration, the overexpression of malate dehydrogenase and isocitrate dehydrogenase genes reported in plants having mycorrhizal association. It concluded that mycorrhizal association improves respiration rate in plants and thus tolerate cold stress. In some plants, the respiration rate decreases in plants and it mainly because of the specificity of interaction between two hosts, colonization stage, and other factors. In the early stages of colonization, AMF decreases the ATP in plants but if it is positive then in mature stages the ATP production increased, and thus respiration also improved and helps plants to tolerate the cold stress (Evelin et al., 2009; Smith et al., 2009).

10.8 IMPROVED NUTRIENTS UPTAKE

Temperature at the root zone directly affects the nutrient's uptake. At low temperature, important macro and micronutrient uptake have been reduced in plants (Tindall et al., 1990). Water and nutrient uptake both reduced by low temperature. Low temperature affects the nutrient uptake by influencing the capacity of ion absorption of plants and also alters the physiology of plant roots. AMF affects the mineral nutrients in plants by helping in the uptake of nutrients from root hairs or hyphae (Smith and Smith, 2011). The fungal hyphae increased the mineral uptake from the soil specifically nitrogen and phosphorous in plants that are in mycorrhizal association (Smith and Read, 2008). Increased uptake of other important nutrients has also been reported. AMF also improved the nutrient uptake and promote growth when plants are subjected to cold stress so that plants can perform better in cold stress. The uptake of phosphorous is controlled by certain different Pi transport systems and it is reported that these Pi transport genes are induced strongly in mycorrhizal plants (Ma et al., 2015, 2018). The high-affinity H^+-Pi cotransporters are belonging to Pht1 gene family and have positive response to the AM association. Other than Pht1 gene family, some other important Pi transporters which involve in internal Pi transport in the plant and are released in subcellular organelles and compartments. These are low affinity H^+-Pi cotransporters and related to Pht2 gene family (Takabatake et al., 1999). When AM plants grow under cold stress, the high expression of genes has been reported. Potassium being important macronutrient plays important role in many physiological processes like photosynthesis, photosynthates translocations, maintaining turgidity, and also activates many enzymes when plants are subjected to stress. The deficiency of potassium leads to the activation of ROS system results in damage to the chlorophyll and membrane. Potassium decreased the production of ROX and increased the antioxidant in plants and helps in alleviating cold stress (Devi et al., 2012). When potassium-mediated cold tolerance increased in plants, it also increased the phospholipids and permeability of the membrane. AMF has the positive effect on the uptake of potassium in plants when grown under cold stress (Liu et al., 2016). Another important nutrient is nitrogen which plays a crucial role in the synthesis of chlorophyll, the uptake of nitrogen also enhanced by mycorrhizal association with plants. The uptake of nitrogen under cold stress is mainly dependent on the host and AMF relationship and specificity, but further studies required to understand host specificity.

10.9 TRANSCRIPTIONAL MODIFICATION

If the AM association is successful, it will modify the transcription levels and improved the overall growth and stress tolerance of plants (Bucher et al., 2014). Because of the beneficial effects of AMF on plants, it gains the attention of many researches to use it against the adverse effects of biotic and abiotic stresses, but the very limited information available about improved transcription level in AM plants. Moreover, no information about the effect of AMF on the alteration of the expression of the gene in plants is available that's why we are unable to understand what molecular mechanisms involved in tolerating cold stress in plants. The first report about transcriptional regulation under chilling stress has been reported by Ma et al. (2018). This report explains that the expression of about 2173 genes has been influenced by cold stress and about 180 genes have been regulated by AMF. *Rhizophagus irregularis* genome consists of 28,232 genes and only 0.007% genes were differentially expressed genes induced by AMF when growing under cold stress thus this very low level gene expression suggests the high level of AMF association required under stress conditions to develop the symbiotic relationship with plants (Tisserant et al., 2012). Studies revealed that these 2173 DEGs are the genes involved in the ROS and ion transport. Moreover, the regulation of mostly genes involved metabolism of glycerol-3-phosphate and oxidation of NADH by AMF-induced DEGs which ultimately improve the redox reaction of plants grown under cold stress. The Kyoto Encyclopedia of Genes and Genomes revealed that DEGs have multiple involvements in different mechanisms like signal transduction of hormones, glycolysis, metabolisms, and biosynthesis of different compounds and molecules. Cold stress also increased the flavonoid biosynthesis which involves in the colonization of AMF (Antunes et al., 2006). So it can be suggested that AMF involve in many transcriptional modifications and also involve in the expression of many important genes which can tolerate the cold stress in plants.

10.10 CONCLUSION

To improve the nutrient uptake efficiency, the tree plants have the ability to utilize the new resources just like the modification of the rhizosphere environment and the symbiotic association with mycorrhizal fungi. AM association helps in the alleviation of the effects of abiotic and biotic stresses. At moderate temperature, AMF also helps in improving photosynthesis, root morphology, nutrient uptake, and growth but the role of mycorrhiza failed at low temperature. AMF involves in many transcriptional modifications and also involves in the expression of very useful genes which can tolerate the cold stress. Using mycorrhiza against cold stress for plants is reported very effective in different plant species. Mycorrhiza plays the significant role in developing tolerance in the plant against the low temperature which includes ROS (reactive oxygen species) system, osmotic adjustment, regulating hydraulic conductance in roots, improving respiration, photosynthesis rate, and nutrient's uptake. The mycorrhizal association removed the osmotic stress in plants because of low temperature by adjusting osmotic pressure by production of sugar, controlling proline, and other osmolytes. The plant–water connection was sustained by the regulation of hydraulic conductance and function of aquaporin. Low temperature-induced oxidative stress also controlled by AMF with the production of oxidant compounds. At the molecular level, AM produced some functional proteins like aquaporin, sugars, and iron transporters. All of the mechanisms were works together to help plants to develop tolerance against cold stress. Major cellular responses such as rigidification of membrane, disturbance in the relationship of plant and water because of reduced hydraulic conductance in root, dehydration at cellular level because of disturbance in osmotic potential, respiration, and photosynthesis are affected in plant due to cold stress alleviated by AMF.

KEYWORDS

- **AMF**
- **cold stress**
- **fungal–plant interaction**
- **hydraulic conductance**
- **osmotic adjustment**
- **plant tolerance**
- **stress tolerance**

REFERENCES

Antunes, P. M.; De Varennes, A.; Rajcan, I.; Goss, M. J. Accumulation of specific flavonoids in soybean (*Glycine max* (L.) Merr.) as a function of the early tripartite symbiosis with arbuscular mycorrhizal fungi and *Bradyrhizobium japonicum* (Kirchner) Jordan. *Soil Biol. Biochem.* **2006**, *38*, 1234–1242.

Bagnall, D. J.; King, R. W.; Farquhar, G. D. Temperature-dependent feedback inhibition of photosynthesis in peanut. *Planta* **1988**, *175*, 348–354.

Bowler, C.; Montagu, M. V.; Inzé, D. Superoxide dismutase and stress tolerance. *Annu. Rev. Plant Biol.* **1992**, *43*, 83–116.

Bucher, M.; Hause, B.; Krajinski, F.; Küster, H. Through the doors of perception to function in arbuscular mycorrhizal symbioses. *New Phytol.* **2014**, *204*, 833–840.

Byun, M. Y.; Lee, J.; Cui, L. H.; Kang, Y.; Oh, T. K.; Park, H.; Lee, H.; Kim, W. T. Constitutive expression of DaCBF7, an Antarctic vascular plant *Deschampsia antarctica* CBF homolog, resulted in improved cold tolerance in transgenic rice plants. *Plant Sci.* **2015**, *236*, 61–74.

Candido, V.; Campanelli, G.; D'Addabbo, T.; Castronuovo, D.; Perniola, M.; Camele, I. Growth and yield promoting effect of artificial mycorrhization on field tomato at different irrigation regimes. *Sci. Hortic.* **2015**, *187*, 35–43.

Chen, S.; Jin, W.; Liu, A.; Zhang, S.; Liu, D.; Wang, F.; Lin, X.; He, C. Arbuscular mycorrhizal fungi (AMF) increase growth and secondary metabolism in cucumber subjected to low temperature stress. *Sci. Hortic.* **2013**, *160*, 222–229.

Chen, X.; Song, F.; Liu, F.; Tian, C.; Liu, S.; Xu, H.; Zhu, X. Effect of different arbuscular mycorrhizal fungi on growth and physiology of maize at ambient and low temperature regimes. *Sci. World. J.* **2014**, 1–7.

Chen, Z, & Gallie, D. R. Dehydroascorbate reductase affects leaf growth, development, and function. *Plant Physiol.* **2006**, *142*, 775–787.

Chu, X. T.; Fu, J. J.; Sun, Y. F.; Xu, Y. M.; Miao, Y. J.; Xu, Y. F.; Hu, T. M. Effect of arbuscular mycorrhizal fungi inoculation on cold stress-induced oxidative damage in leaves of *Elymus nutans* Griseb. *S. Afr. J. Bot.* **2016**, *104*, 21–29.

Chu, X.; Fu, J.; Sun, Y.; Xu, Y.; Miao, Y.; Xu, Y.; Hu, T. Effect of arbuscular mycorrhizal fungi inoculation on cold stress-induced oxidative damage in leaves of *Elymus nutans* Griseb. *S. Afr. J. Bot.* **2016**, *104*, 21–29.

Devi, B. S. R.; Kim, Y. J.; Selvi, S. K.; Gayathri, S.; Altanzul, K.; Parvin, S.; Yang, D. U.; Lee, O. R.; Lee, S.; Yang, D. C. Influence of potassium nitrate on antioxidant level and secondary metabolite genes under cold stress in *Panax ginseng*. *Russ. J. Plant Physiol.* **2012**, *59*, 318–325.

Evelin, H.; Kapoor, R.; Giri, B. Arbuscular mycorrhizal fungi in alleviation of salt stress: A review. *Ann. Bot.* **2009**, *104*, 1263–1280.

Fu, J.; Miao, Y.; Shao, L.; Hu, T.; Yang, P. De novo transcriptome sequencing and gene expression profiling of *Elymus nutans* under cold stress. *BMC Genomics* **2016**, *17*, 870–889.

Gavito, M. E.; Bruhn, D.; Jakobsen, I. Phosphorus uptake by arbuscular mycorrhizal hyphae does not increase when the host plant grows under atmospheric CO_2 enrichment. *New Phytol.* **2002**, *154*, 751–760.

Gavito, M. E.; Olsson, P. A.; Rouhier, H.; Medina-Peñafiel, A.; Jakobsen, I.; Bago, A.; Azcón-Aguilar, C. Temperature constraints on the growth and functioning of root organ cultures with arbuscular mycorrhizal fungi. *New Phytol.* **2005**, *168*, 179–188.

Hajiboland, R.; Joudmand, A.; Aliasgharzad, N.; Tolrá, R.; Poschenrieder, C. Arbuscular mycorrhizal fungi alleviate low-temperature stress and increase freezing resistance as a substitute for acclimation treatment in barley. *Crop. Pasture Sci.* **2019**, *70*, 218–233.

Hannah, M. A.; Heyer, A. G.; Hincha, D. K. A global survey of gene regulation during cold acclimation in *Arabidopsis thaliana. PLoS Genet.* **2005**, *1*, 26–44.

Jan, N.; Andrabi, K. I. Cold resistance in plants: A mystery unresolved. *Electronic Journal of Biotechnology,* **2009**, *12*, 14–15.

Jewell, M. C.; Campbell, B. C.; Godwin, I. D. Transgenic plants for abiotic stress resistance. In: *Transgenic crop plants*; Kole, C., Michler, C. H., Abbott, A. G., Hall, T. C., Eds.; Springer: Berlin, Heidelberg, 2010; pp 67–132.

Karasawa, T.; Hodge, A.; Fitter, A. H. Growth, respiration and nutrient acquisition by the arbuscular mycorrhizal fungus *Glomus mosseae* and its host plant *Plantago lanceolata* in cooled soil. *Plant Cell Environ.* **2012**, *35*, 819–828.

Kaur, G.; Kumar, S.; Thakur, P.; Malik, J. A.; Bhandhari, K.; Sharma, K. D.; Nayyar, H. Involvement of proline in response of chickpea (*Cicer arietinum* L.) to chilling stress at reproductive stage. *Sci. Hort.* **2011**, *128*, 174–181.

Lázaro, J. J.; Jiménez, A.; Camejo, D.; Martí, M. C.; Lázaro-Payo, A.; Barranco-Medina, S.; Sevilla, F. Dissecting the integrative antioxidant and redox systems in plant mitochondria: Effect of stress and S-nitrosylation. *Front. Plant Sci.* **2013**, *460*, 1–20.

Levitt, J. *Responses of plants to environmental stress, chilling, freezing, and high temperature stresses*; Academic Press, 1980; pp 1–497.

Li, C. R.; Liang, D. D.; Li, J.; Duan, Y. B.; Hao, L.; Yang, Y. C.; et al. Unravelling mitochondrial retrograde regulation in the abiotic stress induction of rice alternative oxidase 1 genes. *Plant Cell Environ.* 2013; *36*, 775–788.

Liu, A.; Chen, S.; Wang, M.; Liu, D.; Chang, R.; Wang, Z.; Lin, X.; Bai, B.; Ahammed, G. J. Arbuscular mycorrhizal fungus alleviates chilling stress by boosting redox poise and antioxidant potential of tomato seedlings. *J. Plant Growth Regul.* 2016; *35*, 109–120.

Liu, A.; Wang, B.; Hamel, C. Arbuscular mycorrhiza colonization and development at suboptimal root zone temperature. *Mycorrhiza* **2004**, *14*, 93–101.

Liu, W.; Yu, K.; He, T.; Li, F.; Zhang, D.; Liu, J. The low temperature induced physiological responses of *Avena nuda* L.; a cold-tolerant plant species. *Sci. World J.* **2013**, *6*, 1–7.

Liu, Z.; Li, Y.; Wang, J.; He, X.; Tian, C. Different respiration metabolism between mycorrhizal and non-mycorrhizal rice under low-temperature stress: A cry for help from the host. *J. Agric. Sci.* **2015**, *153*, 602–614.

Locato, V.; De Pinto, M. C.; De Gara, L. Different involvement of the mitochondrial, plastidial and cytosolic ascorbate-glutathione redox enzymes in heat shock responses. *Physiol. Plant.* **2009**, *135*, 296–306.

Luu, D. T.; Maurel, C. Aquaporins in a challenging environment: Molecular gears for adjusting plant water status. *Plant Cell Environ.* **2005**, *28*, 85–96.

Ma, J.; Janoušková, M.; Li, Y.; Yu, X.; Yan, Y.; Zou, Z.; He, C. Impact of arbuscular mycorrhizal fungi (AMF) on cucumber growth and phosphorus uptake under cold stress. *Funct. Plant Biol.* **2015**, *42*, 1158–1167.

Ma, J.; Janouskova, M.; Ye, L.; Bai, L.; Dong, R.; Yan, Y.; Yu, X.; Zou, Z.; Li, Y.; He, C. Role of arbuscular mycorrhiza in alleviating the effect of cold on the photosynthesis of cucumber seedlings. *Photosynthetica* **2019**, *57*, 86–95.

Ma, J.; Sun, C.; Bai, L.; Dong, R.; Yan, Y.; Yu, X.; He, C.; Zou, Z.; Li, Y. Transcriptome analysis of cucumber roots reveals key cold-resistance genes induced by AM fungi. *Plant Mol. Biol. Rep.* **2018**, *36*, 135–148.

Mathur, S.; Jajoo, A. Arbuscular mycorrhizal fungi protects maize plants from high temperature stress by regulating photosystem II heterogeneity. *Ind. Crop. Prod.* **2014**, *143*, 111934.

Matsubara, Y.; Kayukawa, Y.; Fukui, H. Temperature-stress tolerance of *Asparagus* seedlings through symbiosis with arbuscular mycorrhizal fungus. *J. Jpn. Soc. Hortic. Sci.* **2000**, *69*, 570–575.

Maya, M. A.; Matsubara, Y. -I. Influence of arbuscular mycorrhiza on the growth and antioxidative activity in cyclamen under heat stress. *Mycorrhiza* **2013**, *23*, 381–390.

Miransari, M. Hyperaccumulators, arbuscular mycorrhizal fungi and stress of heavy metals. *Biotech. Adv.* **2011**, *29*, 645–653.

Nathalie, D.; Mariama, N.; Pape, I. D.; Dioumacor, F.; Valérie, H.; Sergio, S. Roles of arbuscular mycorrhizal fungi on plant growth and performance: Importance in biotic and abiotic stressed regulation. *Diversity* **2020**, *12*, 370.

Noctor, G.; Mhamdi, A.; Chaouch, S.; Han, Y. I.; Neukermans, J.; Marquez-Garcia, B.; Queval, G.; Foyer, C. H. Glutathione in plants: An integrated overview. *Plant Cell Environ.* **2012**, *35*, 454–484.

Ordoñez, N. M.; Marondedze, C.; Thomas, L.; Pasqualini, S.; Shabala, L.; Shabala, S.; Gehring, C. Cyclic mononucleotides modulate potassium and calcium flux responses to H_2O_2 in *Arabidopsis* roots. *FEBS. Lett.* **2014**, *588*, 1008–1015.

Paredes, M.; Quiles, M. J. The effects of cold stress on photosynthesis in *Hibiscus* plants. *PLoS One* **2015**, *10* (9), 1–13.

Porcel, R.; Redondo-Gómez, S.; Mateos-Naranjo, E.; Aroca, R.; Garcia, R.; Ruiz-Lozano, J. M. Arbuscular mycorrhizal symbiosis ameliorates the optimum quantum yield of photosystem II and reduces non-photochemical quenching in rice plants subjected to salt stress. *J. Plant Physiol.* **2015**, *185*, 75–83.

Rao, A. S. V. C.; Reddy, A. R. Glutathione reductase: A putative redox regulatory system in plant cells. In: *Sulfur assimilation and abiotic stresses in plants*; Khan, N. A., Singh, S., Umar, S., Eds.; Springer: Dordrecht, 2008.

Sanghera, G. S.; Wani, S. H.; Hussain, W.; Singh, N. B. Engineering cold stress tolerance in crop plants. *Curr. Genomics.* **2011**, *12*, 30–43.

Sawers, R. J.; Gutjahr, C.; Paszkowski, U. Cereal mycorrhiza: An ancient symbiosis in modern agriculture. *Trends Plant Sci.* **2008**, *13*, 93–97.

Sharma, S.; Anand, G.; Singh, N.; Kapoor, R. Arbuscular mycorrhiza augments arsenic tolerance in wheat (*Triticum aestivum* L.) by strengthening antioxidant defense system and thiol metabolism. *Front Plant Sci.* **2017**, *8*, 906–927.

Smith, F. A.; Grace, E. J.; Smith, S. E. More than a carbon economy: Nutrient trade and ecological sustainability in facultative arbuscular mycorrhizal symbioses. *New Phytol.* **2009**, *182*, 347–358.

Smith, S. E.; Read, D. J. *Mycorrhizal symbiosis*; Academic Press: San Diego, 1997.

Smith, S. E.; Read, D. J. *Mycorrhizal symbiosis*; Academic Press: London, 2008.

Smith, S. E.; Read, D. J. *Mycorrhizal symbiosis*; Academic Press: London, 2010.

Smith, S. E.; Smith, F. A. Roles of arbuscular mycorrhizas in plant nutrition and growth: New paradigms from cellular to ecosystem scales. *Annu. Rev. Plant Biol.* **2011**, *62*, 227–250.

Takabatake, R.; Hata, S.; Taniguchi, M.; Kouchi, H.; Sugiyama, T.; Izui, K. Isolation and characterization of cDNAs encoding mitochondrial phosphate transporters in soybean, maize, rice and *Arabidopsis*. *Plant Mol. Biol.* **1999**, *40*, 479–486.
Takata, K.; Matsuzaki, T.; Tajika, Y. Aquaporins: Water channel proteins of the cell membrane. *Prog. Histochem. Cytochem.* **2004**, *39*, 1–83.
Theocharis, A.; Clément, C.; Barka, E. A. Physiological and molecular changes in plants grown at low temperatures. *Planta* **2012**, *235*, 1091–1105.
Tindall, J. A.; Mills, H. A.; Radcliffe, D. E. The effect of root zone temperature on nutrient uptake of tomato. *J. Plant Nutr.* **1990**, *13*, 939–956.
Tisserant, E.; Kohler, A.; Dozolme-Seddas, P.; Balestrini, R.; Benabdellah, K.; Colard, A. C. D.; et al. The transcriptome of the arbuscular mycorrhizal fungus *Glomus intraradices* (DAOM 197198) reveals functional tradeoffs in an obligate symbiont. *New Phytol.* **2012**, *193*, 755–769.
Wang, H.; Wang, H.; Shao, H.; Tang, X. Recent advances in utilizing transcription factors to improve plant abiotic stress tolerance by transgenic technology. *Front Plant Sci.* **2016**, *7*, 67–80.
Wani, S. H.; Sandhu, J. S.; Gosal, S. S. Genetic engineering of crop plants for abiotic stress tolerance. In: *Advanced topics in plant biotechnology and plant biology*; Malik, C. P., Kaur, B., Wadhwani, C., Eds.; MD Publications: New Delhi, 2008; pp 149–183.
Winfield, M. O.; Lu, C.; Wilson, I. D.; Coghill, J. A.; Edwards, K. J. Plant responses to cold: transcriptome analysis of wheat. *Plant Biotechnol. J.* **2010**, *8*, 749–771.
Wu, Q. S.; Zou, Y. N. Beneficial roles of arbuscular mycorrhizas in citrus seedlings at temperature stress. *Sci. Hortic.* **2010**, *125*, 289–293.
Xin, Z.; Browse, J. Cold comfort farm: The acclimation of plants to freezing temperatures. *Plant Cell Environ.* **2001**, *23*, 893–902.
Xiong, L.; Schumaker, K. S.; Zhu, J. K. Cell signaling during cold, drought, and salt stress. *Plant Cell* **2002**, *14*, 165–183.
Yang, Y.; Han, X.; Liang, Y.; Ghosh, A.; Chen, J.; Tang, M. The combined effects of arbuscular mycorrhizal fungi (AMF) and lead (Pb) stress on Pb accumulation, plant growth parameters, photosynthesis, and antioxidant enzymes in *Robinia pseudoacacia* L. *PLoS One* **2015**, *10* (12), 1–24.
Zhu, X. C.; Song, F. B.; Liu, F. L.; Liu, S. Q.; Tian, C. J. Carbon and nitrogen metabolism in arbuscular mycorrhizal maize plants under low-temperature stress. *Crop Pasture Sci.* **2015**, *66*, 62–70.
Zhu, X.; Song, F.; Xu, H. Influence of arbuscular mycorrhiza on lipid peroxidation antioxidant enzyme activity of maize plants under temperature stress. *Mycorrhiza* **2010**, *20*, 325–332.

PART IV

MYCORRHIZAL FUNGI FOR SUSTAINABLE AGRICULTURAL DEVELOPMENT, MITIGATING CLIMATE CHANGE, ENSURING FOOD SECURITY, AND GLOBAL ENVIRONMENTAL SUSTAINABILITY

CHAPTER 11

ROLE OF AMF IN SUSTAINABLE AGRICULTURE

SAMIA KHANUM[1], ABDEL RAHMAN MOHAMMAD AL-TAWAHA[2], ABDEL RAZZAQ AL-TAWAHA[3], HIBA ALATRASH[4], ABDUR RAUF[5], ARUN KARNWAL[6], ABHIJIT DEY[7], NUJOUD ALIMAD[8], DEVARAJAN THANGADURAI[9], JEYABALAN SANGEETHA[10], SAHER ISLAM[11], IMRAN[12], AMANULLAH[12], SHAH KHALID[12], PALANI SARANRAJ[13], and ELIF GUNAL[14]

[1]*Department of Botany, University of the Punjab, Lahore, Pakistan*

[2]*Department of Biological Sciences, Al Hussein Bin Talal University, Maan, Jordan*

[3]*Department of Crop Science, Faculty of Agriculture, University Putra Malaysia, Serdang, Selangor, Malaysia*

[4]*General Commission for Scientific Agricultural Research, Syria*

[5]*Department of Chemistry, University of Swabi, Anbar, Khyber Pakhtunkhwa, Pakistan*

[6]*Department of Microbiology, School of Bioengineering and BioSciences, Lovely Professional University, Phagwara, Punjab, India*

[7]*Department of Life Sciences, Presidency University, Kolkata, India*

[8]*Damascus University, Faculty of Agriculture, Damascus, Syria*

[9]*Department of Botany, Karnatak University, Dharwad, Karnataka, India*

Mycorrhizal Technology: Managing Plant Stress and Mitigating Climate Change using Mycorrhizae.
Jeyabalan Sangeetha, Abdel Rahman Mohammad Al-Tawaha, & Devarajan Thangadurai (Eds.)

[10]*Department of Environmental Science, Central University of Kerala, Kasaragod, Kerala, India*

[11]*Institute of Biochemistry and Biotechnology, University of Veterinary and Animal Sciences, Lahore, Pakistan*

[12]*Department of Agronomy, The University of Agriculture, Peshawar, Pakistan*

[13]*Department of Microbiology, Sacred Heart College (Autonomous), Tirupattur, Tamil Nadu, India*

[14]*Department of Soil Science and Plant Nutrition, Faculty of Agriculture, Harran University, Sanliurfa, Turkey*

ABSTRACT

About 80% of terrestrial plants formed mycorrhizal association with AMF. Mycorrhizal association stimulates the physiological and molecular reactions even at subcellular stage leading to modification of the structure of plant community which results in the enhancement of plant resistance against all types of biotic and abiotic stresses. Molecular methodologies have been used to identify the AMF at ecosystem level but for specific plant species only. In a sustainable agricultural system farmers used natural procedures to attain improved food quality and productivity without using fertilizers, decrease input costs, and prevent environmental pollution and its side effects. The importance of using AMF inoculation in the sustainable agriculture system depends on the role of AMF for improving plant growth by increasing nutrients and water uptake and also increasing stability of soil aggregates and enhanced soil fertility. AMF interacts with soil-borne pathogen and reduces their effect on plant by antagonism, mycoparasitism, and antibiosis. Further studies are required for better understanding about ecological importance of AMF at community and ecosystem levels.

11.1 INTRODUCTION

About 80% of terrestrial plants formed mycorrhizal association with AMF (Brundrett, 2002). Arbuscular mycorrhizal association is ubiquitous and one of the most ancient types of symbiotic association between few numbers of AMF and land plants dated back 460 million years (Redecker, 2002). The most important function of this association is the uptake of phosphorous by roots of higher plants and increased uptake of water by hyphae of fungi. Mycorrhizal association also stimulates the physiological and molecular reactions even at subcellular stage leading to modification of the structure of plant community which results in the enhancement of plant resistance against all types of biotic and abiotic stresses. Plants of the same or different species connected with the help of the hyphal network of mycorrhizal association and that is how the transfer of nutrients, releasing signal molecules related to defense proteins like lipochitooligosaccharides and strigolectones. To understand the physiological and taxonomy of fungi recent studies have been carried out (Saito, 2000; Kohout et al., 2014). These are the members of phylum Glomeromycota and obligate in nature (Redecker et al., 2000). Their role for agriculture ecosystems documented very well and can be utilized for the restoration of the forest ecosystem (Solaiman and Abbott, 2003, 2008; Brundrett and Nanjappa, 2013). ECM fungi also distributed extensively, but they only associated with 3% land plants (Smith and Read, 2008) and are members of the phylum Ascomycota and Basidiomycota (Hibbett et al., 2000). AMF rather than ectomycorrhizal fungi depend on plants for Carbon sources, but when they are symbiotically associated, both required 20–40% of C from the plant that is photosynthetically fixed (McNear, 2013).

11.2 CLASSIFICATION AND MOLECULAR IDENTIFICATION OF AMF

Molecular techniques are used to identify the AMF in greenhouse and field circumstances because these organisms are multinuclear in nature so there is limitation in their identification (Schußler et al., 2001; Kohout et al., 2014). Molecular methodologies have been used to identify the AMF at the ecosystem level (Sanders et al., 1995) at the time of seasonal variations when arable crops are colonized by AMF (Daniell et al., 2001), but only for specific plants (Jansa et al., 2003) and coexisting plant species (Gollotte et al., 2004) when influenced by agricultural practices (Brito et al., 2013). Despite the latest studies molecular techniques are not sufficient to identify the AMF because of the low quality and quantity of DNA yield from soil than from roots (Renker et al., 2003). We are still lagging in techniques for precise assessment of the community composition of AMF because of major difference in the quantity of DNA extracted from fungal parts present in the soil for using it as a template DNA strand (Schußler et al., 2001). However,

DNA from multinucleated spores has high quantity than hyphae with fewer nucleuses that are reason of skew composition data of the community (Gamper et al., 2008). This result in possible preference for the estimation of the morphotypes of AMF relative abundance is greater than other morphotypes (Saks et al., 2013). The current advances in sequencing techniques will reveal the detailed explanation of the fungal community with reduced cost for analyzing DNA that will aid in better accessibility. Different primers have been used to analyze four most commonly used AM fungal-specific primers for rDNA, for example SSU, LSU, partial SSU, and ITS and also a new one ITS2 region. These all new research on primers revealed that these primers yield the high rate of fungal and Shannon diversity than SSU primer. The LSU and SSU show the effective bias for *Glomeraceae* and also the composition differences in AMF communities based on Redecker primers. To understand the ecosystem very well, it is significant to study AM mycorrhizal association with higher plants because of the role of AMF for fungal ecology (Willis et al., 2013). Morphological and molecular analyses result in a huge contribution to taxonomic organization of *Glomeromycota* (Oehl et al., 2011). Species from this group form AM association and the phylum consists of 3 classes, 5 orders, 14 families and 29 genera. According to morphological and molecular characteristics, 19 genera, consisting of about 200 species, are classified in AM fungi (Manoharachary et al., 2010). Molecular studies considered the most important to study AMF diversity. Taxonomy of AMF is actually based on morphology of the resting and asexual spores of fungal species, but molecular studies revealed the colonized root diversity (Rosendahl, 2008). Through various experimental studies it is clear that molecular analysis is the key to understanding the phylogeny of AMF which results in improved taxonomy of the fungal species.

11.3 ROLE OF MYCORRHIZA

The clear role of mycorrhiza is not explained clearly yet because under the natural condition association is absence or rare (Merryweather and Fitter, 1995). The establishment of nonmycorrhizal association in a container or field is a difficult challenge because certain treatments of soil like steaming and autoclaving change the soil composition and also disturb the other soil organisms also making it more difficult (Serrasolses et al., 2008). The mycorrhizal association improved the nutrient uptake, especially the elements having low mobility like P, Zn, Cu, etc., and also helped the plants to cope up the stresses and thus help plants to grow better (Sikes et al., 2009). The mycorrhizal association results in the expansion and complementation of the function of the root system in plants. The AMF colonized the inner root area as well as the soil around the root surface, thus uptake the nutrients from the zones which are far from the roots and root hairs of a plant because of the hyphae of the fungal partner which have the ability to grow several centimeters in the soil (Jansa et al., 2005). The uptake and transfer of Phosphorous up to 40 cm via fungal hyphae have been reported (Timonen et al., 1996). Some mycelium-growing pattern is very dense and only extends to few millimeters in the soil (Smith et al., 2004). Through hyphae and mycelium plant nutrient uptake has been improved. ECM fungi are playing an important role in water uptake from the soil (Allen, 2007). This colonization of roots also has an impact on the interaction of the plant with soil-borne pathogens because of several reasons like plant nutrition level changed and changed microbial communities in the rhizosphere (Toljander et al., 2007). The mineralization of certain organic nutrients and bio-weathering of inorganic nutrients by ECM from carbonates, micas, and apatites have also been reported (Blum et al., 2002). Other soil-borne organisms also involve in the mineralization of the nutrients.

The direct involvement of mycorrhizal fungi and other soil microbes in the mineralization of organic P under nonsterile soil situations is still not clear (Finlay, 2008). The AMF also help in the modification of soil like aggregation, wettability and biological activities in the root zone also modulate the competition in plant and soil microbes (Rillig and Mummey, 2006). AMF are heterotrophic and get organic carbon directly from their host plants required for growth, respiration, and biological maintenance. Some mycorrhizal fungi considered limited saprophytic growth (Koide et al., 2008). Mycorrhizal fungi dependence on the plant for carbon source is actually dependence on the availability of nutrients, light, CO_2, etc.; thus, this association varies from beneficial to parasitic (Schroeder and Janos, 2004). The net acquisition of P is less by mycorrhizal association when the limited soil P is available because of low concentration of P or high sorption in weathered soil than the soil in which small amount of P is added (Bolan, 1991). When the P level is very low plants prefer mobilization of P and uptake rather than mycorrhizal association (Lamber et al., 2008). Similarly when the high amount of soluble P fertilizers was added in the soil plant directly gain P from the soil results in no mycorrhizal association (Sorensen et al., 2003). Low or high availability of P, plant species, soil condition, etc., have affected the plant's growth negatively when the mycorrhizal association is observed (Morgan et al., 2005). Nutrient uptake by mycorrhizal fungi when the availability of P is the limited plant growth mycorrhizal association benefits the plant and improved plant growth (Li et al., 2005). The effect of mycorrhizal colonization on the plant is not always specifically positive and required proper isotopic labeling for better understanding of the mechanism. The negative effect of mycorrhizal association has been reported (Smith et al., 2009), but it is studied that the mycorrhizal association has a positive effect on plant growth mostly (Choi et al., 2005). The benefits and the extent of colonization of roots with different fungi mostly depend on environmental conditions rather than the availability of P (Hoeksema et al., 2010).

11.4 NUTRIENTS EXCHANGE MECHANISM OF AM

Nutrients exchange is considered as a very important role of mycorrhizal association. Molecular studies add recent information regarding this (Smith and Smith, 2012). Transfer of P from AMF to plant species is already explained and accepted overall (Smith and Read, 2008). Hyphae absorbed the phosphate solution from the soil with the help of P transporters (Harrison and van Buuren, 1995). Hyphae changed this phosphorous into polyphosphate and then transported it to intraradical hyphae with the help of protoplasmic streaming (Cox et al., 1975; Cooper and Tinker, 1981). It is reported that the considerable amount of P is of fungal origin (Solaiman and Saito, 2001). The efflux of P and decreased poly-P in hyphae can be enhanced by adding glucose and 2-deoxyglucose. The increased rate of P efflux and decreased poly-P are related which explains efflux of P from intraradical hyphae linked with hydrolysis of poly-P. AMF depend on photosynthetic C of the host plant and in exchange fungal partner provides nutrients, especially P, Zn, and N through arbuscules (Harrison, 1999; Balestrini and Bonfante, 2005). There are transporter proteins for N and P transport (Javot et al., 2007b) reported in the colonized roots (Harrison et al., 2002; Javot et al., 2007a). Extraradical hyphal network expands extensively from the root zone of host plant which helps in phosphorous absorption at the greater volume from the soil (Bucher, 2007). High affinity fungal P transporters help in absorption of P from the soil and then it transports to the vacuoles for the polymerization results in polyphosphate and then transports to the intraradical hyphae (Solaiman et al., 1999). Phosphate is released by the hydrolysis of polyphosphate and this phosphate is available for the plant to use. Ammonium, nitrate, and amino acids are also absorbed by hyphae (Hodge et al., 2001), and transport of N by mycorrhiza is well studied (Chalot et al., 2006; Ellerbeck et al., 2013).

We have no proof about the form of N in which it is obtained from the soil; it might be as ammonium or nitrate or in amino acid. Plant carbohydrates absorbed by fungi are hexose (Solaiman and Saito, 1997) via arbuscules. AMF transport P and N via arbuscule to root. C changes into storage lipids inside the intraradical hyphae. Lipids are the source of stored Carbon and transport it from intra to extraradical hyphae and act as the main element of respiratory substrate. AMF increased the nutrient acquisition, developed tolerance against different biotic and abiotic stresses, and also built resistance against different pathogens and toxic compounds present in the soil. When growing under nutrients deficient soil plants colonized with mycorrhizal fungi are showing better growth and development (Janos, 2007). Studies explained the link between acquisition and transport of C and other nutrient cycles between fungi and host plant in temperate regions (Phillips et al., 2013). AMF get photosynthates from the plant and thus improve the N uptake by absorbing the different N sources present in the soil with the help of hyphal networks that result in the increased tolerance in the host plant. The mechanism for the explanation about uptake and transport of N from AMF to the host plant is still unknown. So, further studies are required for better understanding about the role of AMF in N acquisition and ecological importance of AMF at community and ecosystem levels (Hodge and Storer, 2015) (Fig. 11.1).

11.5 INTERACTIONS OF SOIL MICROBIAL POPULATION

The mycorrhizal fungi and pathogens communication and their ability to increase plant nutrition are studied. There is little information about how this interaction affects plant (Schreiner and Bethlenfalvay, 1995). It is studied that plants with mycorrhizal association differ from the plant having no such association in the rhizosphere, due to these interactions, the quality of respiration of root and the quantity of the chemicals secreted from roots altered (Marschner et al., 2001). These interactions between pathogens and fungi are also affecting the growth of fungi. In the presence of bacteria in soil, the hyphae emerging from the germination of fungal spore developed smoothly and form small vesicles and more branched and longer than the hyphae growth in the absence of such bacteria.

AMF changes the mycorrhizosphere by changing the number of such pathogens as well as their activity as a pathogen (Burke et al., 2002). Plants with arbuscular mycorrhizal association with their roots show different impacts

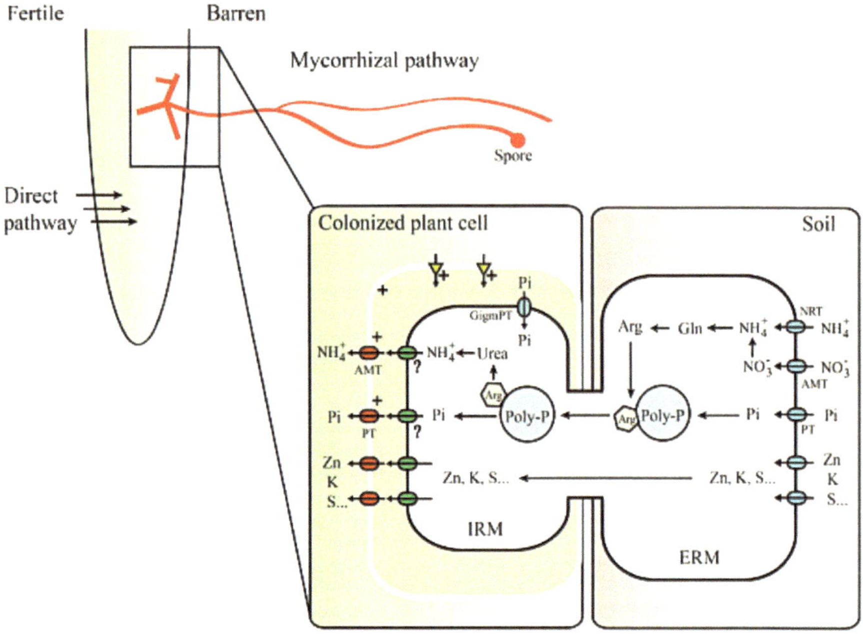

FIGURE 11.1 The direct and mycorrhizal pathways for uptake of Pi, N, and ions in AM symbiosis. In the context of nutrient uptake, the direct pathway mainly occurs at the soil–root interface where plants can directly absorb P/N nutrients from soil. In the mycorrhizal pathway, Pi is absorbed by the extraradical mycelium (ERM) from surrounding soil by fungal phosphate importers. Ammonium (NH_4^+) and/or nitrate (NO_3^-) are taken up by the ERM and assimilated into glutamine (Gln), then into arginine (Arg^+). Assimilation will generate excess H^+ or OH^- with nitrate and ammonium, respectively. Phosphate is mainly transported in the form of polyphosphate granules. The negative charge of these granules makes them possible carriers for arginine and metal ions, which are transported from extraradical hyphae to intraradical hyphae. Efflux mechanisms for Pi^- and NH_4^+ from the intraradical mycelium (IRM) to the interfacial apoplast are still unknown. Metal ions are absorbed by the ERM; metal ion transporters in host plants and AM fungi require further study.

Source: Reproduced with permission from Wanxiao et al. (2017), Copyright © Elsevier (2017).

on microbial composition within rhizosphere. It is studied that many factors affect the efficiency of arbuscular mycorrhizal fungi as a biocontrol agent. These factors are soil moisture, soil content, genotype of host, inoculum of mycorrhiza, inoculation time of mycorrhiza, virulence of mycorrhizal species, inoculation potential of pathogen, microflora of soil (Singh et al., 2000).

It is studied that the protection of plants against all such types of pathogens is totally dependent upon the rate of AMF colonization of plant roots. There are two zones of interaction between pathogens and AMF and these two are mycosphere and rhizosphere. There are some bacteria present in the rhizosphere known as Mycorrhization Helper Bacteria (MHB) which increase the ability of mycorrhizal fungi to colonize plant roots (Tahat et al., 2010).

11.6 PROTECTION AGAINST PATHOGENS

Many agricultural practices are used to control soil-borne pathogens. These techniques include resistant cultivars, seed certification, crop rotation, fumigation of soil, and many chemicals like fungicides. Due to less resistance in plants against pathogen and less chances of control of pathogen inoculum, it is difficult to control plant pathogens (Azcan-Aguilar and Barea, 1997). Due to these reasons, researchers are trying to use other alternatives against these pathogens like the microorganism added to the soil and used fungus inoculum; these organisms increase resistance in plants against pathogen by competing with soil-borne pathogens for nutrients and space by secreting adding antibiotics or act like parasites for those pathogens (Grosch et al., 2005; Berg et al., 2007).

It is studied that using chemicals against soil-borne pathogens also damages plants' growth as well as creating an alarming situation; that is why scientists suggest and use alternative agents for controlling pathogens like fungi and other microbes like bacteria. Many researchers studied that arbuscular mycorrhizal association is very beneficial against soil-borne pathogens. There is a complex interaction between soil pathogen, the plant, and fungal partner which reduced the disease effect caused by pathogens in the host plant which is colonized with arbuscular mycorrhizal fungi (Dehne, 1982).

It is reported that when tomato roots were associated and colonized by *Glomus mosseae*, it reduced the disease effect of pathogen *Phytophthora parasitica* than in nonmycorrhizal tomato roots. It is also reported that when the oil palm plants were colonized by AMF it decreased the disease effect of *Ganoderma boninense* and seedlings of oil palm plant are more resistant to *Ganoderma* (Tahat et al., 2010).

AMF interacts with soil-borne pathogen and reduces their effect on plant by antagonism, mycoparasitism, and antibiosis (Harrier and Watson, 2004). During this association arbuscular mycorrhizal fungus causes certain changes in the host plant, microbial variations in rhizosphere, status of nutrients in soil, changes the anatomy of plant cells, changes morphology of roots, and reduces stress (Hooker et al., 1994).

11.7 AMF ROLE IN SUSTAINABLE AGRICULTURE

In a sustainable agricultural system farmers used natural procedures to attain improved food quality and productivity without using fertilizers, decrease input costs, and prevent pollution and its side effects. Sustainable agriculture is important because of ecofriendly nature. Soil characteristics also contribute in sustainable agriculture by controlling diseases; also improved microbial activity of the soil results in improved parasitism and antagonism at rhizosphere (Jawson et al., 1993; Knudsen et al., 1999). Recent research focused on finding other suitable sources than chemical fertilizers. On the other hand, it is also proved that using the microorganism can also improve the soil quality which results in increased crop productivity. More emphasis has been placed on how we can use already-present soil microbes to increase soil fertility, plant growth, and protection against diseases. AMF study is difficult but considered the important part of the ecosystem other than plant roots only. Recent studies in soil microbiology techniques provoke the use of AMF management in relation to crop nutrition and productivity rate. AMF impact the growth, nutrition, and productivity of the crop even when P is sufficient in the soil (Solaiman and Hirata, 1997; Balzergue et al., 2013). When crops are cultivated in less fertile soil, it is somehow beneficial for AMF in agroecosystem like low nutrient loss in the environment. Suitable strain selection and improved inoculants of AMF help in enhancing the soil fertility, stability and sustainability in agroecosystem (Abbott et al., 1987; Smith et al., 1992; Hamel and Strullu, 2006; Djuuna et al., 2009). Many factors are actually involved to understand the plant response and benefits obtained from mycorrhiza like host dependency for mycorrhizal association, tillage, fertilizers, and potential AMF inoculation application in an agricultural system. The importance of using AMF inoculation in a sustainable agriculture system depends on the role of AMF for improving plant growth by increasing nutrients and water uptake (Auge, 2004) and also increasing stability of soil aggregates and enhanced soil fertility (Rillig, 2004; Rillig et al., 2007) (Table 11.1).

11.8 RESTORATION OF FOREST ECOSYSTEMS

Many important physiochemical and biological soil characteristics have been lost (Skujins and Allen, 1986) because land clearance of terrestrial ecosystem removed millions of hectors in Australia annually (Warren et al.,

TABLE 11.1 A List of Fungal Transporters Described.

Nutrients	Transporters	Species	References
Sugar	GpMST1	Glomeromycotan	Schussler et al. (2006)
	RiMST2	*Glomus* sp.	Helber et al. (2011)
Phosphorus	GvPT	*Glomus versiforme*	Harrison and Vanbuuren (1995)
	GiPT	*Glomus intraradices*	Maldonado-Mendoza et al. (2001)
	GmosPT	*Glomus mosseae*	Balestrini et al. (2007)
	GigmPT	*Gigaspora margarita*	Xie et al. (2016)
Nitrogen	GintAMT	*Glomus intraradices*	Lopez-Pedrosa et al. (2006); Perez-Tienda et al. (2011); Calabrese et al. (2016)
Zinc	GintZnT1	*Glomus intraradices*	Gonzalez-Guerrero et al. (2005)

Source: Reproduced with permission from Wanxiao et al. (2017), Copyright © Elsevier (2017).

1996). The fertility and quality of soil has been determined by these properties which help in productivity and establishment of plants. Restoration of native plants has been decreased because of soil degradation (Agnew and Warren, 1996) which results in increased desertification and soil erosion. The potential of AMF inoculum decreased because of desertification which is the reason for maintaining the nutrient cycle, thus maintaining the vegetative cover of natural habitats. AMF also increase the plant survival during stresses like nutrient's deficiency, drought, or other soil disruptions (Schreiner et al., 1997). Water stable aggregate formation is very important for soil tilth and hyphae play the very important role in it (Jeffries and Barea, 2000). Mycorrhizal propagules can be increased in a damaged ecosystem by revegetation naturally or artificially (Requena et al., 1996). It is studied that during the revegetation process if colonized plants are planted it leads to better establishment (Herrera et al., 1993) and improved soil quality by improving structure of the soil (Carrillo-Garcia et al., 1999). Desertification of the ecosystem can be recovered by planting the plants with already-identified AMF (Azcon-Aguilar et al., 2003). Forest trees depend on the AMF for proper mineral uptake from the soil and in exchange transport the carbon to the AMF. Hyphae are the great source of the carbon for soil microorganism and depend on the plant for energy. In exchange AMF improved the essential mineral nutrient uptake like P, N, S, and Zn by increasing the root system of the plant (Allen, 1991). Many trees connect with one another with hyphal network. Underground biodiversity is very important for balance of forest growth, ecosystem function and also helps in reforestation of damaged land. Mycorrhizal fungi improved water and nutrient uptake and in return get

C from the plant; it also builds resistance against pathogens. Mycorrhizal association helps in the establishment of tree species in new habitat which is difficult in the absence of mycorrhizal fungi (Menkis et al., 2012).

11.9 CONCLUSION

About 80% of terrestrial plants formed mycorrhizal association with AMF. Mycorrhizal association stimulates the physiological and molecular reactions even at subcellular stage leading to modification of the structure of plant community which results in the enhancement of plant resistance against all types of biotic and abiotic stresses. In sustainable agricultural system farmers used natural procedures to attain improved food quality and productivity without using fertilizers, decrease input costs, and prevent environmental pollution and its side effects. On the other hand, it is also proved that using the microorganism can also improve the soil quality that results in increased crop productivity. More emphasis has been placed on how we can use already-present soil microbes to increase soil fertility, plant growth, and protection against diseases. AMF study is difficult but considered the important part of the ecosystem other than plant roots only. Recent studies on soil microbiology techniques provoke the use of AMF management in relation to crop nutrition and productivity rate. AMF impact the growth, nutrition, and productivity of the crop even when P is sufficient in the soil.

KEYWORDS

- **AMF**
- **antibiosis**
- **myco-parasitism**
- **sustainable agriculture**
- **symbiotic interactions**

REFERENCES

Abbott, L. K.; Robson, A. D.; Gazey, C. Selection of inoculant VA mycorrhizal fungi. In: *Mycorrhizas in the next decade, practical applications and research priorities: Proceedings of the 7th North American Conference on Mycorrhizas*; Sylvia, D. M., Hung, L. L., Graham,

J. H., Eds.; Institute of Food and Agricultural Sciences, University of Florida, USA, 1987; pp 10–12.
Agnew, C.; Warren, A. A. framework for tackling drought and land degradation. *J. Arid Environ.* **1996**, *33*, 309–320.
Allen, M. F. Use of mycorrhizae for land rehabilitation. *MIRCEN J. Appl. Microbiol. Biotechnol.* **1986**, *2* (1), 161–176.
Allen, M. F. *The ecology of mycorrhiza*; Cambridge University Press: Cambridge, 1991.
Allen, M. F. Mycorrhizal fungi: Highways for water and nutrients in arid soils. *Vadose Zone J.* **2007**, *6*, 291–297.
Auge, R. M. Arbuscular mycorrhizae and soil/plant water relations. *Can. J. Soil Sci.* **2004**, *84*, 373–381.
Azcon-Aguilar, C.; Barea, J. M. Applying mycorrhiza biotechnology to horticulture: Significance and potentials. *Sci. Hort.* **1997**, *68*, 1–24.
Azcon-Aguilar, C.; Palenzuela, J.; Roldan, A.; Bautista, S.; Vallejo, R.; Barea, J. M. Analysis of the mycorrhizal potential in the rhizosphere of representative plant species from desertification threatened Mediterranean shrub lands. *Appl. Soil. Ecol.* **2003**, *22*, 29–37.
Balestrini, R.; Bonfante, P. The interface compartment in arbuscular mycorrhizae: A special type of plant cell wall? *Plant Biosyst.* **2005**, *139*, 8–15.
Balestrini, R.; Gomez-Ariza, J.; Lanfranco, L.; Bonfante, P. Laser microdissection reveals that transcripts for five plant and one fungal phosphate transporter genes are contemporaneously present in arbusculated cells. *Mol. Plant Microbe Interact.* **2007**, *20*, 1055–1062.
Balzergue, C.; Chabaud, M.; Barker, D. G.; Becard, G.; Rochange, S. F. High phosphate reduces host ability to develop arbuscular mycorrhizal symbiosis without affecting root calcium spiking responses to the fungus. *Front. Plant Sci.* **2013**, *4*, 426.
Berg, G.; Grosch, R.; Scherwinski, K. Risk assessment for microbial antagonists: Are there effects on non-target organisms? *Gesunde Pflanzen* **2007**, *59*, 107–117.
Blum, J. D.; Klaue, A.; Nezat, C. A.; Driscoll, C. T.; Johnson, C. E.; Siccama, T. G.; Eagar, C.; Fahey, T. J.; Likens, G. E. Mycorrhizal weathering of apatite as an important calcium source in base-poor forest ecosystems. *Nature* **2002**, *417*, 729–731.
Bolan, N. S. A critical review on the role of mycorrhizal fungi in the uptake of phosphorus by plants. *Plant Soil* **1991**, *134*, 189–207.
Brito, I.; Carvalho, M.; Goss, M. J. Soil and weed management for enhancing arbuscular mycorrhiza colonization of wheat. *Soil Use Manage.* **2013**, *29*, 540–546.
Brundrett, M. C.; Nanjappa, A. Glomeromycotan mycorrhizal fungi from tropical Australia III. Measuring diversity in natural and disturbed habitats. *Plant Soil* **2013**, *370*, 419–433.
Brundrett, M. C. Coevolution of roots and mycorrhizas of land plants. *New Phytol.* **2002**, *154*, 275–304.
Bucher, M. Functional biology of plant phosphate uptake at root and mycorrhiza interfaces. *New Phytol.* **2007**, *173*, 11–26.
Burke, D. J.; Hamerlynck, E. P.; Hahn, D. Interactions among plant species and microorganisms in salt march sediments. *Appl. Environ. Microbiol.* **2002**, *68*, 1157–1164.
Calabrese, S.; Perez-Tienda, J.; Ellerbeck, M.; Arnould, C.; Chatagnier, O.; Boller, T.; Schussler, A.; Brachmann, A.; Wipf, D.; Ferrol, N. et al. GintAMT3 - a low-affinity ammonium transporter of the arbuscular mycorrhizal *Rhizophagus irregularis. Front. Plant Sci.* **2016**, *7*, 679.

Carrillo-Garcia, A.; Leon de la Luz, J. L.; Bashan, Y.; Bethlenfalvay, G. J. Nurse plants, mycorrhizae, and plant establishment in a disturbed area of the Sonoran Desert. *Restor. Ecol.* **1999**, *7*, 321–335.
Chalot, M.; Blaudez, D.; Brun, A. Ammonia: A candidate for nitrogen transfer at the mycorrhizal interface. *Trends Plant Sci.* **2006**, *11*(6), 263–266.
Choi, D. S.; Quoreshi, A. M.; Maruyama, Y.; Jin, H. O.; Koike, T. Effect of ectomycorrhizal infection on growth and photosynthetic characteristics of *Pinus densiflora* seedlings grown under elevated CO_2 concentrations. *Photosynthetica* **2005**, *43*, 223–229.
Cooper, K. M.; Tinker, P. B. Translocation and transfer of nutrients in vesicular arbuscular mycorrhizas. IV. Effect of environmental variables on movement of phosphorus. *New Phytol. 1981. 88*, 327–339.
Cox, G.; Sanders, F. E.; Tinker, P. B.; Wild, J. A. Ultrastructure evidence relating to host/endophyte transfer in a vesicular–arbuscular mycorrhizae. In: *Endomycorrhizae*; Sanders, F. E., Mosse, B., Tinker, P. B., Eds.; Academic Press: London, 1975; pp 297–312.
Daniell, T. J.; Husband, R.; Fitter, A. H.; Young, J. P. W. Molecular diversity of arbuscular mycorrhizal fungi colonising arable crops. *FEMS Microbiol. Ecol.* **2001**, *36*, 203–209.
Dehne, H. W. Interaction between vesicular-arbuscular mycorrhizal fungi and plant pathogens. *Phytopathology* **1982**, *72*, 1115–1119.
Djuuna, I. A. F.; Abbott, L. K.; Solaiman, Z. M. Use of mycorrhiza bioassays in ecological studies. In: *Symbiotic fungi*; Varma, A., Kharkwal, A. C., Eds.; Springer: Berlin, Heidelberg, 2009; pp 41–50.
Ellerbeck, M.; Schüßler, A.; Brucker, D.; Dafinger, C.; Loos, F.; Brachmann, A. Characterization of three ammonium transporters of the Glomeromycotan fungus *Geosiphon pyriformis. Eukaryotic Cell* **2013**, *12*(11), 1554–1562.
Finlay, R. D. Ecological aspects of mycorrhizal symbiosis: With special emphasis on the functional diversity of interactions involving the extraradical mycelium. *J. Exp. Bot.* **2008**, *59*(5), 1115–1126.
Gamper, H. A.; Young, J. P. W.; Jones, D. L.; Hodge, A. Real-time PCR and microscopy: Are the two methods measuring the same unit of arbuscular mycorrhizal fungal abundance? *Fungal Genet. Biol.* **2008**, *45*(5), 581–596.
Gollotte, A.; Van Tuinen, D.; Atkinson, D. Diversity of arbuscular mycorrhizal fungi colonising roots of the grass species *Agrostis capillaris* and *Lolium perenne* in a field experiment. *Mycorrhiza* **2004**, *14*(2), 111–117.
Gonzalez-Guerrero, M.; Azcon-Aguilar, C.; Mooney, M.; Valderas, A.; MacDiarmid, C. W.; Eide, D. J.; Ferrol, N. Characterization of a *Glomus intraradices* gene encoding a putative Zn transporter of the cation diffusion facilitator family. *Fungal Genet. Biol.* **2005**, *42*, 130–140.
Grosch, R.; Lottmann, J.; Faltin, F.; Berg, G. Use of bacterial antagonists to control diseases caused by *Rhizoctonia solani*. *Gesunde Pflanzen.* **2005**, *57*(8), 199–205.
Hamel, C.; Strullu, D. G. Arbuscular mycorrhizal fungi in field crop production: Potential and new direction. *Can. J. Plant Sci.* **2006,** *86*(4), 941–950.
Harrier, L. A.; Watson, C. A. The potential role of arbuscular mycorrhizal (AM) fungi in the bioprotection of plants against soil-borne pathogens in organic and/or other sustainable farming systems. *Pest Manage. Sci.* **2004**, *60*(2), 149–157.
Harrison, M. J.; van Buuren, M. L. A phosphate transporter from the mycorrhizal fungus *Glomus versiforme*. *Nature* **1995**, *378*(6557), 626–629.

Harrison, M. J.; Vanbuuren, M. L. A phosphate transporter from the mycorrhizal fungus *Glomus versiforme*. *Nature* **1995**, *378*, 626–629.

Harrison, M. J. Biotrophic interfaces and nutrient transport in plant/fungal symbioses. *J. Exp. Bot.* **1999**, 1013–1022.

Helber, N.; Wippel, K.; Sauer, N.; Schaarschmidt, S.; Hause, B.; Requena, N. A versatile monosaccharide transporter that operates in the arbuscular mycorrhizal fungus *Glomus* sp. is crucial for the symbiotic relationship with plants. *Plant Cell* **2011**, *23*, 3812–3823.

Herrera, M. A.; Salamanca, C. P.; Barea, J. Inoculation of woody legumes with selected arbuscular mycorrhizal fungi and rhizobia to recover desertified Mediterranean ecosystems. *Appl. Environ. Microbiol.* **1993**, *59*(1), 129–133.

Hibbett, D. S.; Gilbert, L. B.; Donoghue, M. J. Evolutionary instability of ectomycorrhizal symbioses in basidiomycetes. *Nature* **2000**, *407*(6803), 506–508.

Hodge, A.; Storer, K. Arbuscular mycorrhiza and nitrogen: Implications for individual plants through to ecosystems. *Plant Soil* **2015**, *386*(1), 1–19.

Hodge, A.; Campbell, C. D.; Fitter, A. H. An arbuscular mycorrhizal fungus accelerates decomposition and acquires nitrogen directly from organic material. *Nature* **2001**, *413*(6853), 297–299.

Hoeksema, J. D.; Chaudhary, V. B.; Gehring, C. A.; Johnson, N. C.; Karst, J.; Koide, R. T. et al. Meta-analysis of context-dependency in plant response to inoculation with mycorrhizal fungi. *Ecol. Lett.* **2010**, *13*, 394–407.

Hooker, J. E.; Jaizme-Vega, M.; Atkinson, D. Biocontrol of plant pathogen using arbuscular mycorrhizal fungi. In: *Impact of arbuscular mycorrhizas on sustainable agriculture and natural ecosystems*; Gianinazzi, S., Schhepp, H., Eds.; Birkhauser, Verlag: Basel, Switzerland, 1994; pp 191–209.

Janos, D. P. Plant responsiveness to mycorrhizas differs from dependence upon mycorrhizas. *Mycorrhiza* **2007**, *17*, 75–91.

Jansa, J.; Mozafar, A.; Frossard, E. Phosphorus acquisition strategies within arbuscular mycorrhizal fungal community of a single field site. *Plant Soil* **2005**, *276*(1), 163–176.

Jansa, J.; Mozafar, A.; Kuhn, G.; Anken, T.; Ruh, R.; Sanders, I. R.; Frossard, E. J. E. A. Soil tillage affects the community structure of mycorrhizal fungi in maize roots. *Ecol. App.* **2003**, *13*(4), 1164–1176.

Javot, H.; Penmetsa, R. V.; Terzaghi, N.; Cook, D. R.; Harrison, M. J. A *Medicago truncatula* phosphate transporter indispensable for the arbuscular mycorrhizal symbiosis. *Proc. Natl. Acad. Sci. USA* **2007**, *104*(5), 1720–1725.

Javot, H.; Pumplin, N.; Harrison, M. J. Phosphate in the arbuscular mycorrhizal symbiosis: Transport properties and regulatory roles. *Plant Cell Environ.* **2007**, *30*(3), 310–322.

Jawson, M. D.; Franzluebbers, A. J.; Galusha, D. K.; Aiken, R. M. Soil fumigation within monoculture and rotations: Response of corn and mycorrhizae. *Agron. J.* **1993**, *85*(6), 1174–1180.

Jeffries, P.; Barea, J. M. Arbuscular mycorrhiza: A key component of sustainable plant–soil ecosystems. *Fungal Assoc.* **2012**, 51–75.

Knudsen, I. M.; Debosz, K.; Hockenhull, J.; Jensen, D. F.; Elmholt, S. Suppressiveness of organically and conventionally managed soils towards brown foot rot of barley. *Appl. Soil Ecol.* **1999**, *12*(1), 61–72.

Kohout, P.; Sudová, R.; Janoušková, M.; Čtvrtlíková, M.; Hejda, M.; Pánková, H.; Slavikova, R.; Stajerova, K.; Vosatka, M.; Sykorova, Z. Comparison of commonly used primer sets for evaluating arbuscular mycorrhizal fungal communities: Is there a universal solution?. *Soil Biol. Biochem.* **2014**, *68*, 482–493.

Lambers, H.; Raven, J. A.; Shaver, G. R.; Smith, S. E. Plant nutrient-acquisition strategies change with soil age. *Trends Ecol. Evol.* **2008**, *23*(2), 95–103.

Li, H. Y.; Zhu, Y. G.; Marschner, P.; Smith, F. A.; Smith, S. E. Wheat responses to arbuscular mycorrhizal fungi in a highly calcareous soil differ from those of clover, and change with plant development and P supply. *Plant Soil* **2005**, *277*(1), 221–232.

Lopez-Pedrosa, A.; Gonzalez-Guerrero, M.; Valderas, A.; Azcon-Aguilar, C.; Ferrol, N. GintAMT1 encodes a functional high-affinity ammonium transporter that is expressed in the extraradical of *Glomus intraradices*. *Fungal Genet. Biol.* **2006**, *43*, 102–110.

Maldonado-Mendoza, I. E.; Dewbre, G. R.; Harrison, M. J. A phosphate transporter gene from the extra-radical mycelium of an arbuscular mycorrhizal fungus *Glomus intraradices* is regulated in response to phosphate in the environment. *Mol. Plant Microbe Interact.* **2001**, *14*, 1140–1148.

Manoharachary, C.; Kunwar, I. K.; Tilak, K. V. B. R.; Adholeya, A. Arbuscular mycorrhizal fungi-taxonomy, diversity, conservation and multiplication. *Proc. Natl. Acad. Sci. India Sect. B—Biol. Sci.* **2010**, *80*, 1–13.

Marschner, P.; Crowley, D.; Lieberei, R. Arbuscular mycorrhizal infection changes the bacterial 16S rDNA community composition in the rhizosphere of maize. *Mycorrhiza* **2001**, *11*(6), 297–302.

McNear Jr, D. H. The rhizosphere—roots, soil and everything in between. *Nat. Educ. Knowl.* **2013**, *4*(3), 1.

Menkis, A.; Lygis, V.; Burokiene, D.; Vasaitis, R. Establishment of ectomycorrhiza-inoculated *Pinus sylvestris* seedlings on coastal dunes following a forest fire. *Baltic Forestry* **2012**, *18*(1), 33–40.

Merryweather, J.; Fitter, A. Phosphorus and carbon budgets: Mycorrhizal contribution in *Hyacinthoides non-scripta* (L.) Chouard ex Rothm. under natural conditions. *New Phytol.* **1995**, *129*(4), 619–627.

Morgan, J. A. W.; Bending, G. D.; White, P. J. Biological costs and benefits to plant–microbe interactions in the rhizosphere. *J. Exp. Bot.* **2005**, *56* (417), 1729–1739.

Oehl, F.; Sieverding, E.; Palenzuela, J.; Ineichen, K.; da Silva, G. A. Advances in Glomeromycota taxonomy and classification. *IMA Fungus* **2011**, *2*(2), 191–199.

Perez-Tienda, J.; Testillano, P. S.; Balestrini, R.; Fiorilli, V.; C. Azcon-Aguilar, C.; Ferrol, N. GintAMT2, a new member of the ammonium transporter family in the arbuscular mycorrhizal fungus *Glomus intraradices*. *Fungal Genet. Biol.* **2011**, *48*, 044–1055.

Phillips, R. P.; Brzostek, E.; Midgley, M. G. The mycorrhizal-associated nutrient economy: A new framework for predicting carbon–nutrient couplings in temperate forests. *New Phytol.* **2013**, *199*(1), 41–51.

Redecker, D. Molecular identification and phylogeny of arbuscular mycorrhizal fungi. In: *Diversity and integration in mycorrhizas*; Hans, L., Ed.; Springer: Dordrecht, 2002; pp 67–73.

Redecker, D.; Morton, J. B.; Bruns, T. D. Ancestral lineages of arbuscular mycorrhizal fungi (*Glomales*). *Mol. Phylogenet. Evol.* **2000**, *14*(2), 276–284.

Renker, C.; Heinrichs, J.; Kaldorf, M.; Buscot, F. Combining nested PCR and restriction digest of the internal transcribed spacer region to characterize arbuscular mycorrhizal fungi on roots from the field. *Mycorrhiza* **2003**, *13*(4), 191–198.

Requena, N.; Jeffries, P.; Barea, J. M. Assessment of natural mycorrhizal potential in a desertified semiarid ecosystem. *Appl. Environ. Microbiol.* **1996**, *62*(3), 842–847.

Rillig, M. C.; Mummey, D. L. Mycorrhizas and soil structure. *New Phytol.* **2006**, *171*(1), 41–53.

Rillig, M. C. Arbuscular mycorrhizae, glomalin, and soil aggregation. *Can. J. Soil Sci.* **2004**, *84*(4), 355–363.

Rillig, M. C.; Caldwell, B. A.; Wösten, H. A.; Sollins, P. Role of proteins in soil carbon and nitrogen storage: Controls on persistence. *Biogeochemistry* **2007**, *85*(1), 25–44.

Rosendahl, S. Communities, populations and individuals of arbuscular mycorrhizal fungi. *New Phytol.* **2008**, *178*(2), 253–266.

Saito, M. Symbiotic exchange of nutrients in arbuscular mycorrhizas: Transport and transfer of phosphorus. In: *Arbuscular mycorrhizas: Physiology and Function*; Douds, D. D., Kapunik, Y., Eds.; Kluwer: Dordrecht, The Netherlands, 2000; pp 85–106.

Saks, Ü.; Davison, J.; Öpik, M.; Vasar, M.; Moora, M.; Zobel, M. Root-colonizing and soil-borne communities of arbuscular mycorrhizal fungi in a temperate forest understorey. *Botany* **2014**, *92*(4), 277–285.

Sanders, I. R.; Alt, M.; Groppe, K.; Boller, T.; Wiemken, A. Identification of ribosomal DNA polymorphisms among and within spores of the *Glomales*: Application to studies on the genetic diversity of arbuscular mycorrhizal fungal communities. *New Phytol.* **1995**, *130*(3), 419–427.

Schreiner, R. P.; Bethlenfalvay, G. J. Mycorrhizal interactions in sustainable agriculture. *Crit. Rev. Biotechnol.* **1995**, *15*(3–4), 271–285.

Schreiner, R. P.; Mihara, K. L.; McDaniel, H.; Bethlenfalvay, G. J. Mycorrhizal fungi influence plant and soil functions and interactions. *Plant Soil* **1997**, *188*(2), 199–209.

Schroeder, M. S.; Janos, D. P. Phosphorus and intraspecific density alter plant responses to arbuscular mycorrhizas. *Plant Soil* **2004**, *264*(1), 335–348.

Schüßler, A.; Schwarzott, D.; Walker, C. A new fungal phylum, the Glomeromycota: Phylogeny and evolution. *Mycol. Res.* **2001**, *105*(12), 1413–1421.

Schussler, A.; Martin, H.; Cohen, D.; Fitz, M.; Wipf, D. Characterization of a carbohydrate transporter from symbiotic glomeromycotan fungi. *Nature* **2006**, *444*, 933–936.

Serrasolses, I.; Romanyà, J.; Khanna, P. K. Effects of heating and autoclaving on sorption and desorption of phosphorus in some forest soils. *Biol. Fert. Soils* **2008**, *44*(8), 1063–1072.

Sikes, B. A.; Cottenie, K.; Klironomos, J. N. Plant and fungal identity determines pathogen protection of plant roots by arbuscular mycorrhizas. *J. Ecol.* **2009**, *97*(6), 1274–1280.

Singh, R.; Adholeya, A.; Mukerji, K. G. Mycorrhiza in control of soil borne pathogens. In: *Mycorrhizal biology*; Mukerji, K. G., Chamola, B. P., Singh, J., Eds.; Kluwer Academic/ Plenum Publishers: Netherlands, 2000; pp 173–196.

Smith, F. A.; Grace, E. J.; Smith, S. E. More than a carbon economy: Nutrient trade and ecological sustainability in facultative arbuscular mycorrhizal symbioses. *New Phytol.* **2009**, *182*(2), 347–358.

Smith, S. E.; Read, D. J. *Mycorrhizal symbiosis*; Academic Press, 2008.

Smith, S. E.; Smith, F. A. Fresh perspectives on the roles of arbuscular mycorrhizal fungi in plant nutrition and growth. *Mycologia* **2012**, *104* (1), 1–13.

Smith, S. E.; Robson, A. D.; Abbott, L. K. The involvement of mycorrhizas in assessment of genetically dependent efficiency of nutrient uptake and use. *Plant Soil* **1992**, *146*(1), 169–179.

Smith, S. E.; Smith, F. A.; Jakobsen, I. Functional diversity in arbuscular mycorrhizal (AM) symbioses: The contribution of the mycorrhizal P uptake pathway is not correlated with mycorrhizal responses in growth or total P uptake. *New Phytol.* **2004**, *162*(2), 511–524.

Solaiman, M. Z.; Abbott, L. K. Phosphorus uptake by a community of arbuscular mycorrhizal fungi in jarrah forest. *Plant Soil* **2003**, *248*(1), 313–320.

Solaiman, M. Z.; Hirata, H. Effect of arbuscular mycorrhizal fungi inoculation of rice seedlings at the nursery stage upon performance in the paddy field and greenhouse. *Plant Soil* **1997**, *191*(1), 1–12.

Solaiman, M. Z.; Saito, M. Use of sugars by intraradical hyphae of arbuscular mycorrhizal fungi revealed by radiorespirometry. *New Phytologist* **1997**, *136*(3), 533–538.

Solaiman, M. Z.; Saito, M. Phosphate efflux from intraradical hyphae of *Gigaspora margarita* in vitro and its implication for phosphorus translocation. *New Phytologist* **2001**, *151*(2), 525–533.

Solaiman, M. Z.; Ezawa, T.; Kojima, T.; Saito, M. Polyphosphates in intraradical and extraradical hyphae of an arbuscular mycorrhizal fungus, *Gigaspora margarita. Appl. Environ. Microbiol.* **1999**, *65*(12), 5604–5606.

Solaiman, Z. M.; Abbott, L. K. Influence of arbuscular mycorrhizal fungi, inoculum level and phosphorus placement on growth and phosphorus uptake of *Phyllanthus calycinus* under Jarrah forest soil. *Biol. Fert. Soils* **2008**, *44*(6), 815–821.

Sorensen, J. N.; Larsen, J.; Jakobsen, I. Management strategies for capturing the benefits of mycorrhizas in the production of field-grown vegetables. *Acta. Hortic.* **2003,** *627*, 65–71.

Tahat, M. M.; Kamaruzaman, S.; Othman, R. Mycorrhizal fungi as a biocontrol agent. *Plant Pathol. J.* **2010**, *9*(4), 198–207.

Timonen, S.; Finlay, R. D.; Olsson, S.; Söderström, B. Dynamics of phosphorus translocation in intact ectomycorrhizal systems: Non-destructive monitoring using a β-scanner. *FEMS Microbiol. Ecol.* **1996**, *19*(3), 171–180.

Toljander, J. F.; Lindahl, B. D.; Paul, L. R.; Elfstrand, M.; Finlay, R. D. Influence of arbuscular mycorrhizal mycelial exudates on soil bacterial growth and community structure. *FEMS Microbiol. Ecol.* **2007**, *61*(2), 295–304.

Wanxiao, W.; Jincai, S.; Qiujin, X.; Yina, J.; Nan, Y.; Nutrient exchange and regulation in arbuscular mycorrhizal symbiosis. *Mol. Plant* **2017**, *10*(9), 1147–1158.

Warren, A.; Sud, Y. C.; Rozanov, B. The future of deserts. *J. Arid Environ.* **1996**, *32*(1), 75–89.

Willis, A.; Rodrigues, B. F.; Harris, P. J. The ecology of arbuscular mycorrhizal fungi. *Crit. Rev. Plant Sci.* **2013,** *32*(1), 1–20.

Xie, X. A.; Lin, H.; Peng, X.W.; Xu, C. R.; Sun, Z. F.; Jiang, K. X. et al. Arbuscular mycorrhizal symbiosis requires a phosphate transceptor in the *Gigaspora margarita* fungal symbiont. *Mol. Plant.* **2016**, *9*, 1583–1608.

CHAPTER 12

MITIGATING CLIMATE CHANGE AND ENSURING FOOD SECURITY THROUGH APPLICATION OF MYCORRHIZA IN THE CLIMATE-SMART FARMING SYSTEM

NUJOUD ALIMAD[1], ABDEL RAHMAN MOHAMMAD AL-TAWAHA[2], DEVARAJAN THANGADURAI[3], JEYABALAN SANGEETHA[4], and VEDAVYAS SHIVANAND CHAVAN[3]

[1]*Damascus University, Faculty of Agriculture, Sweidaa Branch, Damascus, Syria*

[2]*Department of Biological Sciences, Al Hussein Bin Talal University, Maan, Jordan*

[3]*Department of Botany, Karnatak University, Dharwad, Karnataka, India*

[4]*Department of Environmental Science, Central University of Kerala, Kasaragod, Kerala, India*

ABSTRACT

Smart farming system is the golden key to driving forward and ensuring food security, plus reserving natural recourses under the presence of massive pressure of difficulties that combined with high consumption of food production that represented with foodborne diseases, fungal toxicities, pests, chemical residue, and so on, especially with huge populations

Mycorrhizal Technology: Managing Plant Stress and Mitigating Climate Change using Mycorrhizae.
Jeyabalan Sangeetha, Abdel Rahman Mohammad Al-Tawaha, & Devarajan Thangadurai (Eds.)

of humans, that increased continuously. The current estimated world population stands at 7.8 billion now, and it is expected to reach 9.7 billion within the next 24 years. Food security is a great issue to solve, it is a critical problem! So all concerted efforts tend to place every possible idea in framing the main solution (smart farming system). Integrated information besides technical communication summarizes the meaning of smart farming system, for best habitude of resource that consists of all agricultural supplement, besides operated GPS robots, and automated watering, plus using microorganisms such as arbuscular mycorrhizal fungi (AMF) for excess plant growth and consequently to raise the yield. However, the high costs, non-accessible of the internet in some regions, and lack of necessary awareness among users affect the proper usage of the technology. Since the old days till now we have struggled and competed all the times, tried to control climate change as much as possible to raise crop yield, and aimed to save lives on the planet, remembering the rule "if the food is yours, be sure the recourses are for all."

12.1 INTRODUCTION

Plants respond to various abiotic stress factors, mainly, high salinity, low or high temperature, ultraviolet radiation, drought, excessive or deficient water, excess light, weighty metals toxicity, nutrient lack. Previous factors threaten plant growth, sustainability, and agriculture industrial development. Producing extra food under less quality soil combined with stress circumstances is a big challenge nowadays. Food security is the upcoming challenge which could be addressed with proper application of eco-friendly approaches. AMF are the better option to combat the abiotic stress and improve the soil quality in natural way (Allen, 1991; Bennett and Bever, 2007). AMF fungi are classified as a spirited function in improving agricultural ecosystems, because they enhance water status and ameliorate plant resistance against natural stresses, plant growth, nutrient plus reproductive capacity, improved soil structure, protection from pathogens, besides representing a standing chemical pesticide alternation in agriculture sphere, especially against nematodes, fungi that are usually the main cause of major product loss (Sharma and Sharma, 2018). Through over AMF, the host plant can activate defense mechanisms by producing operative enzymes of phytoalexins, phenylpropanoids, peroxidases, chitinase, pathogensis-related (PR) proteins, β-1,3-glucanase, hydroxy prolin-rich glycoproteins (HRGP), plus phenolics to act as a biological control besides decreasing N and P fertilizer inputs, also raising plant tolerance to several ecological factors (Duc and Posta, 2018).

The current study discourses how using AMF can give a great hand out in smart farming within climate change circumstances to accomplish good food security and highlight how to plant rejoinder to abiotic tension such as high salinity with toxic effects, drought, heavy metals, via the presence of AMF.

12.2 DROUGHT STRESS

The biggest common abiotic tension that affects growth besides plant development is drought. Plants culminate from osmotic tension under low temperatures, salinity and drought due to minimum water availability for both cellular maintenance and other biological functions. Besides, elongation of the duration of osmotic tension causes the production of highly reactive oxygen spices (ROS) such as (hydrogen peroxide, O_2^-, OH, and IO_2) inside chloroplasts, which may cause serious cellular irreversible damage to lipids, proteins, DNA, and carbohydrates as a result of their high activation (Mahajan and Tuteja, 2005).

Ruizlozano et al. (2012) referred to the significance of symbiosis AMF with plants to neglect and avoid the passive effects induced by dryness. AMF is considered a significant key in rejoinder toward these types of stresses. AMF can design symbiotic combination shape through approximately 80% Tracheophyta, which consist of 90% of whole agricultural plants (Smith and Read, 2008). AMF inoculation provides high advantages to enhance water deficiency of tomato, pepper, wheat, maize, and few varieties of chickpea depending on genotype and phenotype. In this association, the fungus can receive 10–20% of total photosynthetic productivity (Allen, 1991) and lipids from the host plant. Plants are improved out of water and mineral nutrition uptake over the hyphal network of AMF, which enable root steward system to extend via prolonged extraradical hyphae to reach more water, phosphorus (P), nitrogen (N) and other nutrients. Under dryness stress AMF driving plants towards more progressive growth and enable to absorb several mineral nutrient elements that unavailable to uncolonized root (Chen et al., 2017). Considering hyphal thickness is bigger than root intensity in ten-hundred times; consequently, intake superficies of mycorrhizal roots are adequate well enhancement in order superior expansion of mycorrhizal plants compared with unmycorrhizal plant under such stress (Chen et al.,

2017). Several studies illustrated that water scarcity damages chloroplasts and changes photosynthetic pigments. On the contrary, transplantation of AMF mitigates wastage of such parameters which are caused through stress. Research have been conducted that AMF symbioses could protect plants from oxidative tension through minimizing lipid per oxidation level plus strengthening antioxidative enzymes in roots, leaves under moderate or critical drought, therefore increasing root plant biomass by about 17–37% after AMF inoculation so far adding AMF consists of many advantages, mainly increasing plant resistance (Zhao et al., 2015).

12.3 HEAVY METALS

The pollution of heavy metals is considered a significant ecological issue to end plant production, it also threatens our health because of metals' toxicity that affects plants when they present in large concentrations resulting in chlorosis and other symptoms at the end. Intensity of heavy metals curbs the growth of root and shoot; moreover, it minimizes chlorophyll process and respiration. Adding AMF is a convenient form in the sustainable agricultural system to increase plant resistance to heavy metals naturally through planting-resistant cultivars in polluted soils; these cultivars can release many substances containing essential acids like malic, lactic, citric acids, which form metal complexes combination, and therefore minimize soil toxicity (Gomathy et al., 2011). Even though, few reports are available about how tolerant mycorrhizal strains remove heavy metals and clean them up from polluted ecological system (Regvar et al., 2003), by developing an intensive mycelium spread making the roots more expanding and enable absorbing phosphate and other mineral elements through the soil, consequently increasing plant durability to abiotic stress. Besides mycorrhizal management, many agricultural procedures like crop rotation and multicropping can affect positively to increase crop production (Cardoso and Kuyper, 2006) (Table 12.1).

12.4 HIGH SALINITY

Salinity is the essential abiotic ecological factor, affects major plant process such as outgrowth, development, crop productivity, metabolism of lipid, energy production, and protein compilation (Hoorn et al., 2002). Particularly, in arid and semiarid areas, considering that nearly 5-7% of overall land

TABLE 12.1 An Overview of AMF-Assisted Phytoremediation of Various Heavy Metals from Contaminated Sites in Recent Years.

Mycorrhizal species	Host plant	Heavy metals	Mechanisms of HM alleviation	References
Glomus mosseae	*Coreopsis drummondii, Pteris vittata, Lolium perenne, Trifolium repens*	Cu	Phytostabilization	Chen et al. (2007b)
Glomus intraradices, Glomus mosseae, Glomus claroideum and *Glomus geosporum*	*Nicotiana tabacum*	Cd	Phytostabilization	Janouskova et al. (2006)
Acaulospora laevis, Gigaspora margarita, Glomus intraradices	*Astragalus sinicus*	Lanthanum	Phytostabilization	Chen and Zhao (2007)
Glomus intraradices	*Zea mays*	Pb	Phytostabilization	Sudova and Vosatka (2007)
Glomus intraradices	*Agrostis capillaris*	Pb, Zn, Cu, Cd	No effect was noticed	Sudova et al. (2008)
Glomus mosseae	*Trifolium repens*	Cd, Mo, Mn, Zn, Cu, Al, As, Ni	Phytostabilization	Azcon et al. (2009)
Glomus deserticola, Glomus claroideum	*Eucalyptus globulus*	Ar	Phytoextraction	Arriagada et al. (2009)
Glomus intraradices	*Sorghum bicolor*	Cu	Phytostabilization	Toler et al. (2005)
Glomus deserticola	*Eucalyptus globulus*	Zn	Phytoextraction	Arriagada et al. (2010)
Glomus mosseae	*Cajanus cajan*	Cd, Pb	Phytostabilization	Garg and Aggarwal (2012)
Glomus mosseae or *Glomus intraradices*	*Populus alba, Populus nigra*	Zn	Phytostabilization	Lingua et al. (2008)
Glomus mosseae	*Glycine max, Lens culinaris*	Zn, Ni	Phytoextraction	Jamal et al. (2002)
Glomus intraradices	*Helianthus annuus*	Cd	Phytostabilization	De Andrade et al. (2008)

TABLE 12.1 *(Continued)*

Mycorrhizal species	Host plant	Heavy metals	Mechanisms of HM alleviation	References
Glomus mosseae, *Glomus intraradices*	*Capsicum annum*	Cr	Phytostabilization	Ruscitti et al. (2011)
Rhizophagus fasciculatus, *Rhizophagus intraradices*, *Funneliformis mosseae*, *Glomus aggregatum*	*Zea mays*	Cd, Cr, Ni, Pb	Phytoextraction	Singh et al. (2019)
Funneliformis mosseae, *Diversispora spurcum*	*Cynodon dactylon* (bermuda grass)	Pb, Zn, and Cd	AMF changed the HMs content and accumulation characteristics in the plant	Zhan et al. (2019)
Rhizophagus irregularis	*Glycine max* (soybean)	Cd	AM colonization had no effects on Cd accumulation and translocation in both HX3 and HN89 plants	Cui et al. (2019)
Glomus monosporum, *Glomus clarum*, *Gigaspora nigra*	*Trigonella foenum-graecum*	Cd	Phyto-stabilization	Abdelhameed and Metwally (2019)
Glomus aggregatum, *Glomus intraradices*, *Glomus etunicotum*, *Glomus versiforme*	*Medicago sativa*	Cd	Reduction in cadmium uptake of alfalfa grown in Cd polluted soil	Zhang et al. (2019a)
Rhizophagus irregularis	*Phragmites australis*	Cu	Phyto–rhizoremediation	Wu et al. (2020)
Glomus intraradices	*Agrostis capillaris*, *Zea mays*	Pb	*G. intraradices* isolates resulted in decreased Pb concentrations in both shoots and roots in comparison with non-inoculated plants	Malcova et al. (2003)
Glomus mosseae, *Glomus intraradices*	Sun flowers	Cr, Mn, Ni, Cu, Zn, Al, Pb, Co, Mo, Fe and Si	Combined *Glomus mosseae* and sludge amendments resulted with the highest metal uptakes and glomalin contents.	Sayin et al. (2019)

TABLE 12.1 *(Continued)*

Mycorrhizal species	Host plant	Heavy metals	Mechanisms of HM alleviation	References
Funneliformis mosseae, *Rhizophagus intraradices*	*Robinia pseudoacacia*	Pb	AMF symbiosis decreased Pb concentrations in the leaves and promoted the accumulation of biomass as well as photosynthetic pigment contents.	Yang et al. (2015b)
Glomus versiforme, *Rhizophagus intraradices*	*Lonicera japonica*	Cd	Gv and Ri symbiosis alleviated Cd toxicity of *L. japonica* through the decline of shoot Cd concentrations and the improvement of P nutrition.	Jiang et al. (2016)
Rhizophagus irregularis	*Medicago truncatula*	Pb	The water-soluble Pb complex content was decreased by AM inoculation.	Zhang et al. (2019c)
Funneliformis mosseae	*Phragmites australis*	TiO_2NPs	*F. mosseae* changed the distribution of Ti in reeds, increased the accumulation of Ti in roots.	Xu et al. (2019)

Source: Reproduced with permission from Riaz et al. (2021), Copyright © Elsevier B.V. (2020).

is stilted by salinity. Saline soil cause nutrient imbalance and produce many passive outcomes like ethylene, and increase hypertonic solution. Moreover, the photosynthesis process is highly affected by salinity, in addition to the production of reactive oxygen species (ROS). Besides, it is high time to enhance tolerance in plants in order to deal with the salt stress effects and to strongly meet food security issues. Consequently, the toxicity of ion imponderables creates high stress on soil osmosis, by reducing water potency, and minimizing the ability of water uptake as the toxic aggregation of Na^+ inside plant cellular disorganizes nutrient absorption as time progress (Munns and Tester, 2008). Na^+ gathering in shoot cells may affect several processes like photosynthetic ratio, enzymatic reactions, and entrance in K^+ transportation competition. Therefore, salinity negatively affects many physiochemical reactions such as synthesis, respiration, N metabolism, and ion balance.

Ying et al. (2019) found the favorable procedure to enhance plant treatment of saline-polluted soils or metal once, represented by adding some strains of microbes particularly. Nowadays, plant treatment is considered a convenient method if compared with physio-chemical alternatives to remove settled heavy metals out of polluted grounds (Munns and Tester, 2008). Thus reducing harmful effects of salinity problem occurs by adding AMF to get highly chlorophyll continent, and have better plant growth, and therefore better production. Zandavalli et al. (2004) ensured that the presence of mycorrhizae in soils under saline stress gives high defense through better arrival to nutrients. Also, Ying et al. (2019) proposed adding AMF may play a key role in minimizing Na^+ absorption, so as to enhance plant expansion and persistence especially in Nickel-polluted soils under greenhouse-controlled conditions.

Al-Karaki (2002) explained the property mechanism to protect nonhalophytic plants from losing productivity by using AMF, which are energizing root expanding, enhance various elements nutrition, raise synthesis of mycorrhizal plants, in addition to conquest of more P, N, and so on (Smith and Read, 2008). Many researchers considered AM fungi like eco-ameliorators because they can maintain high standards of growth hormones so as to minimize the passive impact of excessive salinity.

12.5 IMPROVING SOIL PROPERTIES

Many authors positively verified AMF capability to relieve the soil pressure and its effect on plant expansion. Soil texture and plant growth can be enhanced under stressed or normal circumstances via AMF (Cho et al., 2006). Moreover, the interactions between microbes and plants may positively affect plant growth, also improve productivity, carbon detention, and nutrition cycle (Kulmatiski et al., 2008). In addition to ameliorating plant responses against abiotic tension (Rodriguez et al., 2008), mycorrhiza network can enhance the property of soil water detention because they produce many chemical substances like polysaccharides, glomalin, mucilages, and hydrophobins which combined soil. Miller and Jastrow (2000) believed in the ability of hyphae as essential bound agents implicated in the microaggregation stability.

AMF serve an important function in both soil fertility and quality through ameliorating nutrient absorption, also raise protection toward fungal soil-borne pathogens, and change soil structure physically and chemically. The second function is improving root morphology via raising more brunching and expanding, consequently more water and minerals absorption abilities. A third great benefit is raising P absorption (Barea, 1991). Also, many studies have been conducted on the ability of AMF in the preservation of soil fertility and erosion. Hyphae can expand several inches into the soil, and therefore reach the nutrient depletion zone, whereas plant cannot access without AMF intermediate (O'keefe and Sylvia, 1992).

Various minerals and nutrients are supplied directly to crops through a hyphal expansion that connects soil, nutrient, and plant (Hart and Forsythe, 2012) (Fig. 12.1). Dorneless et al. (2001) found that hyphal expansion is much more functional than root hairs in ion absorption, especially for immobile elements, like P; therefore, the greater advantage of using AM fungi in some agricultural practices is reducing the application of P fertilizers, with consideration. We cannot get these benefits unless cropping systems save the development of AMF, which is the essential condition for enhanced mycorrhiza. Some families as Chenopodiaceae, Brassicaceae do not form typical arbuscular mycorrhizae, unlike others such as Graminaceae or Solanaceae. Mutually beneficial relationship between roots and soil-borne fungi is served through AMF that coexisted for a long time, about 400 million years (Sieverding, 1991). Plenchette et al. (2005) conducted three factors that weaken the maintenance of AM fungi; they are cropping cycles without mycotrophic, excessive usage of pesticides, and high fertility of P.

AMF represent a form of biological fertilization; they can improve soil quality and structure by producing GRSP (glomalin-related soil proteoglycoprotein) from mycelium and spore walls (Godbold et al., 2006). Sticky GRSP combined tiny soil particles and gathered them with various sizes,

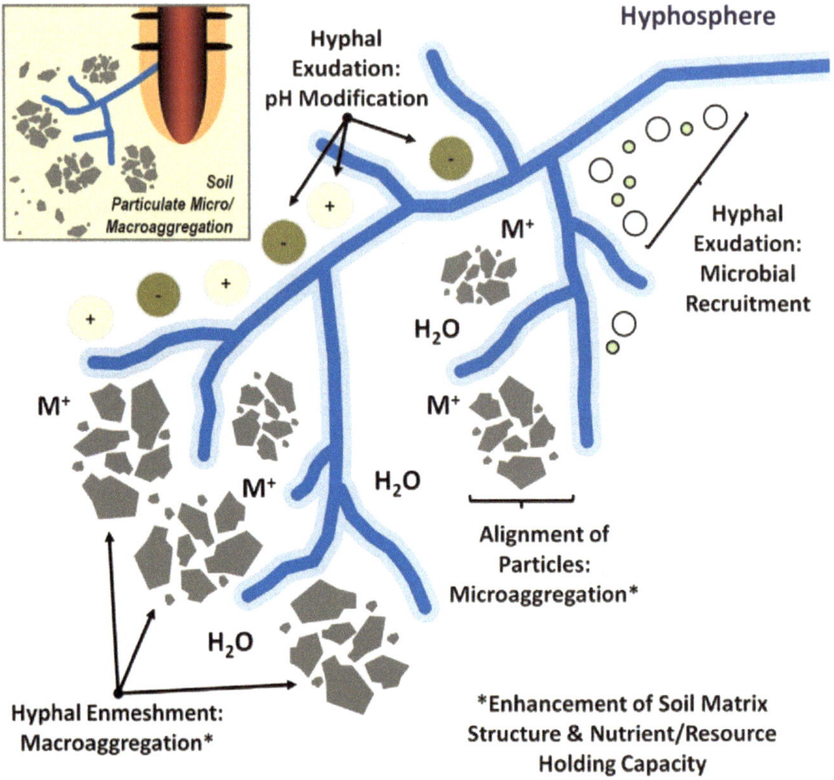

FIGURE 12.1 Generalized depictions of soil particulate micro- and macro aggregation. Enhancement of soil matrix structure and nutrient/resource holding capacity is shown with emphasis on the processes of hyphal exudation (e.g., leading to pH modification and microbial recruitment), alignment of soil particles (e.g., microaggregation), and hyphal enhancement (e.g., microaggregation).

Source: Reproduced with permission from Audet (2014), Copyright © Elsevier Inc. (2014).

so they act as biological glue. The great benefit of glomalin is to raise the ratio of C availability and conservation. Well, combined soil particles can easily resist both water and wind erosion, in addition to filtering water and air with good rates for better growing (Wu et al., 2008). Moreover, Miransari (2010) conducted that inoculation of AMF will enhance enzyme efficacious and raise the functional, also protecting membrane integrity, and stabilizing enzymes/proteins. Moreover, it represents an effective alternative to reduce chemical inputs, saving organic fertility, besides minimizing both biotic and abiotic tension.

12.6 CONCLUSION

Application of AMF in the climate smart farming system helps in mitigation of climate change and ensuring food security, because AMF consists of many advantages, mainly, increase plant resistance, give high defense through better arrival to nutrients, enhance water status, improved soil structure, and ameliorate plant resistance against natural stresses, beside the ability of AMF in preservation of the soil fertility and minimize the erosion. All previous spirited functions of AMF implementation give a hope of ray to improve agricultural ecosystems, especially within the challenge of climate change and food security.

KEYWORDS

- **climate change**
- **food security**
- **mycorrhiza**
- **smart farming system**
- **abiotic stress**

REFERENCES

Abdelhameed, R. E.; Metwally, R. A. Alleviation of cadmium stress by arbuscular mycorrhizal symbiosis. *Int. J. Phytoremed.*. **2019**, *21*, 663–671. https://doi.org/10.1080/15226514.2018.1556584.

Al-Karaki, G. N. Growth of mycorrhizal tomato and mineral acquisition under salt stress. *Mycorrhiza* **2000**, *10*, 51–54.

Allen, M. F.; Swenson, W.; Querejeta, J. I. Egerton-Warburton LM, Treseder KK. Ecology of mycorrhizae: A conceptual framework for complex interactions among plants and fungi. *Annu. Rev. Phytopathol.* **2003**, *41*, 271–303.

Arriagada, C.; Aranda, E.; Sampedro, I.; Garcia-Romera, I.; Ocampo, J. A. Contribution of the saprobic fungi *Trametes versicolor* and *Trichoderma harzianum* and the arbuscular mycorrhizal fungi *Glomus deserticola* and *G. claroideum* to arsenic tolerance of *Eucalyptus globulus*. *Bioresour. Technol.* **2009**, *100*, 6250–6257. https://doi.org/10.1016/j.biortech.2009.07.010.

Arriagada, C.; Pereira, G.; García-Romera, I.; Ocampo, J. A. Improved zinc tolerance in *Eucalyptus globulus* inoculated with *Glomus deserticola* and *Trametes versicolor* or *Coriolopsis rigida*. *Soil Biol. Biochem.* **2010**, *42*, 118–124. https://doi.org/ 10.1016/j.soilbio.2009.10.011.

Audet, P. Arbuscular mycorrhizal fungi and metal phytoremediation: Ecophysiological complementarity in relation to environmental stress. In: *Emerging Technologies and Management of Crop Stress Tolerance*; Ahmad, P., Rasool, S., Eds.; Academic Press: San Diego, 2014; pp 133–160. doi:10.1016/B978–0-12–800875–1.00006–5.

Azcon, R.; Peralvarez, M. del C.; Biro, B.; Roldan, A.; Ruíz-Lozano, J. M. Antioxidant activities and metal acquisition in mycorrhizal plants growing in a heavy-metal multicontaminated soil amended with treated lignocellulosic agrowaste. *Appl. Soil Ecol.* **2009**, *41*, 168–177. https://doi.org/10.1016/j. apsoil.2008.10.008.

Barea, J. M. Vesicular-arbuscular mycorrhizae as modifiers of soil fertility. *Adv. Soil Sci.* **1991**, *15*, 1–40.

Bennett, A. E.; Bever, J. D. Mycorrhizal species differentially alter plant growth and response to herbivory. *Ecology* **2007**, *88*, 210–218.

Cardoso, I. M.; Kuyper, T. W. Mycorrhizas and tropical soil fertility. *Agric. Ecosyst. Environ.* **2006**, *116*(1–2), 72–84.

Chen, B. D.; Zhu, Y. G.; Duan, J.; Xiao, X. Y.; Smith, S. E. Effects of the arbuscular mycorrhizal fungus *Glomus mosseae* on growth and metal uptake by four plant species in copper mine tailings. *Environ. Pollut.* **2007**, *147*, 374–380. https://doi.org/ 10.1016/j. envpol.2006.04.027.

Chen, J.; Zhang, H.; Xinlu Zhang, X.; Ming Tang M. Arbuscular mycorrhizal symbiosis alleviates salt stress in black locust through improved photosynthesis, water status, and K^+/Na^+ homeostasis. *Front Plant Sci.* **2017**, *8*, 1739.

Chen, X. H.; Zhao, B. Arbuscular mycorrhizal fungi mediated uptake of lanthanum in Chinese milk vetch (*Astragalus sinicus* L.). *Chemosphere* **2007**, *68*, 1548–1555. https://doi.org/10.1016/j.chemosphere.2007.02.068.

Cho, K.; Tolera, H.; Leeb, J.; Ownleyc, B.; Stutzd, J. C.; Moorea, J. L.; Augea, R. M. Mycorrhizal symbiosis and response of sorghum plants to combined drought and salinity stresses. *J. Plant Physiol.* **2006**, *163*, 517–528.

Cui, G.; Ai, S.; Chen, K.; Wang, X. Arbuscular mycorrhiza augments cadmium tolerance in soybean by altering accumulation and partitioning of nutrient elements, and related gene expression. *Ecotoxicol. Environ. Saf.* **2019**, *171*, 231–239. https://doi. org/10.1016/j. ecoenv.2018.12.093.

De Andrade, S. A. L.; Da Silveira, A. P. D.; Jorge, R. A.; De Abreu, M. F. Cadmium accumulation in sunflower plants influenced by arbuscular mycorrhiza. *Int. J. Phytoremed.* **2008**, *10*, 1–13. https://doi.org/10.1080/15226510701827002.

Dorneless, M. R. F.; da Silva, C. M and Gomes, A. A. A model for hyphae effects in phosphorus absorption by plants. *Ecol. Model.* **2001**, *142*, 83–89.

Duc, N. H.; Csintalan, Z.; Posta, K. Arbuscular mycorrhizal fungi mitigate negative effects of combined drought and heat stress on tomato plants. *Plant Physiol. Biochem.* **2018**, *132*, 297–307.

Garg, N.; Aggarwal, N. Effect of mycorrhizal inoculations on heavy metal uptake and stress alleviation of *Cajanus cajan* (L.) Millsp. genotypes grown in cadmium and lead contaminated soils. *Plant Growth Regul.* **2012**, *66*, 9–26. https://doi.org/10.1007/s10725–011–9624–8.

Godbold, D. L.; Hoosbeek, M. R.; Lukac, M.; Cotrufo, M. F.; Janssens, L. A, Ceulemans, R.; Polle, A.; Velthorst, E. J.; Scarascia-Mugnozza, G.; Angelis, P.; Miglietta, F.; Peressotti, A. Mycorrhizal hyphal turnover as a dominant process for carbon into soil organic matter. *Plant Soil* **2006**, *281*, 15–24.

Gomathy, M.; Sabarinathan, K. G.; Thangaraju, M.; Subramanian, K. S.; Sivashankari Devi, T.; Ananthi, K. The effect of mycorrhizae inoculated maize root exudates in alleviation of chromium toxicity in chromium polluted environments. *Insight Microbiol.* **2011**, *1*, 20–30.

Hart, M. M.; Forsythe, J. A. Using arbuscular mycorrhizal fungi to improve the nutrient quality of crops; nutritional benefits in addition to phosphorus. *Sci. Hortic.* **2012**, *148*, 206–214.

Hoorn, W.; Katerji, N.; Hamady, A.; Mastroilli, M. Effect of salinity on yield and nitrogen uptake of four-grain legumes and on biological nitrogen contribution from the soil. *Agric. Water Mangt.* **2002**, *51*, 87–98.

Jamal, A.; Ayub, N.; Usman, M.; Khan, A. G. Arbuscular mycorrhizal fungi enhance zinc and nickel uptake from contaminated soil by soybean and lentil. *Int. J. Phytoremed.* **2002**, *4*, 205–221. https://doi.org/10.1080/15226510208500083.

Janouskova, M.; Pavlíkova, D.; Vosatka, M. Potential contribution of arbuscular mycorrhiza to cadmium immobilisation in soil. *Chemosphere* **2006**, *65*, 1959–1965. https://doi.org/10.1016/j.chemosphere.2006.07.007.

Jiang, Q. Y.; Zhuo, F.; Long, S. H.; Di Zhao, H.; Yang, D. J.; Ye, Z. H.; Li, S. S.; Jing, Y. X. Can arbuscular mycorrhizal fungi reduce Cd uptake and alleviate Cd toxicity of *Lonicera japonica* grown in Cd-added soils? *Sci. Rep.* **2016**, *6*. https://doi.org/10.1038/ srep21805.

Kulmatiski, A.; Beard, K. H.; Stevens, J. R.; Cobbold, S. M. Plant-soil feedbacks: A meta-analytical review. *Ecol. Lett.* **2008**, *11*, 980–992.

Lingua, G.; Franchin, C.; Todeschini, V.; Castiglione, S.; Biondi, S.; Burlando, B.; Parravicini, V.; Torrigiani, P.; Berta, G. Arbuscular mycorrhizal fungi differentially affect the response to high zinc concentrations of two registered poplar clones. *Environ. Pollut.* **2008**, *153*, 137–147. https://doi.org/10.1016/j.envpol.2007.07.012.

Mahajan, S.; Tuteja, N. Cold, salinity and drought stress: An overview. *Arch. Biochem. Biophys.* **2005**, *444*, 139–158.

Malcova, R.; Vosatka, M.; Gryndler, M. Effects of inoculation with *Glomus intraradices* on lead uptake by *Zea mays* L. and *Agrostis capillaris* L. *Appl. Soil Ecol.* **2003**, *23*, 55–67.

Miller, R. M.; Jastrow, J. D. Mycorrhizal fungi influence soil structure. In *Arbuscular mycorrhizas: Molecular biology and physiology*; Kapulnik, Y., Douds, D. D., Eds.; Kluwer Academic Publishers: Dordrecht, The Netherlands, 2002; pp 3–18.

Miransari, M. Contribution of arbuscular mycorrhizal symbiosis to plant growth under different types of soil stress. *Plant Biol.* **2010**, *1*, 563–569.

Munns, R.; Tester, M. Mechanisms of salinity tolerance. *Annu. Rev. Plant Biol.* **1992**, *59*, 651–681.

O'keefe, D. M.; Sylvia, D. M. Chronology and mechanisms of P uptake by mycorrhizal sweet potato plants. *New Phytol.* **1992**, *1992*, 651–659.

Plenchette, C.; Clermont-Dauphin, C.; Meynard, J. M.; Fortin, J. A. Managing arbuscular mycorrhizal fungi in cropping systems. *Canadian Journal of Plant Science* **2005**, *85*(1), 31–40.

Regvar, M.; Vogel, K. Colonization of pennycresses (*Thlaspi* spp.) of the Brassicaceae by arbuscular mycorrhizal fungi. *J Plant Physiol.* **2003**, *160*, 615–626.

Riaz, M.; Kamran, M.; Fang, Y.; Wang, Q.; Cao, H.; Yang, G.; Deng, L.; Wang, Y.; Zhou, Y.; Anastopoulos, I.; Wang, X. Arbuscular mycorrhizal fungi-induced mitigation of heavy metal phytotoxicity in metal contaminated soils: A critical review. *J. Hazard. Mater.* **2021**, *15*, 402, 123919. doi: 10.1016/j.jhazmat.2020.123919.

Rodriguez, R. J.; Henson, J.; Van Volkenburgh, E.; Hoy, M.; Wright, L.; Beckwith, F.; Kim, Y. O.; Redman, R. S. Stress tolerance in plants via habitat-adapted symbiosis. *ISME J.* **2008**, *2*, 404–416.

Ruiz-Lozano, J. M.; Porcel, R.; Azcon, R.; Aroca, R. Regulation by arbuscular mycorrhizae of the integrated physiological response to salinity in plants: New challenges in physiological and molecular studies. *J. Exp. Bot.* **2012**, *63*, 4033–4044.

Ruscitti, M.; Arango, M.; Ronco, M.; Beltrano, J. Inoculation with mycorrhizal fungi modifies proline metabolism and increases chromium tolerance in pepper plants (*Capsicum annuum* L.). *Brazilian J. Plant Physiol.* **2011**, *23*, 15–25. https://doi.org/10.1590/S1677–04202011000100004.

Sayin, F. E.; Khalvati, M. A.; Erdinçler, A. Effects of sewage sludge application and arbuscular mycorrhizal fungi (*G. mosseae* and *G. intraradices*) interactions on the heavy metal phytoremediation in chrome mine tailings. *Frontiers in Plant Science*, **2019**, *112*, 217–224.

Sharma, I. P.; Sharma, K. Mycorrhizal colonization and phosphorus uptake in presence of PGPRs along with nematode infection. *Symbiosis* **2018**, *77*, 185–187.

Sieverding, E. *Vesicular-arbuscular mycorrhiza management in tropical agrosystems*; Technical Cooperation, Federal Republic of Germany, Eschborn, 1991; p. 371.

Singh, G.; Pankaj, U.; Chand, S.; Verma, R. K. Arbuscular mycorrhizal fungi assisted phytoextraction of toxic metals by *Zea mays* L. from tannery sludge. *Soil Sediment Contam.* **2019**, *28*, 729–746. https://doi.org/10.1080/15320383.2019.1657381.

Smith, S. E.; Read, D. J. *Mycorrhizal symbiosis*; Academic Press: San Diego, CA, 2008.

Sudova, R.; Vosatka, M. Differences in the effects of three arbuscular mycorrhizal fungal strains on P and Pb accumulation by maize plants. *Plant Soil* **2007**, *296*, 77–83. https://doi.org/10.1007/s11104–007–9291–8.

Sudova, R.; Doubkova, P.; Vosatka, M. Mycorrhizal association of *Agrostis capillaris* and *Glomus intraradices* under heavy metal stress: Combination of plant clones and fungal isolates from contaminated and uncontaminated substrates. *Appl. Soil Ecol.* **2008**, *40*, 19–29. https://doi.org/10.1016/j.apsoil.2008.02.007.

Wu, J. T.; Wang, L.; Zhao, L.; Huang, X. C.; Ma, F. Arbuscular mycorrhizal fungi effect growth and photosynthesis of *Phragmites australis* (Cav.) Trin ex. Steudel under copper stress. *Plant Biol.* **2020**, *22*, 62–69. https://doi.org/10.1111/plb.13039.

Wu, Q. S.; Xia, R. X.; Zou, Y. N. Improved soil structure and citrus growth after inoculation with three arbuscular mycorrhizal fungi under drought stress. *Eur. J. Soil Biol.* **2008**, *44* (1), 122–128.

Xu, Z.; Wu, Y.; Xiao, Z.; Ban, Y.; Belvett, N. Positive effects of *Funneliformis mosseae* inoculation on reed seedlings under water and TiO_2 nanoparticles stresses. *World J. Microbiol. Biotechnol.* **2019**, 35. https://doi.org/10.1007/s11274–019–2656–3.

Yang, Y.; Han, X.; Liang, Y.; Ghosh, A.; Chen, J.; Tang, M. The combined effects of arbuscular mycorrhizal fungi (AMF) and lead (Pb) stress on Pb accumulation, plant growth parameters, photosynthesis, and antioxidant enzymes in *Robinia pseudoacacia* L. *PLoS One* **2015**, *10*. https://doi.org/10.1371/journal.pone.0145726.

Zandavalli, R. B.; Dillenburg, L. R.; Paulo, V. D. Growth responses of *Araucaria angustifolia* (Araucariaceae) to inoculation with the mycorrhizal fungus *Glomus clarum*. *Appl. Soil Ecol.* **2004**, *25*(3), 245–255.

Zhan, F.; Li, B.; Jiang, M.; Li, T.; He, Y.; Li, Y.; Wang, Y. Effects of arbuscular mycorrhizal fungi on the growth and heavy metal accumulation of bermudagrass [*Cynodon dactylon*

(L.) Pers.] grown in a lead–zinc mine wasteland. *Int. J. Phytoremed.* **2019**, *21*, 849–856. https://doi.org/10.1080/ 15226514.2019.1577353.

Zhang, F.; Liu, M.; Li, Y.; Che, Y.; Xiao, Y. Effects of arbuscular mycorrhizal fungi, biochar and cadmium on the yield and element uptake of *Medicago sativa*. *Sci. Total Environ.* **2019**, *655*, 1150–1158. https://doi.org/10.1016/j.scitotenv.2018.11.317.

Zhang, X.; Zhang, H.; Lou, X.; Tang, M. Mycorrhizal and non-mycorrhizal *Medicago truncatula* roots exhibit differentially regulated NADPH oxidase and antioxidant response under Pb stress. *Environ. Exp. Bot.* **2019**, *164*, 10–19. https://doi. org/10.1016/j. envexpbot.2019.04.015.

Zhao, R. B. W.; Wei, B.; Nabi, A.; Guoa, J.; Wanga, L.; Zhaoa, J.; Zhang, J. Arbuscular mycorrhizal fungi affect the growth, nutrient uptake and water status of maize (*Zea mays* L.) grown in two types of coal mine spoils under drought stress. *Appl. Soil Ecol.* **2015**, *88*, 41–49.

CHAPTER 13

IMPACT OF ARBUSCULAR MYCORRHIZAL FUNGI IN GLOBAL SUSTAINABLE ENVIRONMENT

MAHA EL-KADY[1], ABDEL RAHMAN MOHAMMAD AL-TAWAHA[2], MARIYANA LYUBENOVA[3], and JEYABALAN SANGEETHA[4]

[1]*Department of Self-Pollinated Vegetables Crops, Horticulture Research Institute (HRI), Agriculture Research Centre (ARC), Giza, Egypt*

[2]*Department of Biological Sciences, Al Hussein Bin Talal University, Maan, Jordan*

[3]*Sofia University, Department of Ecology and Environmental Protection, 8, Bul. D. Tzankov, Sofia, Bulgaria*

[4]*Department of Environmental Science, Central University of Kerala, Kasaragod, Kerala, India*

ABSTRACT

Although there is an increase in food production and crop yields recently which is considered positive outcome especially for developed countries, but this positive outcome has been accomplished by bushy agrarian practices that lead to massive environmental problems. However, in developing countries there are different kinds of problems that include the shortage of suitable agrarian practices and fertilizers that do not permit intensive production of crops and the yearly overpopulation in these countries leads to a large segment of the population suffering from undernourishment particularly among children. For these reasons, there is a constant need for the

Mycorrhizal Technology: Managing Plant Stress and Mitigating Climate Change using Mycorrhizae.
Jeyabalan Sangeetha, Abdel Rahman Mohammad Al-Tawaha, & Devarajan Thangadurai (Eds.)

unification of sustainable practices of agriculture on the universal level for both developed and developing countries. In the developed world, there is an urgent need to reduce environmental and energy costs, while, in developing countries, sustainable and effective practices are required to permit enough foods production and crops that provide nutrition with suitable cost for the increasing populations. This chapter discusses the impact of arbuscular mycorrhizal fungi (AMF) in global sustainable environment.

13.1 INTRODUCTION

To eliminate the environmental troubles arising from the unwise use of artificial fertilizers and energy costs for manufacturing and at the same time for boosting enough crop yields in the case of lack of resources for having the expensive fertilizers in developing countries, microorganisms can help in fixing these problems and allow more inexpensive and efficient nutrient resources for plants or increase nutrient availability in soils that can give sustainable solutions for the present agricultural practices and in the future (Aggarwal et al., 2011). Since the inception of the earth a relationship has been arose between fungi and plants by mycorrhizas formation which is considered one of the most notable and lasting adaptations to life on earth. Hyphae and spores of glomeromycotan were authenticated since 460 million years of ancient fossils; however, these fossils did not have direct association with plants (Redecker et al., 2000). The direct proof of fungal fossil association with plant came from 407-million years old ancient Rhynie chert (Trewin and Rice, 2004). AM fungi fossils have been reported since 400 million years (Taylor et al., 1995; Phipps and Taylor, 1996). The mycorrhizae formation is founded in 85% of existing plants (Brundrett, 2004; Smith and Read, 2008; van der Heijden et al., 2015; Brundrett and Tedersoo, 2018). Mycorrhiza is a source of regulatory and supporting ecosystem services and directly related to material services through primary productivity. Mycorrhiza (from the Greek μύκης, "mushroom" and ῥίζα, "root") is a mycotrophy in green plants (symbiosis between a mushroom and the roots of a higher plant). It plays an important role in plant physiology, soil biology, and chemistry (Johnson, 1997; Kirk et al., 2001). Over 95% of higher plants can form mycorrhiza, and for some species mycorrhiza is a condition for their existence. Therefore, it is a universal phenomenon that appeared at the beginning of life on land, about 425 million years ago. Only some families, such as Brassicaceae and Chenopodiaceae, do not feed mycotrophically (Fig. 13.1).

Furthermore, mycorrhizal associations take many forms in living species that include different clades of fungi and plants (van der Heijden et al., 2015). Associations including cyanobacteria or green algae form lichen (Selosse and Le Tacon, 1998). While Endomycorrhizas that are considered the most prevailing associations involve Glomeromycotina which form arbuscular mycorrhiza (AM), usually vascular plants roots colonized by these rhizomes and fungi sometimes (Wang and Qiu, 2006; Smith and Read, 2008; Pressel et al., 2016). There hyphae grow between plant cells in the apoplastic space and they perforate cells and form arbuscules, for instance: ramify structures formed in the nutritious exchange between plant and fungus (Bonfante and Genre, 2010). AM (arbuscular mycorrhiza) fungi provided early land plants with nutrition via their hyphae (Dotzler et al., 2009; Bonfante and Selosse, 2010). Since their origin, from 400 million years or more, AM fungi have traveled numerous miles around the earth and faced different ecological conditions. Now, it is recognized clearly that more than 80% of vascular plants are linked with arbuscular mycorrhizal fungi (Smith and Read, 2008; Brundrett, 2009). Associations of arbuscular mycorrhizal fungi are reported from ecological systems all over the world such as moderate to tropical forests, mountains, deserts, sand dunes, grazing lands, agrarian ecosystems, and aquatic plants as well as heavy metals polluted soils (Kehri et al., 2018).

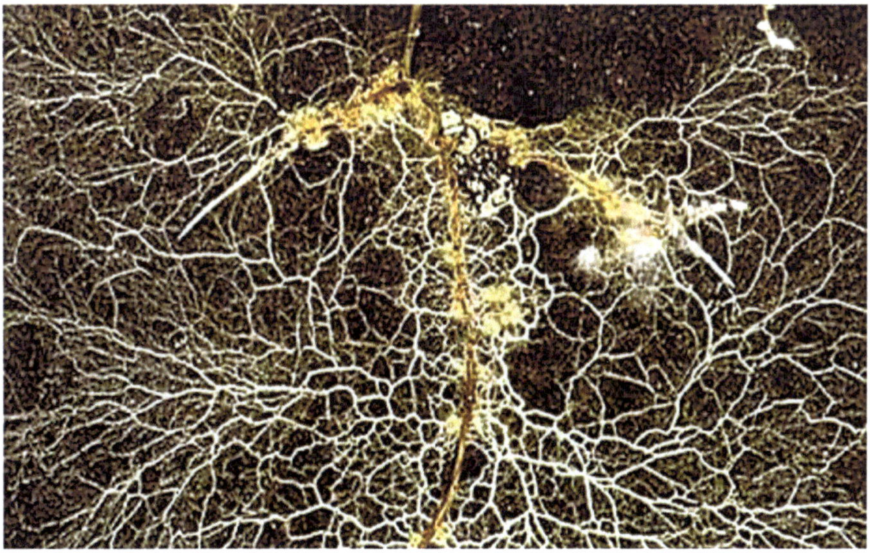

FIGURE 13.1 Root with mycorrhiza.

The benefit for the mycorrhizal fungi from their coexistence with higher plants is the extraction of hydrocarbons from the roots, which has been proven using the radioactive isotope of carbon (^{14}C) (Lyubenova, 2004). Mycorrhiza is highly developed in good light (intensive photosynthesis) and is suppressed in dimming or complete lack of light, also in artificial retention of the movement of assimilates in the roots. In severe soil drought, the mycorrhizal fungi can absorb water from higher plants. When the soil moisture is lower than the coefficient of wilting, the condition of mycorrhiza is completely satisfactory. Among the mycotrophic plants there are species that cannot exist without mycorrhiza and their seeds germinate only if they are infected with symbiont fungi (e.g., species of the family Orchidaceae); others are clearly oppressed or develop in the presence of mycorrhizal agents, and without them, if the conditions are favorable.

Mycorrhizae are divided into ectomycorrhizae and endomycorrhizae. In the first case, the fungal mycelium entangles the young root endings and penetrates only into the intercellular spaces of the integumentary tissues (e.g., *Quercus* L., *Abies* Mill., *Picea* Mill., *Betula* L., *Salix caprea* L.). Ectomycorrhizae consist of a hyphal mantle covering the top of the root and a network of hyphae surrounding the plant cells in the root bark. During the development of this mycorrhiza, the root endings are tightly wrapped with fungal hyphae, and the root hairs die. Endotrophic mycorrhiza is associated with the penetration of fungal hyphae through the cell walls into the parenchymal cells of the roots (species of the family Ericaceae, etc.) (Lyubenova, 2004; Peterson et al., 2004).

Ectomycorrhiza has been found in about 10% of plant families—Betulaceae, Myrtaceae, Fagaceae, Pinaceae, Dipterocarpaceae, Rosaceae, and Orchidaceae, entering into symbiotic relationships with *Basidiomycota* (de Bary, 1866), *Ascomycota* (Cavalier), and *Zygomycota* (Wang, 2006). Some ectomycorrhizal fungi, such as *Leccinum scabrum* (Gray, 1821) and *Suillus luteus* (Roussel, 1796), are symbiotic with only one specific plant species, while other fungi, such as *Amanita*, form mycorrhizae with many different plants. A tree can have 15 or more different mushroom partners at the same time. There are thousands of ectomycorrhizal fungi, which are accepted by over 200 plant families (Rinaldi, 2008). Nutrients can be transported between different plants through the fungal network in the soil. There are records of carbon movement from *Betula papyrifera* Marshall to *Pseudotsuga menziesii* (Mirbel) Franco, contributing to ecosystem succession (Simard et al., 1997). It has been found that the ectomycorrhizal fungus *Laccaria bicolor* (Maire) P.D.Orton (1960) lures and kills soil springtails to obtain nitrogen, some of which can be transferred to the partner plant. According to studies, *Pinus strobus* L. together with *Laccaria bicolor* can take up to 25% of the required nitrogen of the class *Collembola* (Avebury, 1870) — Fungi kill insects and feed host plants (Klironomos et al., 2001). Compared to nonmycorrhizal roots, ectomycorrhizal roots may contain high concentrations of trace elements, including toxic metals (cadmium, silver) or chlorine (Cejpková et al., 2016). An example of ectoendomycorrhiza is arbutoid mycorrhiza. It is a characteristic of plants of the subfamily Arbutoideae and is close to ectomycorrhiza, both functionally and in terms of the participating fungi, but some hyphae penetrate the root cells.

Endotrophic mycorrhizae are largely found in the species of the genera *Quercus* L., *Picea* Mill., *Abies* Mill., *Larix* Mill. under a variety of conditions — different types of soils, with different humidity, different soil acidity (pH from 3.8–8), different altitudes, in plantations or single trees. Endomycorrhiza can take many forms. The most common form is arbuscular mycorrhiza, which occurs in 70% of plant species (Peterson et al., 2004). In this mycorrhiza, the hyphae enter plant cells and form bubbles or two branches. However, the hyphae do not penetrate the protoplast, but invaginate the cell membrane. The structure of the hyphae significantly increases the contact area between the hyphae and the cytoplasm of the cell, thus facilitating the transfer of nutrients between them. Arbuscular mycorrhiza is formed only by fungi of the class Glomeromycetes. Fossil evidence (Remy et al., 1994) and DNA analysis (Simon et al., 1993) show that this mycorrhiza occurred about 400–160 million years ago, when the first plants began to colonize the land. Arbuscular mycorrhiza occurs in 85% of all plant families and is very common in cereals (Wang, 2006). The hyphae of arbuscular mycorrhizal fungi produce the glycoprotein glomalin, which is perhaps one of the largest sources of carbon in the soil (Dimitrova et al., 2014). These fungi have probably been asexual for millions of years, and individuals may contain many genetically distinct nuclei (heterokaryons) (Hijri et al., 2005). Ericoid endomycorrhizae have a simple intra-root phase composed of dense coils of hyphae in the outermost layer of root cells. The outer-root phase is composed of rare hyphae that do not extend very far into the surrounding soil. They can form carpophores (probably in the form of small bowls). Ericoid endomycorrhizae have a significant saprotrophic ability, which allows the intake of nutrients from incompletely decomposed materials (Read et al., 2003).

Monotropoid mycorrhiza occurs in plants of the subfamily Monotropoideae, as well as in several genera of Orchidaceae. These plants parasitize on the fungus, obtaining carbon through mycorrhiza, plant species increase their absorption surface by 75 to 100 times and the suction capacity of roots. In mycorrhizal species the ratio between the biomass of the roots and the aboveground biomass is much smaller in comparison with the nonmycotrophic ones and without mycorrhiza it is not possible to obtain the necessary amounts of water and minerals. Plants receive 10–15 times more nutrients than without mycorrhiza. This is seven to eight times more than what plants can get from best agricultural practices and fertilization methods. Mycorrhiza not only provides many times more nutrients and water to plants, but also increases the speed of delivery of mineral elements and balances them by elements. Symbiont fungi have been reported to be able to decompose some complex organic compounds inaccessible to higher plants. By their powerful enzymes, hyphae decompose the water-insoluble forms of phosphorus, potassium, and nitrogen, but also copper, zinc, and other trace elements supplying them to the host plant. In addition to the necessary minerals, the mycorrhiza provides everything necessary for plant development—vitamins, enzymes, biostimulators, hormones, and other biologically active substances. A rich and balanced diet is directly related to the yields. The most important ability of mycorrhiza from a practical point of view is that it supplies the water necessary for the development of plants. It acts as a powerful water pump, transferring significant amounts of water to the host plant, from quite large distances. And here it is not a question of a liter or 2 or 10 L, but of quantities measured in tons for one season. Therefore, where there is mycorrhiza, the plants never need water. This is true even for the driest places. The vitamins and other growth substances released by the symbiotic fungi stimulate the development of the roots.

In the processes of the nutrient cycle, fungi release not only enzymes but also various toxins that block the development of a number of soil pathogens such as *Pythium*, *Phytophthora*, *Fusarium*, and others. The peculiarity of mycorrhiza is that it can connect several plants in one common communication network. This mechanism is directly related to the generic functions of plants. Mycorrhiza is a characteristic of almost all cultivated plants in fruit growing, vegetable production, viticulture, landscaping, forestry, and field crops. However, there is the question of the economic efficiency of its artificial import. Since in nature mycorrhizal symbiosis is a spontaneously arising relationship, the question arises why it is necessary to infect plants with mycorrhizal fungi artificially? Won't this happen on its own? Unfortunately, the intensive cultivation of the land, the increasing use of pesticides, and the production of sterile planting material have greatly reduced the number of mycorrhizal fungi living in the soil; therefore, their number needs to be restored. And although with mycorrhiza we will not be able to completely stop the use of chemical fertilizers, they can be significantly reduced. If we take into account the ever-increasing prices of fertilizers and the restrictions imposed on the use of nitrates and pesticides, then the use of mycorrhiza becomes more than desirable. Mycorrhiza as a process provides the following regulatory and supporting ecosystem services:

1. Regulation and maintenance of the nutrient regime and the soil moisture regime—increasing the intake of nutrients and water from the plants.
2. Regulation of flowering and production—earlier and simultaneous flowering, therefore better yields.
3. Maintaining good physiological status of plants.
4. Increasing the resistance of plants to various types of stress (drought, heat, uneven distribution of precipitation, the appearance of pests and diseases, transplanting, etc.).
5. Blocking the development of some pathogenic microorganisms that cause significant damages of cultures.
6. Reduction of costs for watering, the use of fertilizers and pesticides, due to the reduced needs of cultivated plants.

13.2 TAXONOMY OF ARBUSCULAR MYCORRHIZAL FUNGI

The arbuscular mycorrhizal fungi (AMF) taxonomy is based on the morphological attributes of spores produced in soil, including cell wall layers, color, and size (Morton, 1988; Morton and Benny, 1990). There are some difficulties that appear while making AMF taxonomy and these difficulties include, the spores collected from soils are not in their authentic conditions and a lot of AMF species do not produce spores under controlled status. These morphological identification problems of AMF make species characterization only determined by experts, in the next pages we show some important stations of arbuscular mycorrhizal taxonomy. Arbuscular mycorrhizal fungi (AMF) have been placed in order Glomales (Zygomycota) within another six genera using both morphological characteristics and phylogenetic analysis (Morton and Benny, 1990). The first characterization of the AM fungi has been published by Tulsane and Tulsane (1845). They described the two species of *Glomus* (*G. macrocarpus* and *G. microcarpus*), according to its spores formation. Berkeley and Broome (1873) created the genus *Sclerocystis* for involving all species which form spores in teeny sporocarps. The classification of both genera was in order Mucorales, family Endogonaceae that was proposed for the first time by Fries in 1849. The Endogonaceae family was amended later by Thaxter (1922) and included four genera (*Endogone*, *Glaziella*, *Sclerocystis*, and *Sphaerocreas*) and all genera were from sporocarpic fungi that produce both zygospores and chlamydospores. Later in 1953 Mosse discovered that adding the sporocarps isolated from the mycorrhizal strawberry roots of *Endogone* species to sterilized soil with strawberry shoots resulted in an ideal mycorrhizal colonization. This fungus is later described as *Endogone mosseae* (= *Glomus mosseae*) in honor of its discoverer. In 1963, Gerdemann and Nicolson discovered moisten sieving and decanting method for nonsporocarpic and sporocarpic fungal isolation which created an interest in AM fungi taxonomy significantly among researchers. In 1968, the first key of endogonaceous spores identification was prepared by Mosse and Bowen who classified seven genera (*Glomus*, *Sclerocystis*, *Acaulospora*, *Gigaspora*, *Endogones*, *Glaziella*, and *Modicella*) of vesicular arbuscular mycorrhizal fungi which belong to Endogonaceae family, order Mucorales, and phylum Zygomycota. Endogonaceae family was amended and classified later by Gerdemann and Trappe (1974); they described 44 species subdivided into seven genera. Twelve new species and two new genera (*Acaulospora* and *Gigaspora*) were described. The genus *Glomus* was split up from the *Endogone* (as merged previously by Tulasane brothers into *Endogone*) and established as independent genus. The genus *Glomus* included 19 species and they all produce spores blastically and genus *Sclerocystis* included four species. There are different types of spores according to species ranging from compact to loose aggregates of spores without or with peridial wall. Thereafter, as *Modicella* and *Glaziella* genera not forming arbuscular mycorrhizal association, Trappe (1982) transferred *Modicella* to the family Mortierellaceae and Gibson et al. (1986) transferred *Glaziella* to the family Ascomycota. This classification presented a fundamental regulation for the AM fungi taxonomy and was used as reference for many years later. A lot of new species have been discovered during the period from 1975 to 1989, after 12 years of Gerdemann and Trappe (1974) monograph, the AM fungi species jumped to 77 (Trappe, 1982) and after another 6 years Schenck and Perez (1988) registered 126 species. In this period of time AM fungi were being classified only on confirmed subcellular structures of spores and morphological characters. A new terminology was proposed by Walker (1983) for describing the AM fungi. He proposed (walls) expression for the AM fungi identification, which can be grouped according to wall types, specified by spore's phenotype whether it is broken or intact, also he proposed the (murograph) expression for walls description. The murograph is a graphical exemplification of spore's wall layer differentiation. Walker defined wall types as unit, evanescent, laminated, and membranous. When new species were added, more mycorrhizal wall types were introduced: amorphous (Morton, 1986), expanding (Berch and Koske, 1986), notched (Koske and Gemma, 1995), coriaceous (Walker and Sanders, 1986), and germinal (Spain et al., 1989). Later, Berch (1986) criticized this terminology, he had argument that each wall layer should describe the spore characters and be based on the knowledge of each wall layer origin. The Berch view has opened the road to study the development of spore. Morton (1990) analyzed 57 AM fungi species, 27 phenotypic mycorrhiza, and spores characters. Morton hypothesized that all AM fungi included a monophyletic association by participation nature of combined symbiosis and arbuscules formation with roots. Later, Morton and Benny (1990) put the arbuscule fungi in a new order named Glomales (now Glomerales). They divided the order glomales into three families, Glomeraceae, Gigasporaceae, and Acaulosporaceae. Every family included two genera, Glomeraceae family consisted of *Sclerocystis* and *Glomus*, Acaulosporaceae family included *Entrophospora* and *Acaulospora*, and family Gigasporaceae included *Scutellospora* and *Gigaspora*. The difference between families Glomeraceae and Gigasporaceae is fundamentally by the existence of vesicles for the initial one, which are absent in Gigasporaceae family. This classification was based on the spore development, cladistic analysis, and germination mode of spore. The AM fungi was placed into the order Endogonales by Gerdemann and Trappe (1974) and placed into order Glomerales by Morton and Benny (1990). But, both of them did not obviously state the AM fungi class. Then, Cavalier-Smith (1998) placed fungi which formed arbuscular mycorrhizal group with plants in the new class Glomeromycetes and the new phylum Archeomycota. Microscopic checking of spores was the only way to differentiate AM fungi before using molecular techniques (Kehri et al., 2018). The first try for using molecular techniques was made by Morton and Redecker (2001) who instituted in the order Glomales (now Glomerales) two new families, which are Paraglomaceae (now Paraglomeraceae) containing single genus, *Paraglomus* and Archaeosporaceae which contained single genus, *Archaeospora*, with three species forming *acaulospora*, *gerdemannii*, and *leptoticha*. Then, ribosomal RNA sequence analysis cleared that all arbuscular mycorrhizal fungi species are involved in a monophyletic clade and excluded from other groups of fungi. Besides that, all arbuscular mycorrhizal fungi species group were diverse so they set into a new monophyletic phylum Glomeromycota and this separation gives it natural and proper status, the most widely used region was rDNA small subunit (SSU) which is used in phylogenetic sequence analysis for AMF (Schüßler et al., 2001). In addition, Schüßler et al. (2001) moved Glomarales to phylum level; moreover, they established three new orders Paraglomerales, Archaeosporales, and Diversisporales. In early June 2004, Oehl and Sieverding added four new species, redescribed and placed two species from *Glomus* to new genus *Pacispora* and this new genus belongs to the family glomeraceae. Later in 2006, Sieverding and Oehl modified five species of *Entrophospora*, according to differences in root infection characteristics and spore morphologies and they kept only *Entrophospora baltica* and *Entrophospora infrequens* in the genus *Entrophospora* in the new family Entrophosporaceae.

The other three species that were excluded from the genus *Entrophospora* were organized in two new genera. *Kuklospora* genus was placed in family Acaulosporaceae with two species *Kuklospora colombiana* and *K. kentinensis* and *Intraspora* genus was included in Archaeosporaceae family with only one species *Intraspora schenckii*. In 2006, a new genus has been described by Spain et al. named *Appendicispora* in Archaeosporaceae family with three species (*Ap. gerdemannii, Ap. jimgerdemannii*, and *Ap. appendicula*), but *Ap. gerdemanii* moved later to new genus *Ambispora*. Walker et al. (2007) established and described a new genus *Ambispora* that contains three species (*A. fennica*, *A. leptoticha*, and *A. gerdemannii*) on the morphological evidences basis and using techniques of molecular ITS and SSU regions of rDNA, and these three species are known to produce both acualosoroid and glomoid spores, while another species *A. callosa* is known to produce glomoid spores only. The new genus was set into new family named *Ambisporaceae* along with its four species in the order *Archaeosporales*. Oehl et al. (2011a) proposed a classification based on both morphological and genetic characters, genetic characters include SSU and LSU rRNA and partial sequences of β-tubulin, morphological characters include shape, color, and thickness, etc. Their classification divided Glomeromycota Phylum into three classes Glomeromycetes, Paraglomeromycetes, and Archaeosporomycetes, class Glomeromycetes includes three orders Glomerales, Gigasporales, and Diversisporales. In the same year, Oehl et al. (2011d) proposed a new genus namely *Orbispora* in the family Glomeromycota (Scutellosporaceae), *Orbispora* spores are formed eventually with three spore walls on sporogenous cells. Furthermore, the same authors Oehl et al. (2011f) divided the Phylum Glomeromycota into 3 classes, 5 orders, 14 families, and 29 genera. They found that the spore formation in 10 genera was exclusively glomoid, seven were scutellosporoid, four were entrophosporoid, two were acaulosporoid, one was pacisporoid, and one was gigasporoid. The rest four genera, three have bimorphic spore formation, and one genus *Geosiphon* are related to cyanobacteria. A new classification is based on morphological and molecular studies suggested by Goto et al. (2012) who established the new family Intraornatosporaceae which is placed in the order Gigasporales which has new two species (*P. bahiana* and *P. maritime*) and *Intraornatospora* which has one new species *I. intraornata*. Błaszkowski et al., in the year 2013, described two new species within *Septoglomus* genus as *S. furcatum* and *S. fuscum*, based on ITS, LSU, and SSU of nrDNA sequences phylogenetic analyses. Moreover, the authors used *Plantago lanceolata* as a host plant with single species culture and found that both species formed arbuscular mycorrhizae. Later, three new species were discovered by Symanczik et al. (2014) from the dried sandy plain of the Arabian Peninsula, based on morphological and molecular analyses of ITS, LSU, ITS, and SSU of rDNA sequences. The new species namely *Diversispora omaniana* and *Rhizophagus arabicus* were isolated from the native dried habitat and *Septoglomus nakheelum* was isolated from the nearby area planted with date palm. Also, in the same year, Błaszkowski et al. (2014) established two new species and placed them in *Septoglomus* genus (*S. turnauae* and *S. jasnowskae*) of Glomeromycota. Not long ago, Sieverding et al. (2015) described new genus, *Rhizoglomus* that was typified by *Rhizoglomus intraradices* which belongs to Glomeraceae family. The new genus phylogenetic analysis showed that it forms a separate clade within family Glomeraceae. The new genera consisted of 13 species *R. intraradices*, *R. antarcticum*, *R. microaggregatum*, *R. arabicum*, *R. custos*, *R. clarum*, *R. fasciculatum*, *R. irregulare*, *R. invermaium*, *R. manihotis*, *R. natalense*, *R. aggregatum*, and *R. proliferum*. In the same year, two new species (*Dominikia difficilevidera* and *Dominikia duoreactiva*) have been reported by Błaszkowski et al. (2015) and placed within recently established genus *Dominikia* (Glomeromycota) on the phylogenetic analyses basis (ITS, LSU, and SSU sequences of rRNA) and morphology analyses. Moreover, the authors reported that *D. duoreactiva* is distributed in specific places, while *D. difficilevidera* has a distribution worldwide.

On the same line, Błaszkowski et al. (2015) established a new species *Glomus tetrastratosum*, they proved that *Glomus tetrastratosum* belongs to the family Glomeraceae with the basis of morphological and phylogenetic analyses of ITS, LSU, and SSU of rRNA sequences. In single species culture with *Plantago lanceolata*, *G. tetrastratosum* was found to form structures like vesicles and arbuscules.

There is another kind of taxonomy which describes the relation types between mycorrhiza fungi and the host plant roots which change in the root morphology and fungi (Harley and Smith, 1983); this classification consists of the following:

1. *Ectomycorrhizae* are differentiated by the existence of hyphal plait arising among cortical cells of root and producing a net or structure called "Hartig net." An ectomycorrhizae fungal associate in the most cases belongs to the Basidomycota, but in some cases, it belongs to an Ascomycete.
2. *Ectendomycorrhiza* was first discovered by Melin in 1923. It is defined as an intermediate association type between Ecto and Endo-mycorrhizae and has the features of both the types (Harley, 1989). It has intercellular as well as intracellular hyphae in the cortical region of roots.
3. *Endomycorrhizae* are differentiated by the fungal hyphae type which breaks through the host plant roots cortical cells. The hyphae have intercellular as well as intracellular associations within roots and they do not have external sheath. This association type takes place in all crops species and basically consists of three categories according to Wilcox (1991): (1) Ericaceous mycorrhizae, (2) Orchidaceous mycorrhizae, (3) Arbuscular mycorrhizal fungi (AMF).
4. *Ericaceous mycorrhizae*: This type of association is found in order Ericales plants. It does not form arbuscules. Also it is divided into three types: (1) Arbutoid mycorrhizae, (2) Ericoid mycorrhizae, and (3) Monotropoid mycorrhizae.
5. *Orchidaceous mycorrhizae*: Form intracellular hyphal coils called "peloton coil" by penetrating the host root cells and majority of it are members in Basidiomycota, such as *Rhizoctonia*, *Tulasnella*, *Armillaria*, etc.
6. *Arbuscular mycorrhizal fungi (AMF)*: They are differentiated by the existence of intracellular hyphae in the primary cortex that form arbuscular and vesicles. Arbuscules had been named by Gallaud (1905) like that because they resemble with trees shape. Vesicles are with thick-walls, thin-walled globose or subglobose or with asymmetrical shape structures. They are found in all angiospermic families especially cereals and legumes (Newman and Reddell, 1987) as well as in various tree crops (Cuenca and Meneses, 1996; Graham et al., 1997; Mamatha et al., 2002). Exclude some families such as Urticaceae, Commelinaceae, Betulaceae, Polygonaceae, and

Cyperaceae. Earlier, since some groups do not produce vesicles, so AMF expression is preferred (Friberg, 2001). Some important genera names are *Glomus*, *Acaulospora*, *Scutellospora*, *Sclerocystis*, *Enterophospora*, and *Gigaspora*.

13.3 THE ROLE IN IMPROVING CROPS NUTRITION UPTAKE

In the modern earthly ecosystems, cyanobacteria, plants, and algae develop fundamental symbiotic associations with fungi in the soil and these relations are recognized to be old (Strullu-Derrien et al., 2017). In association with plant roots, AM fungi produce arbuscules, vesicles, spores, and hyphae inside the roots skin and hyphae, spores, and vesicle out of the roots (Kehri et al., 2018). Carbon from the host plant is usually received by fungi, which in return is provided with fundamental soil components such as phosphorus (P) and nitrogen (N); other benefits include crops development under abiotic and biotic stresses which have negative effects on crops production such as toxic compounds, drought, pests, and immoderate use of pesticides and fertilizers as well as encouraging crops development in normal conditions (Alqarawi et al., 2014a; Alqarawi et al., 2014b; Navarro et al., 2014). AMF is able to improve soil fertility, soils inoculated with arbuscular mycorrhizal fungi composed of more steady masses and numerous hyphal mycelium which promotes the water-holding capacity in soil. Moreover, it is considered as natural growth regulators for a large number of terrestrial plants, bio-inoculants, and bio-fertilizers in sustainable productivity of crops (Barrow, 2012).

There are some growth-related functions that are affected by AMF inoculation, for instance, stomatal conductance, leaf water potential, relative water content (RWC), CO_2 assimilation, and PSII efficiency (He et al., 2017; Chandrasekaran et al., 2019). AMF enhances the uptake of mineral nutrition by plants even under stress conditions. Zhao et al. (2015) inoculated soil with *Zea mays* and exposed it to drought conditions with *Rhizophagus intraradices* strain BGCBJ09 that enhanced the uptake of P, N, K and increased plant dry mass, water use efficiency, and Mg concentration in shoots (Fig. 13.2).

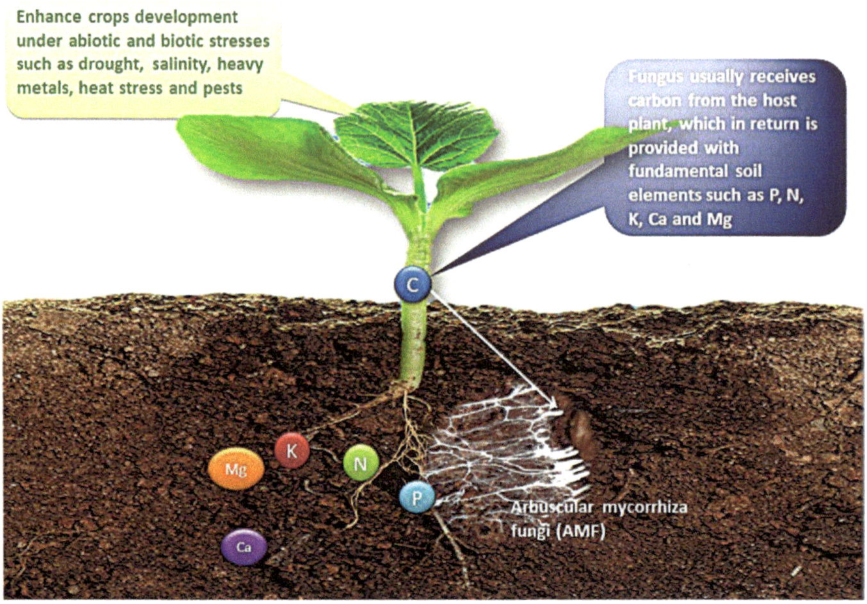

FIGURE 13.2 Arbuscular mycorrhizal fungi role in mineral nutrition.

13.4 ARBUSCULAR MYCORRHIZA FUNGI AS BIO-FERTILIZERS

Bio-fertilizers are a mix of natural substances that are used to enhance soil fertility and crops production (Sadhana, 2014). Numerous studies cleared the countless benefits of arbuscular mycorrhiza fungi on both crop productivity and soil health during the past two decades (Begum et al., 2019). Thus, arbuscular mycorrhizal fungi (AMF) could be considered a benefit replacement of nonorganic fertilizers in near future, because using it can successfully reduce chemical fertilizer quantities that are used in agriculture particularly phosphorus (Ortas, 2012). It is known that persistent use of nonorganic fertilizers, fungicides, and herbicides has caused numerous disturbances to plants, soil, animal, and human health by their damaging effects on soil, water, and air systems and food production quality through different food chains (Yang et al., 2004). AM fungal inoculation was made in a greenhouse experiment on "Rodade" tomatoes; the results showed superior seedlings performance according to their high shoot/root ratio, higher shoot fresh weight, higher root growth rate, and higher root biomass (Oseni et al., 2010). In addition, Ziane et al. (2017) reported the AM fungi importance in easing fertilizer exploitation to the most to achieve satisfying yield and growth of the tomato crop. In addition, they proposed that the AM fungi usage could recompense the decreasing of chemical fertilizers, consequently introducing a further sustainable agricultural system that was suitable for the ecosystem. Organic farms depend on AMF as bio-fertilizer, the degree of AMF colonization with roots, species diversity, and spore density of AMF are reported as the highest in the organic farms (Limonard and Ruissen, 1989). Samples taken by Eason et al. (1999) from grassland areas in organic and nonorganic farms showed that AMF colonization of spore density and pasture roots were higher for both of them, for the organic sites. The higher plenty of AMF in organic farms is generally ascribed to the absence of easily dissoluble P fertilizers or the lack of availability in the soil, the lower biocides usage, and more diversity in agricultural rotations (Ryan and Tibbett, 2009).

Inoculation of arbuscular mycorrhiza fungi stimulates plants' nutrient uptakes and then boosts macro and micro nutrient concentrations significantly, which leads to increasing photosynthesis percent and biomass aggregation (Chen et al., 2017; Mitra et al., 2019). Furthermore, AMF is capable of increasing inorganic nutrients uptake in plants, particularly phosphate fertilizers (Smith et al., 2003; Nell et al., 2010) can increase the concentration of phosphorus significantly in both shoots and roots (Al-Hmoud and Al-Momany, 2017) and also it is very efficient in helping plants to uptake nutrients from the poor nutrients soil (Kayama and Yamanaka, 2014). Besides, Smith and Read (1997) reported that AMF association can boost the availability of some micronutrients such as copper and zinc and improve the capability of surface absorbing of host roots (Bisleski, 1973). Inoculating soil with AMF can enhance plant growth such as in experimental trials on tomato plants planted in that soil, the results showed increase in nitrogen, calcium, leaf area, potassium, and phosphorus contents (Balliu et al., 2015). Nitrogen translocation inside seeds can promote from beginning to maturity, after AMF symbiosis setting that produces extraradical mycelia underground extending from the roots up to the surrounded rhizosphere and that helps in increasing nutrients absorption particularly nitrogen (Battini et al., 2017).

Arbuscular mycorrhiza fungi provides the mineral nutrition to plants even under unsuitable conditions; some examples for that unsuitable statuses, Zhao et al. (2015) inoculated soil planted with *Zea mays* and exposed to drought conditions with strain BGCBJ09 and *Rhizophagus intraradices* and that increased uptake of P, N, K, plant dry weight, water use efficiency, and Mg concentration in shoots. While Amiri et al. (2015) planted *Pelargonium graveolens* with drought conditions and inoculated soil with *Funneliformis mosseae* and *Rhizophagus intraradices*, that increased plant biomass, improved nutrient concentration, and increased essential oil content and glomalin proteins in related soil (GRSP). Also, Amiri et al. (2017) reported significant increase in P, N, and Fe concentrations in *Pelargonium graveolens* L. under the conditions of drought stress and mycorrhizal inoculation. Planting *Acacia nilotica* in soil with salinity conditions and inoculated with *Glomus fasciculata*, boosted P, Cu, Zn contents and improved shoot and root biomass (Giri et al., 2007). Under the conditions of limited phosphorus, AMF can increase phosphorus uptakes of the host plants infected roots (Bucher, 2007); for instance, phosphorus absorption rate was increased significantly in maize plants colonized with AMF (Garcés-Ruiz, 2017). Improvement of photosynthetic rate and else leaf functions are related to improved growth frequency directly by AMF colonization that related directly to the N and P absorption under lower and higher phosphorus levels with diverse irrigation systems (Liu et al., 2014; Liu et al., 2018). In addition, Gomez-Bellot et al. (2015) noticed positively increased levels of P, K, and Ca in *Euonymus japonica* under the conditions of salinity stress due to attachment of instant fungi. Under low fertilizer levels, arbuscular mycorrhiza fungi mediated improved shoot biomass distribution to grains and panicles through increase in redistribution of N and P, especially to panicles that had been demonstrated by Zhang et al. (2018a).

13.5 ARBUSCULAR MYCORRHIZA FUNGI ROLE IN ENHANCING CROPS QUALITY CHARACTERS

AMF role not only limited for improving the nutrient crops uptake, but also can improve the quality of crop characters. For instance, strawberry planted in soil wealthy with AMF presented increased secondary metabolites levels leading to improving antioxidant properties (Castellanos-Morales et al., 2010). Arbuscular mycorrhiza fungi can improve the crops' characters quality by affecting volatile compounds and carotenoids production (Hart et al., 2015). Beneficial effects of arbuscular mycorrhiza fungi on tomatoes quality had been observed by Bona et al. (2017). In another study, enhanced citrus fruit quality had been reported by Zeng et al. (2014) through increasing organic acids contents, sugars, vitamin C, minerals, and flavonoids due to *Glomus versiforme* (arbuscular mycorrhizal fungi). AMF improved lettuce growth and increased anthocyanins, Cu, Fe, carotenoids, and phenolics concentrations in mycorrhizal plants than in non-mycorrhizal plants (Baslam et al., 2011). Furthermore, Dasgan et al. (2008) determined mycorrhizal inoculation effects on tomato plants under hydroponic greenhouse system. The fruits' size and fruit yield significantly increased with mycorrhizal inoculation. Moreover, mycorrhiza formed a superior nutritional condition by increasing the soluble sugars and ascorbic acid amounts by solubilizing phosphorus in tomato plants (Kowalska et al., 2015) (Fig. 13.3).

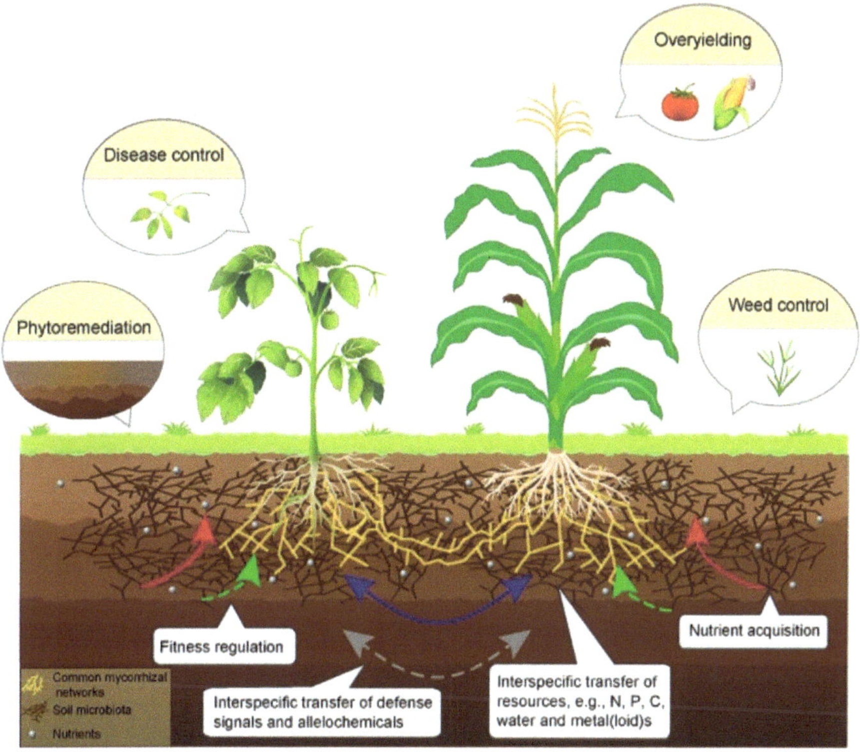

FIGURE 13.3 The potential benefits of arbuscular mycorrhizal (AM) fungi in intercropping systems, for example, overyielding, weed and disease control, and phytoremediation. Arrows and boxes show proposed mediation of AM fungi for uptake and/or interspecific transfer of metabolic resources, infochemicals and defense signals below ground.

Source: Reproduced with permission from Li et al. (2021), Copyright © Springer Nature (2021).

13.6 ARBUSCULAR MYCORRHIZA FUNGI AND ABIOTIC STRESSES (DROUGHT, SALINITY, AND HEAVY METALS)

Drought stress affects plant life in different ways such as follows: drought reduces induced oxidative stress as well as transpiration rate (Impa et al.,

2012; Hasanuzzaman et al., 2013). Furthermore, drought stress causes harmful effects on plant growth by affecting ion uptake, nutrient assimilation, and enzyme activity (Ahanger and Agarwal, 2017; Ahanger et al., 2017a). There is a vigorous proof of AMF reduction effect of drought stress in many crops such as soybean, maize, onion, strawberry, wheat, and barley (Mena-Violante et al., 2006; Ruiz-Lozano et al., 2015; Yooyongwech et al., 2016; Moradtalab et al., 2019). Such a symbiotic relationship regulates some plant processes such as controlling ABA metabolism that controls stomatal conductance (Duan et al., 1996), increasing osmotic adjustment (Kubikova et al., 2001), raising glutathione level (Rani, 2016), and enhanced proline aggregation (Ruiz-Sanchez et al., 2010; Yooyongwech et al., 2013). Symbiotic association of AMF with many plants improves plant biomass under drought conditions, root size, and leaf area index (Al-Karaki et al., 2004; Gholamhoseini et al., 2013). Furthermore, AMF symbiosis with host plant enhances gas exchange, leaf water relationships, stomatal regulation, and the rate of transpiration (Morte et al., 2000; Mena-Violante et al., 2006). A study made by Li et al. (2019) experienced the enhancement effect in photosynthesis and growth made by AMF association with C4 (*Hemarthria altissima*) and C3 (*Leymus chinensis*) plant species out of antioxidant system regulation.

Salinity stress is suppressing plant's growth by affecting shoots development and the absorption rate performed to reduction in yield productivity (Hasanuzzaman et al., 2013; Ahanger et al., 2017a). In addition, it also increases the extreme reactive oxygen species generation (Ahmad et al., 2010; Ahanger and Agarwal, 2017; Ahanger et al., 2017b; Ahanger et al., 2018). Efforts are made to enhance crop production under the conditions of salinity stress, using AMF is one of the potential methods that relieve the salinity harmful effects on plants (Santander et al., 2019). The useful effects of AMF association on physiological processes under saline regimes such as photosynthesis rate, stomatal conductance, and leaf water relations were reported by Ait-El-Mokhtar et al. (2019). Another example of beneficial role of AMF-inoculation to mitigate salinity stress when associating with *Allium sativum* plants presented growth traits improvement such as leaf area index and dry and fresh biomass under salinity conditions (Borde et al., 2010). Furthermore, Santander et al. (2019) have shown that the mycorrhizal lettuce plants under stress conditions had higher biomass weight, increased N uptake, proline percent, and remarkable ionic relations changes especially reducing Na^+ accumulation, than those in uninoculated plants. The key growth regulator levels can also be regulated by AMF inoculation. Talaat and Shawky (2014) and Hameed et al. (2014) observed that AMF-colonized plants improved the concentration of growth regulator cytokinin resulting a significantly increase in photosynthesis translocation under salinity stress condition. In addition, arbuscular mycorrhiza fungi-mediated plants with salinity stress conditions can reduce the oxidative stress by inhibiting lipid membrane peroxidation (Abdel Latef and Chaoxing, 2014; Talaat and Shawky, 2014). Also, AMF symbiosis was studied by Sheng et al. (2011) to enhance *Zea mays* plants accumulation or synthesis of organic acids which grow in saline soil, resulting in inducing increased production of betaine organic acid, ratifying the circuitous role of arbuscular mycorrhiza fungi in plant osmoregulation under the salinity stress conditions.

There is wide belief that AMF can support plants' arise in contaminated soils with heavy metals, because of their ability to support defense system of the AMF associated with plants growing in that contaminated soils. Heavy metals come from different sources such as manufacturing processes and sewage water can contaminate soils and reach and accumulate in fruits, grains, and vegetables causing numerous health problems (Liu et al., 2013; Yousaf et al., 2016). The fungal hyphae can accumulate heavy metals (Ouziad et al., 2005) inside their cell wall, vacuole and with other cytoplasm substances (Punamiya et al., 2010), thus reduces metal toxicity in plants. The effect on plant growth of AMF under acute stressful conditions is frequent because of the fungi ability to increase physiological and morphological processes that lead to increasing biomass of plants and then boosting important nutrients uptake such as Zn, P, and Cu (Kanwal et al., 2015; Miransari, 2017). The heavy metals absorption into plants can be interrupted by mycorrhizae and disturb their moving from the root area to the shoots (Dong et al., 2008; Li et al., 2015). AMF-colonized with wheat plants increase nutrient absorption positively under aluminum stress (Aguilera et al., 2014). There is a relationship between low Cd toxicity and movability with arbuscular mycorrhiza fungi by raising soil pH (Shen et al., 2006), storing cadmium in the extraradical mycelium (Janouškova and Pavlikova, 2010), and associating Cd to glomalin, which is glycoprotein. For instance, arbuscular mycorrhiza fungi are very effective in lowering the Cd levels in cell wall as well as the vacuoles, which lead to Cd detoxification in rice crop (Li et al., 2016a). AMF inoculated with alfalfa (*Medicago sativa* L.) improved tolerance to Cd perhaps as a result of the alteration in chemical forms of Cd in various plant tissues which were observed by Wang et al. (2012).

Heat stress affects plant growth significantly by suppressing seed germination, plant strength loss, slowing the growth rate, reducing biomass production, and affecting reproductive organs. Also, it affects leaves with burning and wilting, discoloring, damage in fruits, cell death, yield decreasing and raising oxidative stress (Wahid et al., 2007; Hasanuzzaman et al., 2013). Plants associated with arbuscular mycorrhiza fungi show better growth than the non-arbuscular mycorrhiza fungi associated plants under the heat stress conditions (Gavito et al., 2005). The association of AMF (*Glomus fasciculatum*) species leads to making positive changes in plant growth at high temperatures (Maya and Matsubara, 2013). Furthermore, a lot of reports declare that AMF inoculation with plants at low temperature leads to better growing and plants become more tolerant to cold stress than noninoculated plants with AMF (Zhu et al., 2010b; Abdel Latef and Chaoxing, 2011b; Chen et al., 2013; Liu et al., 2013). In addition, Zhu et al. (2010a) reported that AMF can also hold moisture in host plants, while Abdel Latef and Chaoxing (2011b) observed that AMF can boost plant secondary metabolites which lead to supporting the defense system of plants, and also supporting plants resistance to cold stress by improving protein content. Also, symbiotic relationship of AMF enhances water with plants and increases osmotic adjustment and gas exchange (Zhu et al., 2012); otherwise, it can significantly enhance many metabolites concentrations resulting from improving the synthesis of chlorophyll in plants that undergo cold stress (Zhu et al., 2010a; Abdel Latef and Chaoxing, 2011b). In addition, Abdel Latef and Chaoxing (2011b) reported the role of AMF through cold stress to modify protein content in tomato and other vegetables.

KEYWORDS

- **arbuscular mycorrhizal fungi**
- **biofertilizers**
- **drought**
- **heavy metals**
- **global sustainable environment**
- **salinity**
- **taxonomy**

REFERENCES

Abass, M.; Alyemeni, M. N.; Wijaya, L.; Alamri, S. A.; Alam, P.; Ashraf, M.; Ahmad, P. Potential of exogenously sourced kinetin in protecting *Solanum lycopersicum* from NaCl-induced oxidative stress through up-regulation of the antioxidant system, ascorbate–glutathione cycle glyoxalase system. *PLoS One* **2018**, *13*(9), e02020175.

Abdel Latef, A. A.; Chaoxing, H. J. Arbuscular mycorrhizal influence on growth, photosynthetic pigments, osmotic adjustment and oxidative stress in tomato plants subjected to low temperature stress. *Acta Physiol. Plant.* **2011**, *33*, 1217–1225.

Abdel Latef, A. A.; Chaoxing, H. J. Does the inoculation with *Glomus mosseae* improve salt tolerance in pepper plants? *Plant Grow. Regul.* **2014**, *33*, 644–653.

Aggarwal, A.; Kadian, N.; Tanwar, A.; Yadav, A.; Gupta, K. Role of arbuscular mycorrhizal fungi (AMF) in global sustainable development. *J. Appl. Natural Sci.* **2011**, *3*(2), 340–351.

Ahanger, M. A.; Agarwal, R. M. Potassium up-regulates antioxidant metabolism and alleviates growth inhibition under water and osmotic stress in wheat (*Triticum aestivum* L.). *Protoplasma* **2017**, *254*(4), 1471–1486.

Ahanger, M. A.; Tittal, M.; Mir, R. A.; Agarwal, R. M. Alleviation of water and osmotic stress-induced changes in nitrogen metabolizing enzymes in *Triticum aestivum* L. cultivars by potassium. *Protoplasma* **2017a**, *254* (5), 1953–1963.

Ahanger, M. A.; Tomar, N. S.; Tittal, M.; Argal, S.; Agarwal, R. M. Plant growth under water/salt stress: ROS production; antioxidants and significance of added potassium under such conditions. *Physiol. Mol. Biol. Plants.* **2017b**, *23*(4), 731–744.

Ahmad, P.; Jaleel, C. A.; Salem, M. A.; Nabi, G.; Sharma, S. Roles of enzymatic and non-enzymatic antioxidants in plants during abiotic stress. *Crit. Rev. Biotechnol.* **2010**, *30*, 161–175.

Ait-El-Mokhtar, M.; Laouane, R. B.; Anli, M.; Boutasknit, A.; Wahbi, S.; Meddich, A. Use of mycorrhizal fungi in improving tolerance of the date palm (*Phoenix dactylifera* L.) seedlings to salt stress. *Sci. Hort.* **2019**, *253*, 429–438.

Al-Hmoud, G.; Al-Momany, A. Effect of four mycorrhizal products on squash plant growth and its effect on physiological plant elements. *Adv. Crop. Sci. Tech.* **2017**, *5*, 260.

Al-Karaki, G.; Mcmichael, B.; Zak, J. Field response of wheat to arbuscular mycorrhizal fungi and drought stress. *Mychorrhiza* **2004**, *14*, 263–269.

Alqarawi, A. A.; Abd-Allah, E. F.; Hashem, A. Alleviation of salt induced adverse impact via mycorrhizal fungi in *Ephedra aphylla* Forssk. *J. Plant. Interact.* **2014a**, *9*(1), 802–810.

Alqarawi, A. A.; Hashem, A.; Abd_Allah, E. F.; Alshahrani, T. S.; Huqail, A. A. Effect of salinity on moisture content, pigment system, and lipid composition in *Ephedra alata* Decne. *Acta Biol. Hung.* **2014b**, *65*(1), 61–71.

Amiri, R.; Ali, N.; Nematollah, E.; Mohammad, R. S. Nutritional status, essential oil changes and water-use efficiency of rose geranium in response to arbuscular mycorrhizal fungi and water deficiency stress. *Symbiosis* **2017**, *73*, 15–25.

Amiri, R.; Nikbakht, A.; Etemadi, N. Alleviation of drought stress on rose geranium *Pelargonium graveolen* L. Herit. in terms of antioxidant activity and secondary metabolites by mycorrhizal inoculation. *Sci. Hort.* **2015**, *197*, 373–380.

Balliu, A.; Sallaku, G.; Rewald, B. AMF inoculation enhances growth and improves the nutrient uptake rates of transplanted, salt-stressed tomato seedlings. *Sustainability* **2015**, *7*, 15967–15981.

Barrow, C. J. Biochar potential for countering land degradation and for improving agriculture. *App. Geogr.* **2012**, *34*, 21–28.

Barzana, G.; Aroca, R.; Paz, J. A.; Chaumont, F.; Martinez-Ballesta, M. C.; Carvajal, M.; Ruiz-Lozano, J. M. Arbuscular mycorrhizal symbiosis increases relative apoplastic water flow in roots of the host plant under both well-watered and drought stress conditions. *Ann. Bot.* **2012**, *109*(5), 1009–1017.

Baslam, M.; Garmendia, I.; Goicoechea, N. Arbuscular mycorrhizal fungi (AMF) improved growth and nutritional quality of greenhouse grown lettuce. *J. Agric. Food Chem.* **2011**, *59*, 5504–5515.

Battini, F.; Gronlund, M.; Agnolucci, M.; Giovannetti, M.; Jakobsen, I. Facilitation of phosphorus uptake in maize plants by mycorrhizosphere bacteria. *Sci. Rep.* **2017**, *7*, 4686.

Berch, S. M.; Koske, R. E. *Glomus pansihalos*, a new species in the Endogonaceae, Zygomycetes. *Mycologia* **1986**, *78*(5), 832–836.

Berch, S. M. Endogonaceae: Taxonomy, specificity, fossil record, phylogeny. *Front. Appl. Microbiol.* **1986**, *2*, 161–188.

Berkeley, M. J.; Broome, C. E. Enumeration of the Fungi of Ceylon. Part II: Containing the remainder of the Hymenomycetes, with the remaining established tribes of Fungi. *Bot. J. Linn. Soc.* **1973**, *14*(73), 65–140.

Bisleski, R. L. Phosphate pools, phosphate transport, and phosphate availability. *Annu. Rev. Plant Physiol.* **1973**, *24*, 225–252.

Błaszkowski, J.; Chwat, G.; Góralska, A.; Bobrowska-Chwat, A. *Glomus tetrastratosum*, a new species of arbuscular mycorrhizal fungi (Glomeromycota). *Mycoscience* **2015**, *56*(3), 280–286.

Błaszkowski, J.; Chwat, G.; Góralska, A.; Ryszka, P. & Kovács, G. M. Two new genera, *Dominikia* and *Kamienskia*, and *D. disticha* sp. nov. in Glomeromycota. *Nova Hedwigia* **2015**, *100*(1–2), 225–238.

Błaszkowski, J.; Chwat, G.; Góralska, A.; Ryszka, P.; Orfanoudakis, M. *Septoglomus jasnowskae* and *Septoglomus turnauae*, two new species of arbuscular mycorrhizal fungi (Glomeromycota). *Mycol. Progr.* **2014**, *13*(4), 985.

Błaszkowski, J.; Chwat, G.; Kovács, G. M.; Gáspár, B. K.; Ryszka, P.; Orłowska, E.; Buscot, F. *Septoglomus fuscum* and *S. furcatum*, two new species of arbuscular mycorrhizal fungi (Glomeromycota). *Mycologia* **2013**, *105*(3), 670–680.

Bona, E.; Cantamessa, S.; Massa, N.; Manassero, P.; Marsano, F.; Copetta, A.; et al. Arbuscular mycorrhizal fungi and plant growth-promoting pseudomonads improve yield, quality and nutritional value of tomato: A field study. *Mycorrhiza* **2017**, *27*, 1–11.

Bonfante, P.; Selosse, M. A. A glimpse into the past of land plants and of their mycorrhizal affairs: from fossils to evo-devo. *New Phytol.* **2010**, *186*(2), 267–270.

Bonfante. P.; Genre, A. Mechanisms underlying beneficial plant–fungus interactions in mycorrhizal symbiosis. *Nat. Commun.* **2010**, *1*, 48.

Borde, M.; Dudhane, M.; Jite, P. K. AM fungi influences the photosynthetic activity, growth and antioxidant enzymes in *Allium sativum* L. under salinity condition. *Not. Sci. Biol.* **2010**, *2*, 64–71.

Brundrett, M. C. Diversity and classification of mycorrhizal associations. *Biol. Rev.* **2004**, *78*, 473–495.

Brundrett, M. C. Mycorrhizal associations and other means of nutrition of vascular plants: understanding the global diversity of host plants by resolving conflicting information and developing reliable means of diagnosis. *Plant Soil* **2009**, *320* (1–2), 37–77.

Brundrett, M.; Tedersoo, L. Evolutionary history of mycorrhizal symbiosis and global host plant diversity. *New Phytol.* **2018**, *220*, 1108–1115.

Bucher, M. Functional biology of plant phosphate uptake at root and mycorrhizae interfaces. *New Phytol.* **2007**, *173*, 11–26.

Castellanos-Morales, V.; Villegas, J.; Wendelin, S.; Vierheiling, H.; Eder, R.; Cardenas-Navarro, R. Root colonization by the arbuscular mycorrhizal fungus *Glomus intraradices* alters the quality of strawberry fruit (*Fragaria ananassa* Duch.) at different nitrogen levels. *J. Sci. Food Agric.* **2010**, *90*, 1774–1782.

Cavalier-Smith, T. A revised six-kingdom system of life. *Biol. Rev.* **1998**, *73*(3), 203–266.

Cejpková, J.; Gryndler, M.; Hršelová, H.; Kotrba, P.; Řanda, Z.; Synková, I.; Borovička, J. Bioaccumulation of heavy metals, metalloids, and chlorine in ectomycorrhizae from smelter-polluted area. *Environ. Pollut.* **1998**, *218*, 176–185.

Chandrasekaran, M.; Chanratana, M.; Kim, K.; Seshadri, S.; Sa, T. Impact of arbuscular mycorrhizal fungi on photosynthesis, water status, and gas exchange of plants under salt stress—a meta-analysis. *Front. Plant Sci.* **2019**, *10*, 457.

Chen, S.; Jin, W.; Liu, A.; Zhang, S.; Liu, D.; Wang, F.; Lin, X.; He, C. H. Arbuscular mycorrhizal fungi (AMF) increase growth and secondary metabolism in cucumber subjected to low temperature stress. *Sci. Hort.* **2013**, *160*, 222–229.

Chen, S.; Zhao, H.; Zou, C.; Li, Y.; Chen, Y.; Wang, Z.; Jiang, Y.; Liu.; A, Zhao, P.; Wang, M.; Ahammed, G. J. Combined inoculation with multiple arbuscular mycorrhizal fungi improves growth, nutrient uptake and photosynthesis in cucumber seedlings. *Front. Microbiol.* **2017**, *8*, 25–16.

Cuenca, G.; Meneses, E. Diversity patterns of arbuscular mycorrhizal fungi associated with cacao in Venezuela. *Plant Soil* **1996**, *183*, 315–322.

Dasgan, H. Y.; Kusvuran, S.; Ortas, I. Responses of soilless grown tomato plants to arbuscular mycorrhizal fungal (*Glomus fasciculatum*) colonization in re-cycling and open systems. *Afr. J. Biotechnol.* **2008**, *7*, 3606–3613.

Dimitrova, V.; Zhianska, M.; Lyubenova, M. *Role of underground phytomass in the annual fixation of* CO_2 *by representative forest ecosystems in Bulgaria*; Sofia, LTU Press, **2014**, p 112.

Dong, Y.; Zhu, Y. G.; Smith, F. A.; Wang, Y.; Chen, B. Arbuscular mycorrhiza enhanced arsenic resistance of both white clover *Trifolium repens* L. and ryegrass *Lolium perenne* L. plants in an arsenic-contaminated soil. *Environ. Pollut.* **2008**, *155*, 174–181.

Dotzler, N.; Walker, C.; Krings, M.; Hass, H.; Kerp, H.; Taylor, T. N.; Agerer, R. Acaulosporoid glomeromycotan spores with a germination shield from the 400-million-year-old Rhynie chert. *Mycol. Progr.* **2009**, *8*, 9–18.

Duan, X.; Neuman, D. S.; Reiber, J. M.; Green, C. D.; Arnold, M.; Saxton, A. M.; Auge, R. M. Mycorrhizal influence on hydraulic and hormonal factors implicated in the control of stomatal conductance during drought. *J. Exp. Bot.* **1996**, *47*(303), 1541–1550.

Eason, W. R.; Scullion, J.; Scott, E. P. Soil parameters and plant responses associated with arbuscular mycorrhizas from contrasting grassland management regimes. *Agric. Ecosyst. Environ.* **1999**, *73*, 245–255.

Friberg, S. Distribution and diversity of arbuscular mycorrhizal fungi in traditional agriculture on Niger Inland delta, Mali, West Africa. *CBN Skriftserie*. **2001**, *3*, 53–80.

Fries, E. M. Summa Vegetabiliun Scandinaveae **1849**, *2*, 261–572.

Gallaud, I. Etudes surles mycorrhizes endophytes. *Rev. Gen. Bot.* **1905**, *17*, 5.

Garcés-Ruiz, M.; Calonne-Salmon, M.; Plouznikoff, K.; Misson, C.; Navarrete-Mier, M.; Cranenbrouck, S.; Declerck, S. Dynamics of short-term phosphorus uptake by intact mycorrhizal and non-mycorrhizal maize plants grown in a circulatory semi-hydroponic cultivation system. *Front. Plant Sci.* **2017**, *8*(1471), 1–8.

Gavito, M. E.; Olsson, P. A.; Rouhier, H.; Medinapeoafiel, A.; Jakobsen, I.; Bago, A. Temperature constraints on the growth and functioning of root organ cultures with arbuscular mycorrhizal fungi. *New Phytol.* **2005**, *168*, 179–188.

Gerdemann, J. W.; Trappe, J. M. The Endogonaceae in the Pacific Northwest. *Mycologia Memoir.* **1974**, *5*, 1–76.

Gholamhoseini, M.; Ghalavand, A.; Dolatabadian, A.; Jamshidi, E.; Khodaei-Joghan, A. Effects of arbuscular mycorrhizal inoculation on growth, yield, nutrient uptake and irrigation water productivity of sunflowers grown under drought stress. *Agric. Water Manag.* **2013**, *117*, 106–114.

Gibson, J. L.; Kimbrough, J. W.; Benny, G. L. Ultrastructural observations on Endogonaceae (Zygomycetes). II. Glaziellales ord. nov. and Glaziellaceae fam. nov.: New taxa based upon light and electron microscopic observations of *Glaziella aurantiaca. Mycologia.* **1986**, *78*, 941–954.

Giri, B.; Kapoor, R.; Mukerji, K. G. Improved tolerance of *Acacia nilotica*, to salt stress by arbuscular mycorrhiza, *Glomus fasciculatum*, may be partly related to elevated K/Na ratios in root and shoot tissues. *Microbiol. Ecol.* **2007**, *54*, 753–760.

Gomez-Bellot, M. J.; Ortuoo, M. F.; Nortes, P. A.; Vicente-Sanchez, J.; Boon, S.; Sanchez Blanco, M. J. Mycorrhizal euonymus plants and reclaimed water: Biomass, water status and nutritional responses. *Sci. Hort.* **2015**, *186*, 61–69.

Goto, B. T.; Silva, G. A.; Assis, D.; Silva, D. K.; Souza, R. G.; Ferreira, A. C.; Oehl, F. Intraornatosporaceae (Gigasporales), a new family with two new genera and two new species. *Mycotaxon* **2012**, *119*(1), 117–132.

Graham, J. H.; Duncan, L. W.; Eissenstat, D. M. Carbohydrate allocation patterns in citrus genotypes as affected by phosphorus nutrition, mycorrhizal colonisation and mycorrhizal dependency. *New Phytol.* **1997**, *135*, 335–343.

Hameed, A.; Dilfuza, E.; Abd-Allah, E. F.; Hashem, A.; Kumar, A.; Ahmad, P. Salinity stress and arbuscular mycorrhizal symbiosis in plants. In: *Use of microbes for the alleviation of soil stresses*; Miransari, M., Ed.; Springer Science+Business Media: New York, 2014; pp 139–159.

Harely, J. L.; Smith, S. E. *Mycorrhiza symbiosis*; Academic Press: London, 1983; p 483.

Harley, J. L. The significance of mycorrhiza. *Mycol. Res.* **1989**, *92*, 129–139.

Hart, M.; Ehret, D. L.; Krumbein, A.; Leung, C.; Murch, S.; Turi, C.; Franked, P. Inoculation with arbuscular mycorrhizal fungi improves the nutritional value of tomatoes. *Mycorrhiza* **2015**, *25*, 359–376.

Hasanuzzaman, M.; Gill, S. S.; Fujita, M. Physiological role of nitric oxide in plants grown under adverse environmental conditions, In: *Plant acclimation to environmental stress*; Tuteja, N., Gill, S. S., Eds.; Springer Science+Business Media: New York, 2013; pp 269–322.

He, F.; Sheng, M.; Tang, M. Effects of *Rhizophagus irregularis* on photosynthesis and antioxidative enzymatic system in *Robinia pseudoacacia* L. under drought stress. *Front. Plant Sci.* **2017**, *8*, 183.

Hijri, M.; Sanders, I. R. Low gene copy number shows that arbuscular mycorrhizal fungi inherit genetically different nuclei. *Nature* **2005**, *433*(7022), 160–163.

Impa, S. M.; Nadaradjan, S.; Jagadish, S. V. K. Drought stress induced reactive oxygen species and anti-oxidants in plants. In *Abiotic stress responses in plants: Metabolism, productivity and sustainability*; Ahmad, P., Prasad, M. N. V., Eds.; Springer Science + Business Media: New York, 2012; pp 131–147.

Janouškova, M.; Pavlikova, D. Cadmium immobilization in the rhizosphere of arbuscular mycorrhizal plants by the fungal extraradical mycelium. *Plant Soil* **2010**, *332*, 511–520.

Johnson, N. C.; Graham, J-H.; Smith, F. A. Functioning of mycorrhizal associations along the mutualism–parasitism continuum. *New Phytol.* **1997**, *135*(4), 575–585.

Kanwal, S.; Bano, A.; Malik, R. N. Effects of arbuscular mycorrhizal fungi on metals uptake, physiological and biochemical response of *Medicago sativa* L. with increasing Zn and Cd concentrations in soil. *Am. J. Plant Sci.* **2015**, *6*, 2906–2923.

Kayama, M.; Yamanaka, T. Growth characteristics of ectomycorrhizal seedlings of *Quercus glauca*, *Quercus salicina*, and *Castanopsis cuspidata* planted on acidic soil. *Trees* **2014**, *28*, 569–583.

Kehri, H. K.; Akhtar, O.; Zoomi, I.; Pandey, D. Arbuscular mycorrhizal fungi: taxonomy and its systematics. *Int. J. Life Sci. Res.* **2018**, *6*(4), 58–71.

Kirk, P. M.; Cannon, P. F.; David, J. C. *Ainsworth and Bisby's Dictionary of the Fungi*; CAB International: Wallingford, 2001.

Klironomos, J. N.; Hart, M. Animal nitrogen swap for plant carbon. *Nature* **2001**, *410*(6829), 651–652.

Koske, R. E.; Gemma, J. N. *Scutellospora hawaiiensis*: A new species of arbuscular mycorrhizal fungus from Hawaii. *Mycologia* **1995**, *87*, 678–683.

Kowalska, I.; Konieczny, A.; Gąstoł, M.; Sady, W.; Hanus-Fajerska, E. Effect of mycorrhiza and phosphorus content in nutrient solution on the yield and nutritional status of tomato plants grown on Rockwool or coconut coir. *Agric. Food Sci.* **2015**, *24*, 39–51.

Kubikova, E.; Moore, J. L.; Ownlew, B. H.; Mullen, M. D.; Augé, R. M. Mycorrhizal impact on osmotic adjustment in *Ocimum basilicum* during a lethal drying episode. *J. Plant Physiol.* **2001**, *158*, 1227–1230.

Li, H.; Chen, X. W.; Wong, M. H. Arbuscular mycorrhizal fungi reduced the ratios of inorganic/organic arsenic in rice grains. *Chemosphere* **2015**, *145*, 224–230.

Li, H.; Luo, N.; Zhang, L. J.; Zhao, H. M.; Li, Y. W.; Cai, Q. Y.; Wong, M. H.; Mo, C. H. Do. Arbuscular mycorrhizal fungi affect cadmium uptake kinetics, subcellular distribution and chemical forms in rice. *Sci. Total Environ.* **2016**, *571*, 1183–1190.

Li, J.; Meng, B.; Chai, H.; Yang, X.; Song, W.; Li, S.; Lu, A.; Zhang, T.; Sun, W. Arbuscular mycorrhizal fungi alleviate drought stress in C3 (*Leymus chinensis*) and C4 (*Hemarthria altissima*) grasses via altering antioxidant enzyme activities and photosynthesis. *Front. Plant Sci.* **2019**, *10*, 499.

Li, M.; Hu, J.; Lin, X. The roles and performance of arbuscular mycorrhizal fungi in intercropping systems. *Soil Ecol. Lett.* 2021. https://doi.org/10.1007/s42832–021–0107–1.

Limonard, T.; Ruissen, M. A. The significance of VA-mycorrhiza to future arable farming in the Netherlands. *Neth. J. Plant. Path.* **1989**, *95* (1), 129–135.

Link, H. F. Observations in ordine plantarum naturales. *Ges Naturforsch Freunde Berl Mag.* **1809**, *3*, 3–42.

Liu, C.; Ravnskov, S.; Liu, F.; Rubek, G. H.; Andersen, M. N. Arbuscular mycorrhizal fungi alleviate abiotic stresses in potato plants caused by low phosphorus and deficit irrigation/ partial root-zone drying. *J. Agric. Sci.* **2018**, *156*, 46–58.

Liu, L. Z.; Gong, Z. Q.; Zhang, Y. L.; Li, P. J. Growth, cadmium uptake and accumulation of maize *Zea mays* L. under the effects of arbuscular mycorrhizal fungi. *Ecotoxicology* **2014**, *23*, 1979–1986.

Liu, X.; Song, Q.; Tang, Y.; Li, W.; Xu, J.; Wu, J.; Wang, F.; Brookes, P. H. Human health risk assessment of heavy metals in soil–vegetable system: A multi-medium analysis. *Sci. Total. Environ.* **2013**, *463*, 530–540.

Lyubenova, M. *Phytoecology*; Acad. M. Drinov Publishing House: Sofia, 2004; p 574 (In Bulgarian).

Mamatha, G.; Bagyaraj, D. J.; Jaganath, S. Inoculation of field-established mulberry and papaya with arbuscular mycorrhizal fungi and a mycorrhiza helper bacterium. *Mycorrhiza* **2002**, *12*, 313–316.

Maya, M. A.; Matsubara, Y. Influence of arbuscular mycorrhiza on the growth and antioxidative activity in Cyclamen under heat stress. *Mycorrhiza* **2013**, *23*(5), 381–390.

Melin, E. Experimentelle Untersuchungen iiber die Konstitution und Okologie der Mykorrhizen von *Pinus silvestris* L. and *Picea abies* (L.) Karst. *Mykol Untersuch* **1923**, *2*, 73–331.

Mena-Violante, H. G.; Ocampo-Jimenez, O.; Dendooven, L.; Martinez-Soto, G.; Gonzalez-Castafeda, J.; Davies, F. T.; Portugal, V. O. Arbuscular mycorrhizal fungi enhance fruit growth and quality of Chile ancho *Capsicum annuum* L. cv San Luis plants exposed to drought. *Mycorrhiza* **2006**, *16*(4), 261–267.

Midgley, D. J. Cairney, S. M.; Cairney, J. W. G. Spatial distribution of fungal endophyte genotypes in a *Woollsia pungens* (Ericaceae) root system. *Aust. J. Bot.* **2002**, *50*(5), 559–565.

Miransari, M. Arbuscular mycorrhizal fungi and heavy metal tolerance in plants. In: *Arbuscular mycorrhizas and stress tolerance of plants*; Wu. Q. S., Ed.; Springer Nature: Singapore, 2017; p 174.

Mitra, D.; Navendra, U.; Panneerselvam, U.; Ansuman, S.; Ganeshamurthy, A. N.; Divya, J. Role of mycorrhiza and its associated bacteria on plant growth promotion and nutrient management in sustainable agriculture. *Int. J. Life Sci. Appl. Sci.* **2019**, *1*, 1–10.

Moradtalab, N.; Roghieh, H.; Nasser, A.; Tobias, E. H.; Günter, N. Silicon and the association with an arbuscular-mycorrhizal fungus (*Rhizophagus clarus*) mitigate the adverse effects of drought stress on strawberry. *Agronomy* **2019**, *9*(41), 1–22.

Morte, A.; Lovisolo, C.; Schubert, A. Effect of drought stress on growth and water relations of the mycorrhizal association *Helianthemum almeriense–Terfezia claveryi*. *Mycorrhiza* **2000**, *10*, 115–119.

Morton, J. B.; Benny, G. L. Revised classification of arbuscular mycorrhizal fungi (Zygomycetes): A new order, Glomales, two new suborders, Glomineae and Gigasporineae, and two new families, Acaulosporaceae and Gigasporaceae, with an emendation of Glomaceae. *Mycotaxon* **1990**, *37*, 471–491.

Morton, J. B.; Redecker, D. Two new families of Glomales, Archaeosporaceae and Paraglomaceae, with two new genera *Archaeospora* and *Paraglomus*, based on concordant molecular and morphological characters. *Mycologia* **2001**, *93*(1), 181–195.

Morton, J. B. Three new species of *Acaulospora* (Endogonaceae) from high aluminum, low-pH soils in West Virginia. *Mycologia* **1986**, *78*, 641–648.

Morton, J. B. Taxonomy of VA mycorrhizal fungi: Classification, nomenclature, and identification. *Mycotaxon* **1988**, *32*, 267–324.

Morton, J. B. Evolutionary relationships among arbuscular mycorrhizal fungi in the Endogonaceae. *Mycologia* **1990**, *82*, 192–207.

Mosse, B.; Bowen, G. D. A key to the recognition of some *Endogone* spore types. *Trans. Br. Mycol. Soc.* **1968**, *51*, 469–483.
Mosse, B. Fructifications associated with mycorrhizal strawberry roots. *Nature* **1953**, *171*(4361), 974–974.
Navarro, J. M.; Perez-Tornero, O.; Morte, A. Alleviation of salt stress in citrus seedlings inoculated with arbuscular mycorrhizal fungi depends on the root stock salt tolerance. *J. Plant Physiol.* **2014**, *171*(1), 76–85.
Nell, M.; Wawrosch, C.; Steinkellner, S.; Vierheilig, H.; Kopp, B.; Lossl, A. Root colonization by symbiotic arbuscular mycorrhizal fungi increases sesquiterpenic acid concentrations in *Valeriana officinalis* L. *Planta Med.* **2010**, *76*, 393–398.
Newman, E. I.; Reddell, P. The distribution of mycorrhizas among families of vascular plants. *New Phytol.* **1987**, *106*, 745–751.
Oehl, F.; Sieverding, E. *Pacispora,* a new vesicular arbuscular mycorrhizal fungal genus in the Glomeromycetes. *J. Appl. Bot. Angew. Botanik* **2004**, *72*, 72–82.
Oehl, F.; Alves, A.; Silva, G.; Goto, B. T.; Costa M. L.; Sieverding, E. Glomeromycota: Two new classes and a new order. *Mycotaxon* **2011a**, *116*(1), 365–379.
Oehl, F.; Silva, G.; Goto, B.T.; Sieverding, E. Glomeromycota: Three new genera and glomoid species reorganized. *Mycotaxon* **2011b**, *116*(1), 75–120.
Oehl, F.; Silva, G.; Sánchez-Castro, I.; Goto, B. T.; Maia, L. C.; Vieira, H.; Palenzuela, J. Revision of Glomeromycetes with entrophosporoid and glomoid spore formation with three new genera. *Mycotaxon* **2011c**, *117*(1), 297–316.
Ortas, I. The effect of mycorrhizal fungal inoculation on plant yield, nutrient uptake and inoculation effectiveness under long-term field conditions. *Field Crops Res.* **2012**, *125*, 35–48.
Oseni, T. O.; Shongwe, N. S.; Masarirambi, M. T. Effect of arbuscular mycorrhiza (AM) inoculation on the performance of tomato nursery seedlings in vermiculite. *Int. J. Agric. Biol.* **2010**, *12*, 789–792.
Ouziad, F.; Hildebrandt, U.; Schmelzer, E.; Bothe, H. Differential gene expressions in arbuscular mycorrhizal-colonized tomato grown under heavy metal stress. *J. Plant Physiol.* **2005**, *162*, 634–649.
Peterson, R. L.; Massicotte, H. B.; Melville, L. H. *Mycorrhizas: Anatomy and cell biology*; NRC Press: Canada, 2004.
Phipps, C. J.; Taylor, T. N. Mixed arbuscular mycorrhizae from the Triassic of Antarctica. *Mycologia* **1996**, *1*, 707–714.
Pressel, S.; Bidartondo, M. I.; Field, K. J.; Rimington, W. R.; Duckett, J. G. Pteridophyte fungal associations: Current knowledge and future perspectives. *J. System. Evol.* **2016**, *54*, 666–678.
Punamiya, P.; Datta, R.; Sarkar, D.; Barber, S.; Patel, M.; Da, P. Symbiotic role of *Glomus mosseae* in phytoextraction of lead in vetiver grass *Chrysopogon zizanioides* L. *J. Hazard. Mater.* **2010**, *177*, 465–474.
Rani, B. Effect of arbuscular mycorrhiza fungi on biochemical parameters in wheat *Triticum aestivum* L. under drought conditions; PhD Thesis, CCSHAU, Hisar, 2016.
Read, D. J.; Perez-Moreno, J. Mycorrhizas and nutrient cycling in ecosystems—A journey towards relevance? *New Phytol.* **2003**, *157*(3), 475–492.
Redecker, D.; Kodner, R.; Graham, L. E. Glomalean fungi from the Ordovician. *Science* **2000**, *289*, 1920–1921.

Remy, W.; Taylor, T. N.; Hass, H.; Kerp, H. 4 hundred million year old vesicular-arbuscular mycorrhizae. *Proc. Natl. Acad. Sci. USA.* **1994**, *91*(25), 11841–11843.

Rinaldi, A. C.; Comandini, O.; Kuyper, T. W. Ectomycorrhizal fungal diversity: Separating the wheat from the chaff. *Fungal Diversity* **2008**, *33*, 1–45.

Ruiz-Lozano, J. M.; Aroca, R.; Zamarreoo, M.; Molina, S.; Andreo-Jiménez, B.; Porcel, R.; Garcia-Mina, J. M.; Ruyter-Spira, C.; Lopez-Raez, J. A. Arbuscular mycorrhizal symbiosis induces strigolactone biosynthesis under drought and improves drought tolerance in lettuce and tomato. *Plant Cell Environ.* **2015**, *39*(2), 441–452.

Ruiz-Sanchez, M.; Aroca, R.; Muooz, Y.; Polon, R.; Ruiz-Lozano, J. M. The arbuscular mycorrhizal symbiosis enhances the photosynthetic efficiency and the antioxidative response of rice plants subjected to drought stress. *J. Plant Physiol.* **2010**, *167*, 862–869.

Ryan, M. H.; Tibbett, M. The role of arbuscular mycorrhizas in organic farming. In: *Organic crop production, ambitions and limitations*; Kirchmann, H., Bergstrom, L., Eds.; Springer, 2009; pp 189–229.

Sadhana, B. Arbuscular mycorrhizal fungi (AMF) as a biofertilizers—A review. *Int. J. Curr. Microbiol. App. Sci.* **2014**, *3*(4), 384–400.

Santander, C.; Sanhueza, M.; Olave, J.; Borie, F.; Valentine, C.; Cornejo, P. Arbuscular mycorrhizal colonization promotes the tolerance to salt stress in lettuce plants through an efficient modification of ionic balance. *J. Soil Sci. Plant Nutr.* **2019**, *19*(2), 321–331.

Schenk, N. C.; Perez, Y. *Manual for identification of vesicular arbuscular mycorrhizal fungi*; Synergistic Publications: Florida, 1988; pp 1–255.

Schüßler, A.; Walker, C. Evolution of the plant-symbiotic fungal phylum, Glomeromycota. In: *Evolution of fungi and fungal-like organisms, the mycota XIV*; Poggeler, S., Wostemeyer. Eds.; Springer Verlag: Berlin, Heidelberg, 2011; pp 163–185.

Schüβler, A.; Schwarzott, D.; Walker, C. A new fungal phylum, the Glomeromycota: Phylogeny and evolution. *Mycol. Res.* **2001**, *105*(12), 1413–1421.

Selosse, M. -A.; Le Tacon, F. The land flora: A phototroph–fungus partnership? *Trends Ecol. Evol.* **1998**, *13*, 15–20.

Sheng, M.; Tang, M.; Zhang, F.; Huang, Y. Influence of arbuscular mycorrhiza on organic solutes in maize leaves under salt stress. *Mycorrhiza* **2011**, *21*, 423–430.

Sieverding, E.; Oehl, F. Revision of *Entrophospora* and description of *Kuklospora* and *Intraspora*, two new genera in the arbuscular mycorrhizal Glomeromycetes. *J. Appl. Bot. Food Qual.* **2006**, *80* (1), 69–81.

Sieverding, E.; da Silva, G. A.; Berndt, R.; Oehl, F. *Rhizoglomus*, a new genus of the Glomeraceae. *Mycotaxon* **2015**, *129*(2), 373–386.

Simard, S.; Perry, D. A.; Jones, M. D.; Myrold, D. D.; Durall, D. M.; Molina, R. Net transfer of carbon between ectomycorrhizal tree species in the field. *Nature* **1997**, *388*(6642), 579–582.

Simon, L.; Bousquet, J.; Leactevesque, R. C.; Lalonde, M. Origin and diversification of endomycorrhizal fungi and coincidence with vascular land plants. *Nature* **1993**, *363*(6424), 67–69.

Smith, S. E.; Read, D. J. *Mycorrhizal symbiosis*; Academic Press: San Diego, 1997; p 607.

Smith, S. E.; Read, D. J. *Mycorrhizal symbiosis*; Academic Press: Cambridge, 2008.

Smith, S. E.; Smith, F. A.; Jakobsen, I. Mycorrhizal fungi can dominate phosphate supply to plants irrespective of growth responses. *Plant Physiol.* **2003**, *133*, 16–20.

Spain, J. L.; Sieverding, E.; Oehl, F. *Appendicispora*: A new genus in the arbuscular mycorrhiza forming Glomeromycetes, with a discussion of the genus *Archaeospora*. *Mycotaxon* **2006**, *97*, 163–182.

Spain, J. L.; Sieverding, E.; Schenck, N. C. *Gigaspora ramisporophora*: A new species with novel sporophores from Brazil. *Mycotaxon* **1989**, *34*, 667–677.

Spatafora, J. W.; Aime, M. C.; Grigoriev, I. V.; Martin, F.; Stajich, J. E.; Blackwell, M. The fungal tree of life: From molecular systematics to genome-scale phylogenies. *Microbiol. Spectrum* **2017**, 5, FUNK-0053–2016.

Strullu-Derrien, C.; Spencer, A.; Goral, T.; Dee, J.; Honegger, R.; Kenrick, P.; Longcore, J.; Berbee, M. New insights on the evolutionary history of Fungi from a 407-million-year-old Blastocladiomycota-like fossil showing a complex thallus. *Phil. Trans. R. Soc. Lond. B* **2017**, *373*.

Symanczik, S.; Błaszkowski, J.; Chwat, G.; Boller, T.; Wiemken, A.; Al-Yahyaei, M. N. Three new species of arbuscular mycorrhizal fungi discovered at one location in a desert of Oman: *Diversispora omaniana*, *Septoglomus nakheelum* and *Rhizophagus arabicus*. *Mycologia* **2014**, *106*(2), 243–259.

Talaat, N. B.; Shawky, B. T. Protective effects of arbuscular mycorrhizal fungi on wheat (*Triticum aestivum* L.) plants exposed to salinity. *Environ. Exp. Bot.* **2014**, *98*, 20–31.

Taylor, T. N.; Remy, W.; Hass, H.; Kerp, H. Fossil arbuscular mycorrhizae from the early Devonian. *Mycologia* **1995**, *87*, 560–573.

Thaxter, R. A revision of the Endogoneae. *Proc. Am. Acad. Arts Sci.* **1922**, *57*(12), 291–350.

Trappe, J. M. Synoptic keys to the genera and species of zygomycetous mycorrhizal fungi. *Phytopathology* **1982**, *72*(8), 1102–1108.

Trewin, N. H.; Rice, C. M. *The Rhynie Hot-Spring System: Geology, Biota and Mineralisation. Trans. R. Soc. Edinburgh Earth Sci.* **2004**, *94*, 283–521.

Tulasne, L. R.; Tulasne, C. Fungi *Nonnulli hypogaei*, noviv. minus cogniti. *Giornale Botanico Italiano* **1845**, *2*(7–8), 55–63.

Van der Heijden, M.; Martin, F. M.; Selosse, M. A.; Sanders, I. R. Mycorrhizal ecology and evolution: The past, the present, and the future. *New Phytol.* **2015**, *205*, 1406–1423.

Wahid, A.; Gelani, S.; Ashraf, M.; Foolad, M. R. Heat tolerance in plants: An overview. *Environ. Exp. Bot.* **2007**, *61*, 199–223.

Walker, C. H.; Sanders, F. E. Taxonomic concepts in the Endogonaceae. III. The separation of *Scutellosopra* gen. nov. from *Gigaspora* Gerd. and Trappe. *Mycotaxon* **1986**, *27*, 169–182.

Walker, C. Taxonomic concepts in the Endogonaceae; Spore wall characteristics in species descriptions. *Mycotaxon* **1983**, *18*, 443–455.

Walker, C.; Vestberg, M.; Demircik, F.; Stockinger, H.; Saito, M.; Sawaki, H. & Schüßler, A. Molecular phylogeny and new taxa in the Archaeosporales (Glomeromycota): *Ambispora fennica* gen. sp. nov.; Ambisporaceae fam. nov. and emendation of *Archaeospora* and Archaeosporaceae. *Mycological Research* **2007**, *111*(2), 137–153.

Wang, B.; Qiu, Y. L. Phylogenetic distribution and evolution of mycorrhizas in land plants. *Mycorrhiza* **2006**, *16*, 299–363.

Wang, B. Phylogenetic distribution and evolution of mycorrhizas in land plants. *Mycorrhiza* **2006**, *16*(5), 299–363.

Wang, Y.; Jing, H.; Gao, Y. Arbuscular mycorrhizal colonization alters subcellular distribution and chemical forms of cadmium in *Medicago sativa* L. and resists cadmium toxicity. *PLoS One* **2012**, *7*, 3161–3164.

Wilcox, H. *Mycorrhizae*. In: *The plant root: The hidden half*; Waisel, Y., Eshel, A., Kafkati, U., Eds.; Marcel Dekker: New York, 1991; pp 731–765.

Yang, S.; Li, F.; Malhi, S. S.; Wang, P.; Dongrang, S.; Wang, J. Long term fertilization effects on crop yield and nitrate nitrogen accumulation in soil in Northwestern China. *Agron. J.* **2004**, *96*, 1039–1049.

Yooyongwech, S.; Phaukinsang, N.; Cha-Um, S.; Supaibulwatana, K. Arbuscular mycorrhiza improved growth performance in *Macadamia tetraphylla* L. grown under water deficit stress involves soluble sugar and proline accumulation. *Plant Growth Regul.* **2013**, *69*, 285–293.

Yooyongwech, S.; Samphumphuang, T.; Tisarum, R.; Theerawitaya, C.; Chaum, S. Arbuscular mycorrhizal fungi (AMF) improved water deficit tolerance in two different sweet potato genotypes involves osmotic adjustments via soluble sugar and free proline. *Sci Hort.* **2016**, *198*, 107–117.

Yousaf, B.; Liu, G.; Wang, R.; Imtiaz, M.; Zia-Ur-Rehman, M.; Munir, M. A. M.; Niu, Z. Bioavailability evaluation, uptake of heavy metals and potential health risks via dietary exposure in urban-industrial areas. *Environ. Sci. Pollut. Res.* **2016**, *23*(22), 22443–22453.

Zeng, L.; JianFu, L.; JianFu, L.; MingYuan, W. Effects of arbuscular mycorrhizal (AM) fungi on citrus quality under nature conditions. *Southwest China J. Agric. Sci.* **2014**, *27*, 2101–2105.

Zhang, F.; Jia-Dong, H. E.; Qiu-Dan, N. I.; Qiang-Sheng, W. U.; Zou, Y. N. Enhancement of drought tolerance in trifoliate orange by mycorrhiza: changes in root sucrose and proline metabolisms. *Not. Bot. Hort. Agrobot. Cluj-Napoca.* **2018**, *46*(1), 270.

Zhao, R.; Guo, W.; Bi, N.; Guo, J.; Wang, L.; Zhao, J.; Zhang, J. Arbuscular mycorrhizal fungi affect the growth, nutrient uptake and water status of maize (*Zea mays* L.) grown in two types of coal mine spoils under drought stress. *Appl. Soil Ecol.* **2015**, *88*, 41–49.

Zhu, X. C.; Song, F. B.; Xu, H. W. Arbuscular mycorrhizae improve low temperature stress in maize *via* alterations in host water status and photosynthesis. *Plant Soil* **2010a**, *331*, 129–137.

Zhu, X. C.; Song, F. B.; Xu, H. W. Effects of arbuscular mycorrhizal fungi on photosynthetic characteristics of maize under low temperature stress. *Acta Ecol. Sin.* **2010b**, *21*, 470–475.

Zhu, X.; Song, F.; Liu, S.; Liu, T.; Zhou, X. Arbuscular mycorrhizae improve photosynthesis and water status of *Zea mays* L. under drought stress. *Plant Soil Environ.* **2012**, *58*, 186–191.

Ziane, H.; Meddad-Hamza, A.; Beddiar, A.; Gianinazzi, S. Effects of arbuscular mycorrhizal fungi and fertilization levels on industrial tomato growth and production. *Int. J. Agric. Biol.* **2017**, *19*, 341–347.

INDEX

B

C

E

F

G

H

I

J

M

O

P

R

S

T

V

W